READERS TELL US THAT *CHOOSE T*[illegible]

"How do you begin to thank someone [illegible] change your life? I went from a size 26-28 to an 8! My blood pressure is now 110/70. My cholesterol is now about 180. I am enclosing some pictures so you can see the dramatic change in my appearance. What you can't see is the change in my attitude and outlook on life. I'm told by my coworkers and friends that I always have a smile on my face. I'm full of energy and life and the days are not long enough to accomplish all that I want to do."

Mary Miller, New York City

"This program is not about losing weight, it's about living healthy and eating a healthy, low-fat, high-fiber diet and lowering cholesterol. . . . I am now a much happier, healthier person, thanks to *Choose to Lose.* I'm finally in control of my life now and I FEEL GREAT!!"

Barbara Moschgat, Johnstown, Pennsylvania

"I can do it! I am in control."

Marla Summers, Lakeland, Florida

"*Choose to Lose* has given me such a high level of motivation and commitment. There is no comparison between how I felt on my previous diets and how motivated and confident I am now with my new eating lifestyle. After my heart attack I've truly been given another chance, and what a way to live my life now because I 'Choose to Lose.'"

Joanne Nesbitt, Narragansett, Rhode Island

"About four years ago I chose to lose weight using your book *Choose to Lose.* Your book made it simple to follow, but most of all it's been great fun. I've managed to lose about 150 lbs and am having no problem keeping it off. As a matter of fact, sometimes I have to work at keeping it on."

Jeanne Bennett, Stowe, Pennsylvania

"Your approach has made total sense to me, and I have actually enjoyed losing weight and getting into shape. I have lost about 15 lbs in 5-6 weeks and feel great. I have even started enjoying foods such as fish and vegetables that I previously scorned."

Jon Maren, Andover, Massachusetts

"Thank you for changing my life. Your book should be called *Choose to Live*!"

Carolyn Applegate, West Palm Beach, Florida

"I now feel that I can manage my life as well as my food intake. This truly has been a life-changing experience. I have tried many things and have always failed, but in *Choose to Lose,* failure is not an option."

Joie Wagler, R.N., Washington, Indiana

"I wrote to you to let you know how great I think your books are and that I'm living a low-fat lifestyle. I have a lot of weight to lose, but I AM getting there. I've had to buy new pants in a smaller size, and I can even tell just by looking in the mirror that I'm losing weight. Best of all I feel good about myself. The hardest part is accepting that I can eat without feeling guilty."

Linda Lape, Pittsburgh, Pennsylvania

"*Choose to Lose* makes you conscious of your eating habits, and when you are aware you can change your behavior. Being aware gives you control and control gives you freedom."

Leslie Ayvazian, Leonia, New Jersey

"I am about 120, down from 132, and I can't believe it because I eat so well, so often, and never feel starved or deprived. It's a pleasure."

Zola Bryen, Cherry Hill, New Jersey

"I want to thank you for *Choose to Lose.* I'm a Brazilian girl who has always fought against the scale. I used to starve in order to get thin, but it didn't work. I was unhappy because I thought I would never be able to eat carbohydrates (which I love), that every single slice of bread or pasta would make me fat. I went to America, and fortunately my friend introduced me to this wonderful book. I lost 20 pounds and I feel great now, because I CAN EAT. Since then my family adopted the low-fat diet and everyone feels great. Right now, I'm so used to eating in a healthy way that it's been years since I've eaten a hamburger — I'd rather have a turkey sandwich instead!!!! Thank you very much!"

Mariana Correa, Brazil

"With this program, it feels like I am constantly eating — which has always been a problem in the past. The only difference now is that I know what to eat. So far I've lost 40 pounds in nine months, and my goal weight is within sight (25 pounds away)."

Steve Becker, Two Rivers, Wisconsin

"I want to share with you my success, as I have recently passed the 100-pounds-lost milestone. I lost 12 inches in my waist from 54" to 42". I thank you for the book that has initiated such a major improvement in my life. This has become a lifestyle change rather than a diet, and I really believe it has lengthened (or, should I say, saved) my life."

Dan King, Springville, California

"I am amazed that I am not hungry as I have been with other diets. I have never lost weight before while having more food than I can possibly eat. Another thing that astounds me is that it is incredibly easy to follow and I can work "treats" every now and then and not feel guilty. Your book has given me a new sense of hope."

LaRhonda Hunt, Trenton, Tennessee

PHYSICIANS RECOMMEND *CHOOSE TO LOSE:*

"I find your book very well written, easily understood, and extremely informative. Not only do I recommend it to my patients, I also find it an excellent gift for family and friends."

Michael E. Rubin, M.D.

"As you may recall, I received your book on October 19, 1993, and began making changes on October 20, 1993. Since that time I have lost 66 pounds using the information from your book as well as entirely changing my exercise routine. I have recommended your book to hundreds of patients, some of whom have made significant changes in their diet and weight."

Kathy Sturino, M.D., Waukesha, Wisconsin

"I have found *Choose to Lose* to be tremendously helpful both personally and professionally. You have my strongest endorsement and my *heartfelt* thanks."

Larry Saladino, M.D., Vienna, Virginia

"I came across *Choose to Lose* when it was first out. I have used the information in the book extensively in patient counseling. I have referred many patients to get the book, and I have seen excellent results in food behavior change and better health in these patients. I have also encouraged people to get *Eater's Choice* as a companion since it teaches people how to make more healthy foods. All this to say — thank you for the information put in a practical way for patients."

Carolyn V. Brown, M.D., M.P.H.

Choose to Lose

Also by Dr. Ron Goor and Nancy Goor

Eater's Choice: A Food Lover's Guide to Lower Cholesterol

Eater's Choice Low-Fat Cookbook

Choose to Lose

A Food Lover's Guide to Permanent Weight Loss

Dr. Ron Goor and Nancy Goor

THIRD EDITION

HOUGHTON MIFFLIN COMPANY
Boston New York

Library of Congress Cataloging-in-Publication Data

Goor, Ron.
Choose to lose : a food lover's guide to permanent weight loss /
Ron Goor and Nancy Goor. — 3rd ed.
p. cm.
Includes bibliographical references and index.
ISBN 0-395-97097-0
1. Low-fat diet. 2. Food — Fat content—Tables.
I. Goor, Nancy. II. Title.
RM237.7.G66 1999
613.2′5 — dc21 99-18106 CIP

Printed in the United States of America

AGM 10 9 8 7 6 5 4 3 2 1

Before beginning this or any diet, you should consult your physician to be sure it is safe for you.

In loving memory of Charles Goor

Contents

Acknowledgments

Ron and Nancy Goor wish to thank the many people who have contributed to the development of this book: Sharon Bergner, Robert R. Betting, Shari Bilt, Rebecca Carter, Letitia Cornelius, J.P. Flatt, Alex Goor, Dan Goor, Jeanette Goor, Anita Hamel, Kimra Hawk, Lynn Karnitz, Genevieve Kazdin, Dan King, Sandra Leonard, Suzanne Lieblich, Rolland Lippoldt, Jennifer Martin, Suzanne McGovern, Helen Miller, Ted Mummery, Nancy Ochsner, Margo Sipiora, Kendra Stemple Todd, Dawn Weddle, Paula Wickstrom, and all the *Choose to Lose* Leaders.

A Note on the Third Edition

In the four years since *Choose to Lose* was revised, we have witnessed the explosion of fat-free processed foods, the reintroduction of ineffective and dangerous approaches to weight loss — including diet drugs — and the growing prevalence of eating out rather than preparing food at home. Because these issues have such a profound effect on everyone who wants to eat healthfully, we have addressed them all in the third edition of *Choose to Lose.*

Don't worry. *Choose to Lose* is basically the same book — it's just better! The basic principles of *Choose to Lose* have stayed the same. The method, which reflects those principles, has not changed because it works. It has stood the test of time. We have always promoted three strategies for healthy eating and weight loss: reduce fat, eat lots of low-fat, nutrient dense/fiber-rich foods, and exercise aerobically.

In the first edition of *Choose to Lose,* we thought reducing fat would be a major stumbling block, so we focused primarily on fat. After a short while, our readers and *Choose to Lose* Program Leaders told us that people didn't have a problem lowering their fat intake — they had a problem eating enough food.

In the second edition we told our readers to Eat! Eat! Eat! to counteract their fear of eating. The U.S. food system of 1995 allowed that recommendation without too many caveats.

But the U.S. food supply didn't stay the same. Responding to people's desire to have sweets but not fat, food manufacturers produced a panoply of nonfat cookies, cakes, ice cream, frozen yogurt, chips, and crackers. Consumers went berserk. They plundered the supermarket shelves of fat-free products. They even mobbed the SnackWell's trucks before they were unloaded. But fat-free, nutrition-free, and fiber-free empty calories were not really the answer.

In the third edition we have modified our message to Eat! Eat! Eat!

nutrient-dense/fiber-rich foods by tacking on a strong warning. You may Eat! Eat! Eat! *only* if you are eating nutrient-dense/fiber-rich foods because their bulk makes them self-limiting (and, of course, healthy). You must eat highly processed empty carbohydrates rarely and with extreme care because they will stall your weight loss and also force out nutritious food. We are repeating the warning so often in this edition that you'll probably find yourself murmuring it in your sleep.

In this edition we correct a lot of misinformation promoted by diet writers and the media. Two such myths are that dietary carbohydrates turn to fat and that the production of insulin in reaction to eating carbohydrates is dangerous and fattening.

Since the second edition was published, diet drugs appeared, became wildly popular, caused death and injury, and were pulled from the market. Since the drug industry knows that overweight people would kill (or die) for a weight loss drug, they are furiously working on creating new ones (one is being introduced to the market as we write these words). As it is only a matter of time before diet pills become popular again, we discuss the problems inherent in all diet drugs.

This new edition brings two other changes. We have included more psychological issues in the new chapter, "It's Up to You." In addition, we have updated the Food Tables because thousands of new foods have become available. We also discuss the current trends in foods and hype on packaging.

As we assured you in the beginning of this note, *Choose to Lose* is the same book, but even better! We hope you enjoy it thoroughly, take it to heart, follow it to a tee, and that it brings joy and good health into your life.

P.S. We have gotten so much pleasure from hearing of your successes. We recently received an e-mail from Chooser to Lose Dan King to alert us that he had reached the 100-pound loss mark. One hundred pounds ago, his weight made walking difficult. Now he rises at 5:45 A.M. every day to walk three miles! Chooser to Lose Juanita Brooks lost 36 pounds and has kept it off for more than four years so far. Even suffering serious injury in a car accident hasn't made her backtrack. Recently she wrote, "My doctor told me something I never thought I'd hear.

'Juanita,' he said, 'I do *not* want you to lose any more weight.' I now weigh 128 lbs and I feel great."

We would love for you to tell us how YOU are doing. Call us. E-mail us. (Our e-mail address is goor@choicediets.com; our website is http://www.choicediets.com.) Send us before and after or before and in process pictures.

We'd also like to thank the talented, bright, caring, creative Choose to Lose Program Leaders around the country who do a phenomenal job teaching Choose to Lose in their communities and worksites. They are an inspiration to the many participants they teach and motivate.

Introduction

Choose to Lose is an outgrowth of our book *Eater's Choice: A Food Lover's Guide to Lower Cholesterol.* Soon after *Eater's Choice* was published we began receiving letters from ecstatic readers. Not only were they lowering their cholesterol, they were losing weight — typically 10 to 15 pounds and some up to 40 or more pounds. This was not the intention of the cholesterol-lowering plan, but it was a side effect that was appreciated as much (or more!).

Readers tell us that they find the *Eater's Choice* method simple, direct, and flexible. It gives them the tools and knowledge to take control of their diet for the first time. And it is positive. All foods are allowed. No exchanges, no two-week meal plans to follow for the rest of your life. *You* determine which foods you eat.

Choose to Lose uses essentially the same approach that was successfully utilized in *Eater's Choice.* But while *Eater's Choice* limits only saturated fat intake, *Choose to Lose* limits *total* fat consumption. Saturated fat is the chief culprit in the diet that raises blood cholesterol, but all three types of fat — saturated, monounsaturated, and polyunsaturated — in the foods you eat make you gain weight. In recent years scientific research has shown that dietary fat, not carbohydrates or total calories, is the primary cause of weight gain. Furthermore, researchers note that weight loss can be achieved simply by limiting fat intake. That is why *Choose to Lose* uses a Fat Budget and the total fat content of foods to help you make food choices that satisfy your palate and help you lose weight.

Like *Eater's Choice, Choose to Lose* helps you adopt a new eating lifestyle. After all, permanent weight loss, like permanent cholesterol lowering, requires a new lifelong eating pattern, not just a two-week crash diet. Whatever changes are made to lower cholesterol and weight must be continued. The diet plan must therefore be palatable and practical.

Not only is *Choose to Lose* easy to follow, it is also enjoyable. Most people eat a limited number of foods and are probably bored with their diet. This new low-fat way of eating will broaden your culinary horizons and add spice and variety to your life. You will explore new foods and new food preparations. To discover a world of new taste treats, try the mouth-watering recipes in *Choose to Lose,* as well as 300 more in *Eater's Choice Low-Fat Cookbook* — recipes for real foods, such as *Tortilla Soup, Cajun Chicken, Broiled Ginger Fish, Cinnamon Sweet Cakes, Honey Whole-Wheat Bread, Calzone, Divine Buttermilk Pound Cake,* and *Deep-Dish Pear Pie,* to mention just a few favorites.

Any change in eating patterns takes commitment and motivation. *Choose to Lose* provides you with the tools and knowledge to make food choices to keep you lean for a lifetime. You will find you are eating more food and enjoying it more. And most important, you will also lose weight.

Good luck and good health.

Choose to Lose

1

Fat's the One

"As for myself, I'd like to tell you that *Choose to Lose* has given me back my smile."

— Maggie Hall, Dunedin, Florida

TAKE A LOOK in the mirror — a good look. Is the person staring out at you a little pudgier than you would like? Has the muscle turned to flab? Are the seams and buttons on the verge of splitting and popping? Don't turn away in disappointment. Give yourself a smile. Already you are making a big change in *you*. You are starting *Choose to Lose*. Soon you will see a trim version of what you see now. The trim version will be beaming. No suffering will have made the pounds melt away. Only good, abundant food, including your all-time favorites. *Choose to Lose* is not a miracle diet. In fact, it's not even a diet. It's a new way of eating that is easy, healthy, and delicious. And it works.

The key to *Choose to Lose* is FAT.

FAT MAKES FAT

The word is FAT. FAT is what you don't want to be and FAT is what you ate to become FAT. So the way to become NOT FAT is to reduce fat in your diet. Don't focus on total calories . . . or sugar . . . or starch. Focus on FAT.

Until recently, most diet formulators based their weight loss methods on reducing *total calories*. Sugar and starch were always forbidden. These approaches pinpointed the wrong culprit. The reason that foods containing sugar make you gain weight is not because they contain sugar. It is because they contain *fat* and sugar. It is the fat in the cheesecake, not the sugar, that puts dimples in your knees. The reason that foods containing starch make you gain weight is not because they contain starch. It is because they contain *fat* and starch. It is the fat in the sour cream, not the starch in the potato, that rounds your belly.

EAT TO LOSE

It is not *how much* you eat that gets you into trouble — it is *what* you eat that makes you fat or prevents you from getting thin. As long as you are eating nutrient-dense/fiber-rich carbohydrates (fruits, vegetables, whole grains), low-fat meats, and nonfat and low-fat dairy in sensible amounts, eat up and enjoy.

In fact, you *must* eat lots of nutritious carbohydrates to lose weight. These foods help maintain your metabolism so that your body burns fat. Eating lots of food keeps you happy and satisfied so that you don't feel hungry and binge, and it gives you energy so that you stay vital all day. That is why *Choose to Lose* doesn't focus on reducing total calories; it focuses on reducing the nutrient that is making you fat — FAT.

You may find this recommendation hard to accept. After all, you have probably been trying to cut *total calories* and limit carbohydrates for years. Please. Don't just accept this advice on faith. Read the following section, which explains simply how the fats and carbohydrates you eat affect your weight. The science behind *Choose to Lose* is fascinating, illuminating — and motivating!

It is always difficult to replace old ideas with new ones. The mistaken concept that all calories affect the body in the same way is imbedded in the public consciousness. It is repeated over and over in magazines and on TV so that it becomes accepted as fact. We are lucky that although our readers often begin with the false impression that they must limit all food to lose weight, they soon learn that they can eat a lot of food (albeit low-fat, nutrient-dense, fiber-rich food) and still lose. In fact, one of the most common comments we hear is, "I can't believe how much food I can eat and still lose!"

FOOD — MORE THAN JUST PLEASURE

This is a book about fat, food, and you. Ice cream sundaes, chicken soup, T-bone steaks, potato chips — for most people food offers more than just gustatory satisfaction. Food may represent love (you want to please those you love, so you feed them); status (with your raise you can afford to eat filet mignon and caviar); comfort (you feel lonely or frustrated, so you console yourself with a bowl of ice cream); nostalgia (the aroma of hot dogs conjures up visions of baseball games and family picnics); or something just plain wonderful to eat. Food fulfills a different need in each person, but for everyone, no matter how thin, fat, young, old, rich, poor — food is energy.

ENERGY EXPENDITURE

BMR

Your body needs a certain amount of energy to function. The amount of energy you use when you are completely at rest is called your basal metabolic rate, BMR. This is the energy needed to power your heart, lungs, brain, kidneys, and other organs and keep them in good repair and, for children, the energy to grow.

Physical Activity

You also need energy for physical activity — walking, running, moving. This is an expandable amount depending on how active you are.

So the total amount of energy you expend is equal to the sum of your BMR and your physical activity.

ENERGY INTAKE

Food — Source of Energy

Where does your body get this energy? Energy is stored in fats, proteins, and carbohydrates in the foods you eat. More calories are stored in fat than in carbohydrates or proteins. One gram of fat has 9 calories (some scientists say 11). One gram of carbohydrate or protein has

What Is a Calorie?

A calorie is a unit of energy. Energy is stored in fats, proteins, and carbohydrates in units called calories. Total calories are the sum of the fat calories, protein calories, and carbohydrate calories you eat. Alcohol calories may also be part of the total.

Fat calories
Carbohydrate calories
Protein calories
+ Alcohol calories (if consumed)
Total Calories

The pie chart below shows the caloric composition of a daily total caloric intake of 2000 calories.

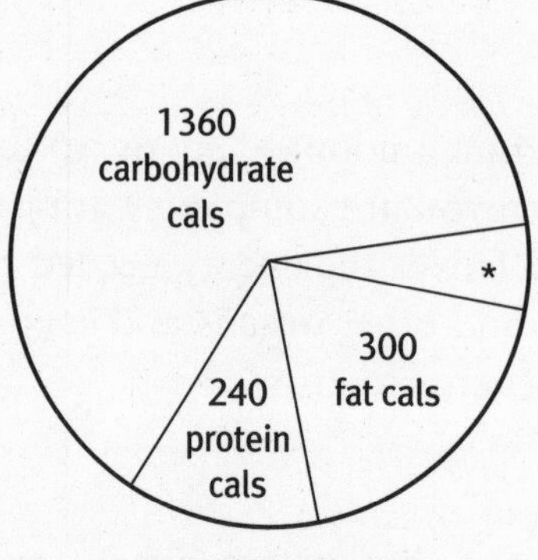

2000 Total Calories

*If alcohol is consumed, it is included in the total.

4 calories. Fat is so densely caloric that when you eat a little, you've eaten a lot of calories. For example, when you eat 1½ ounces of fat-laden potato chips you are consuming the same number of calories as when you eat 12 ounces of potatoes (two medium-large potatoes).

To release the energy stored in fats, carbohydrates, and protein, they must be burned. Picture a stack of logs blazing in a fireplace. The burning of nutrients and the burning of wood in a fireplace are similar chemical processes. In both cases the chemicals are combined with oxygen (burned or oxidized) and the energy stored in them is released. In the case of wood, the energy is released all in one step as heat. In the body, the oxidation goes on at a lower temperature (98.6°F) in small steps controlled by enzymes. In this way, most of the energy stored in the food is captured for growth, maintenance, repair, and physical activity, and little is wasted as heat.

All Calories Are Not Equal

You may have thought that everything you ate ended up as those rolls of fat rippling down your belly. You may even have been avoiding carbohydrates in the belief that they are fattening. But you were misguided.

Many people believe that all calories contribute equally to weight gain. But scientific evidence has shown that fats, carbohydrates, and proteins are metabolized (burned and stored) differently. Fat calories are stored as fat, while carbohydrate and protein calories (consumed in reasonable amounts) are burned off. In fact, eating carbohydrates can help you lose weight. To understand why, read on.

CARBOHYDRATES

In order to produce the energy it needs, the body burns a mixture of carbohydrate, fat, and protein derived from the foods you eat. The primary fuel, carbohydrate, normally accounts for over 50 percent of the calories you consume. Carbohydrates provide most of the energy to fuel the muscles and most other bodily functions and are the sole source of energy for the brain. Carbohydrates include both simple sugars (you know a sugar by its sweet taste) and starches (complex carbohydrates) such as potatoes, other vegetables, rice, pasta, and bread.

Each day you consume about 1000 calories of carbohydrate. Most of it is burned within a few hours of intake. Some is stored as glycogen — long chains of the simple sugar glucose — in the muscles and in the liver. Glycogen in the muscles is used for short bursts of intense activity, as in competitive sports and in fight or flight responses. Glycogen in the liver provides a constant supply of glucose to the brain. The brain can burn only the simple sugar glucose. The brain is a fuel hog and burns about 30 percent of the energy you use each day. Without a constant supply of glucose to the brain you would lose consciousness.

The body has a limited capacity to store carbohydrates (about 800–1000 calories, or approximately one day's intake). When you eat a meal, a small amount of carbohydrate tops up the glycogen stores, which have been partially depleted between meals. The rest of the carbohydrate (the majority) is burned within a few hours of consumption and fuels your physical activity and internal functions.

Insulin: You Need It

Don't let anyone tell you that eating carbohydrates is dangerous because it makes your insulin rise. Insulin is essential for good health. Insulin is released from the pancreas to move carbohydrate out of the blood and into the cells where it is burned or stored. A rise in insulin in response to eating carbohydrates is a *normal* reaction — *not* a symptom leading to diabetes. Insulin production in response to eating carbohydrates is as normal as is an increased pulse in response to exertion. In the absence of insulin, carbohydrate (glucose) would remain in your blood and thus be unavailable to your cells for energy. In essence, your cells would be starving to death in the midst of plenty.

Running on Empty

What happens if you eat too few carbohydrates? If you are following a low-calorie/low-carbohydrate and/or high-protein diet and consuming too few carbohydrates, your glycogen stores never get filled up. Result: fatigue and loss of stamina and endurance. This may be why you felt so tired and cranky on every other diet you tried.

Natural Regulation of Carbohydrates

What prevents you from consuming too much carbohydrate?

If you are truly eating nutrient-dense/fiber-rich foods such as fruits, vegetables, and whole grains, it is almost impossible to overeat them. They are so filling, they limit themselves. Imagine eating five potatoes at one sitting. You couldn't do it. Selecting nutrient-dense/fiber-rich foods is the best protection from overeating carbohydrates.

What happens to the carbohydrate you consume in excess of your needs?

If you eat more carbohydrate (within reasonable limits) than necessary for your immediate needs — BMR + physical activity — it is burned, and the energy is converted to heat. Under normal circumstances, none is converted to fat. This process of producing heat from burning carbohydrate is called the thermogenic effect of food. When you feel warm during or after a meal, you're feeling the thermogenic effect.

Won't I Gain Weight If I Overdose on Carbohydrates?

Within normal limits, the body burns off almost all the carbohydrate you eat. Only if you were to eat more than 2200 calories of pure carbohydrate over your normal daily total caloric intake for five to six days in a row might the excess carbohydrates possibly turn to fat. This is called glycogen loading, and it isn't easy to do. If the sources of carbohydrates you are eating are mainly fruits, vegetables, and whole grains, it is difficult, even impossible, to overeat them.

So you can scratch the idea that carbohydrates turn to fat. It doesn't happen. But you can push the system and affect your weight loss progress.

Empty Carbohydrates Provide No Brakes

When man evolved there were no fat-free, processed foods. The bulk inherent in the natural foods available — vegetables, fruits, and whole grains — limited their consumption. Today we are subjected to a cornucopia of fiber-free/nutrient-free processed foods. You can gobble up fat-free cookies, ice cream, and cakes. You can inhale nonfat chips and crackers. These foods have no bulk, so you can easily consume hundreds and hundreds of empty calories without feeling in the least bit full.

Overloading on Empty Carbohydrate Calories Stalls Weight Loss

Overeating nonfat, high-sugar, high-calorie foods will *not* make you gain weight, because as we explained earlier, carbohydrates do not turn to fat under normal circumstances. But because the body can't store carbohydrates, it must get rid of them, so it burns them. But instead of a limited number of carbohydrate calories, there are hundreds or even thousands of empty carbohydrate calories to be burned. Consuming fiber-free foods slows down and even stops weight loss because the body prefers to burn these hundreds of empty calories instead of stored fat.

You might ask, "What about high-fiber carbohydrates? Won't they be burned in preference to stored fat, too?" This doesn't happen, because high-fiber foods are self-limiting so you don't overeat them. Remember those five potatoes. They contain 500 total calories. You might eat one (100 calories), but you couldn't possibly eat all of them. However, you could easily inhale a whole box of twenty Entenmann's fat-free oatmeal cookies (1000 total calories); see page 123.

The Gray Area

Of course, it is always possible to push an eating plan beyond its limits. There are processed foods, such as pasta, bagels, breads, and some cereals, which you can certainly include in a healthy diet, but since they are not high in fiber, it is easy to overeat them. Just like overdosing on empty calories, accumulating too many calories from these processed foods will slow down or stall your weight loss. In other words, we are not recommending that you stuff yourself with half loaf of whole-wheat bread or three cups of pasta or four cups of cereal. You have to use sense.

Don't Blame Carbohydrates

We hear people complain that they are eating only carbohydrates yet gaining weight. There are three reasons for this.

1. They don't know what carbohydrates are and/or the carbohydrates they are eating are laced with fat. Do they call croissants carbohydrates? Or fettuccine Alfredo? They look a lot like fat to us.
2. If they are not writing down and analyzing everything they eat, do they really know what and how much they are eating? They may

think they are eating only carbohydrates or they may *think* they eat little or no fat.

3. The carbohydrates they are eating may be high-sugar, nonfat sweets and snacks with little or no bulk to prevent overconsumption. Eating too many empty calories will stall fat loss because the body will burn them in preference to burning stored fat. Under normal circumstances overeating carbohydrates will not cause weight gain.

FAT AND WHY *CHOOSE TO LOSE* FOCUSES ON IT

In contrast to carbohydrates, the fat that you eat is not burned right away but is immediately added to the fat stores, where it promptly becomes the soft, squishy blubber that pads your body. It's as if you wadded the ice cream, pizza, cheeseburger, French fries, cookies directly onto your hips, belly — *you* know where — except it is happening from the inside.

The adipose tissue contains enormous stores of fat. A normal lean person stores about 140,000 calories of fat, but there is no upper limit. Thus, a person who weighs 300 pounds may be storing 200 pounds of fat.

Unlike carbohydrate consumption, eating fat does not cause the body to burn fat. Since there are no mechanisms to limit fat consumption and since high-fat foods have little bulk, it is easy to overeat fat day after day. And since almost all the fat is stored, when you overeat fat day after day you become fatter and fatter. Given the high-fat food system in the U.S., that is exactly what we Americans do.

The Secret to Weight Loss: Raiding the Fat Stores

Now you know why the carbohydrates you eat don't become fat on your body and the fats you eat do. But what can you do about the fat that already rounds your edges? How can you make your body mobilize the fat out of the fat stores so that you can lose weight?

Each day fat from the foods you eat is added to the fat stores. Some is removed for burning to furnish energy not supplied by the carbohydrate and protein you eat. Your weight is determined by how much fat you add to the fat depots versus how much you remove.

If you eat just the amount of fat needed to furnish the energy not supplied by the carbohydrate and protein, your weight will remain the same. If you eat more fat than is needed, the excess will go into the fat stores and you will gain weight. If you eat less fat than is required to satisfy your energy needs, then the body will have to make up the deficit by burning additional fat. And where does this fat come from? The fat is removed from the fat stores and **you lose weight.**

It all makes sense. You became fat because you ate *more* fat than you burned. To lose weight, you have to eat *less* fat than you burn so that your body will use up the excess fat luxuriating in your adipose tissues to help supply your energy needs.

THREE STRATEGIES TO SHRINK THE FAT STORES

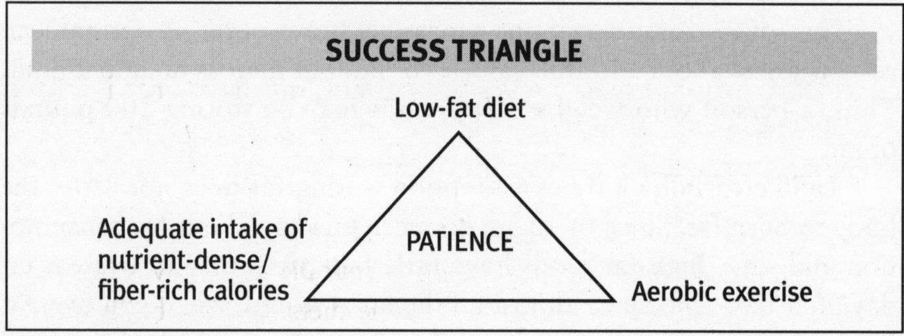

Strategy #1: Low-fat Diet — Put Less Fat into the Fat Stores

Obviously, the less fat you add to the fat stores, the less you will need to take out later. The way to do this is to eat a low-fat diet. *Choose to Lose* provides a powerful tool, the Fat Budget, and an effective method (see Chapter 2, "The *Choose to Lose* Plan") to help you follow a low-fat diet.

Maximizing Fat Loss. Eating less fat is essential for weight loss, but you can't stop there. To shrink the fat stores faster, you must not only add less fat to them, you need to remove fat that is already there. *Choose to Lose* has two strategies aimed at removing stored fat. Strategy #2 you are not going to believe.

Strategy #2: Adequate Intake of Nutrient-Dense/Fiber-Rich Calories — You Have to Eat to Lose Weight

Can you imagine a weight loss program insisting that you eat an abundance of food to lose weight? It's not a dream — only a dream program! You need to eat enough calories to keep your BMR chugging along at a maximum rate so that you will burn fat at a maximum rate and thus help reduce your fat stores. Of course, you must eat a lot of the right foods. This does not mean a bowl of fat-laden peanuts or a box of fat-free cookies. It does mean replacing much of the fat in your diet with nutritious low-fat food — fruits, vegetables, whole grains, non-fat dairy, poultry, fish, seafood and so on.

The Benefits of Eating Nutrient-Dense/Fiber-Rich Foods

1. Maximizing your metabolic rate
2. Ensuring an adequate intake of vitamins, minerals, and fiber* for long-term good health
3. Keeping you full, satisfied, and content
4. Increasing your energy and stamina

Eating Too Little Can Be Hazardous to Your Diet. Have we convinced you? After years of following low-calorie diets, you probably have a mindset against eating. You may still think that a nifty way to force your body to burn fat from its fat stores is to reduce your total energy intake drastically. How about an 800-calorie-a-day crash diet for three weeks to turn you into a string bean? The problem is that not only are 800-calorie-a-day diets dangerous and sometimes even fatal, deficient in vital nutrients, and impossible to follow for more than a few weeks, but starvation diets depress your basal metabolic rate. If you eat fewer than about 1500 total calories a day, you will find that your weight loss

*See page 115 for a discussion of fiber.

will *slow down.* Your body doesn't know that you are starving just to lose weight. It reacts to the intake of too few calories by lowering the BMR to conserve stored energy and thus burning food more slowly. That's a great help if food is really scarce, but if you are trying to lose weight, burning stored energy (fat) more slowly means losing weight more slowly. Conclusion: Starvation makes you hungry, dissatisfied, unhealthy, and eventually fatter.

Important Message #1

Although you just read that you can eat lots of total calories as long as you limit fat and empty calories (and use sense), we know in your heart of hearts you don't *really* believe it. You will limit your fat, but you will still be afraid to eat a lot of food — low-fat, nutritious food. DON'T BE AFRAID! Limiting total calories has not been successful for you in the past, at least not in the long-term. You can't lose anything (but weight) by trying a different approach. So TRY IT! (But be sure the foods you eat in abundance are nutritious, low-fat foods.)

Important Message #2

In case you misread the previous message to mean you can eat *any* total calories in abundance, let us clarify.

1. Eating lots of nonfat, processed fluff like fat-free cakes and nonfat cookies, chips made with olestra, fat-free ice cream, fat-free frozen yogurt, and fat-free cheese will stall or stop your weight loss. These foods contribute so many calories that your body will burn them instead of the fat from your fat stores. They offer little or nothing of value to your body. Treat them like a high-fat splurge.
2. Use sense. Eating enormous quantities of even relatively healthy foods will slow your weight loss. For example, 1½ to 2 cups (340 total calories) of raisin bran cereal is acceptable, but more is overdoing it. Two cups of spaghetti (420 total calories) is acceptable, but more is too much. You don't need all those calories of processed white flour. Don't push the system. Use sense.

Strategy #3: Aerobic Exercise — Build Muscle to Burn Fat

This third strategy to maximize fat removal may not be your favorite. But it is absolutely essential and you may learn to enjoy, even love, it. The third strategy is daily aerobic exercise. Plain, old-fashioned walking is just fine. You don't have to strain and sweat to make it work. **The main value of aerobic exercise is to build and preserve muscle. Muscle burns fat.** The more muscle you have the more fat-burning capacity you have. Muscle is metabolically active (burns energy at a high rate) and the more you have of it, the higher your BMR and the faster you burn everything, including fat. Not only will aerobic exercise help you lose weight, it will help you keep it off. That doesn't mean you have to exercise until you look like Arnold Schwarzenegger. Thirty minutes of walking each day is fine. It doesn't even have to be fast, just non-stop. See Chapter 11, "Exercise: Is It Necessary?" for more information on exercise.

Preserving Muscle: Another Good Reason to Exercise. You think of your muscle as being permanent, but it isn't. When you don't use your muscle, it breaks down and the protein is burned. (Have you ever had your leg in a cast for five or six weeks? When the cast comes off, the muscle is withered and wasted from disuse. Much of the protein has been burned for energy. But as soon as you start walking again, the muscle begins rebuilding and within a short time your leg looks normal.) When you use your muscles, some of the protein you eat is used to build them up again.

When you are in energy deficit, that is, you are using more energy than you are taking in (as in a semi-starvation or calorie-restricted diet), you must make up for the deficit by using energy stored in the body. The two major sources of stored energy are fat in the adipose tissues and protein in the muscle. Wouldn't it be great if the body automatically burned fat from the fat stores? No such luck. Unfortunately, fat is the stored energy of last resort. The body prefers to burn protein from the muscle — if you are *not* exercising. YOU DO NOT WANT YOUR BODY TO BURN PROTEIN INSTEAD OF FAT. **To protect your muscle from being broken down and burned for energy and to force your body to burn stored fat instead, you must exercise aero-**

bically* (not in intensive bursts and spurts but steadily and continuously).

Yo-Yoing: Starvation Diets Make You Fatter

Almost all other diets are a combination of starvation and no exercise. Now that you know about the importance of exercise, you'll understand why this duo can be so destructive to your weight loss goals.

People who go on low-calorie diets rarely exercise. They believe the ads that show a beautiful, slender young woman sipping a diet shake as she languishes on the beach. If this were reality, the next picture would show her hungry, bored, and fatigued, leaving her chaise longue to find a hamburger stand. The immediate and natural response to a week or two of deprivation is to overeat. After all, so much suffering deserves a reward. But now, because of a lowered BMR and less muscle mass, the reward sits on hips and thighs instead of being burned off. And since the dieters haven't learned new low-fat eating habits and still don't exercise, the combination of a lower BMR and their old high-fat diet results in new pounds of fat being added back at a faster rate. This is the famous yo-yo effect. With each cycle of weight loss and gain, you lose more muscle, your weight comes off more slowly and is regained more quickly, and you end up fatter (fat being a higher percentage of your body weight) and heavier. Starvation plus no exercise is a system destined to fail. (You will find a discussion of other cockamamie diet ideas in Chapter 15, "Frequently Asked Questions.")

Breaking the Cycle

If you are a former yo-yoer, don't be discouraged. You don't have to be trapped in the cycle. By eating a diet low in fat and high in nutrient-dense/fiber-rich calories and doing a moderate amount of aerobic exercise, you will build new muscle and get your metabolism going again.

---

*Aerobic exercise is nonstop, uses large muscle mass, and requires a steady supply of oxygen. Example: walking.

A Related Word about Dieting and Guilt

Eating only 500–1000 calories a day is like holding your breath for 20 minutes. It's impossible. After a few minutes of not breathing you gasp for breath. You don't feel guilty. Why should you? But if you quit your diet after a few weeks of starving, you think you have failed — you are bad. Don't blame yourself. Blame the diet. What you need is an eating plan that works, and *Choose to Lose* is it. With *Choose to Lose* you will not starve. Rather you will eat more than you ever ate before. You will be full and satisfied. You will learn to make choices that will make and keep you lean. You will be in control. And you will be able to follow *Choose to Lose* forever.

Patience Is at the Center

You'll notice that the word *patience* lies at the center of the triangle on page 10. Patience is the byword of *Choose to Lose.* People lose weight at different rates. You may be one of the fortunate people whose fat melts away like butter in the summer sun. Or you may find that fat clings to you like a wet tee-shirt. In either case, if you follow *Choose to Lose,* you will lose weight. One reader wrote to us after she had lost 45 pounds, "My goal is to lose another 50 lbs. I don't worry about losing weight any longer, because I know that living this lifestyle will put me there in the next year or so."

Telling It Like It Is

That's it. You are overweight —

not because you eat too much food,
  but because you eat too much fat (more than you are burning) and the excess is going directly into your fat stores;
not because you eat too much carbohydrate,
  but because you eat too many empty carbohydrates and too little nutrient-dense/fiber-rich carbohydrates;
not because you are not a marathon runner,

but because you are not exercising regularly to preserve and build new muscle and force your body to burn fat from the fat stores.

So to lose weight you must eat less fat and more nutrient-dense/fiber-rich carbohydrates, and also exercise aerobically every day, which includes walking.

Throw Away Your Scale!

The only weight loss that is meaningful and permanent and has long-term health benefits is loss of fat. If you reduce your intake of fat, eat adequate nutritious calories, and exercise aerobically every day to build and preserve your muscle, you will lose fat. Don't weigh yourself every half-hour. In fact, THROW AWAY YOUR SCALE. Give it to your worst enemy. *Choose to Lose* is not about weight loss. It is about fat loss. (Rest assured, if you lose fat, you will lose weight.)

The scale does not distinguish water loss*, muscle loss, and fat loss. A more accurate and positive way to measure your progress is to take note of the way you fit into your clothes. Fitting into a smaller pair of pants or dress indicates that you are losing fat and building lean muscle mass, which is just as important as losing pounds. In fact, it is more important because the greater your muscle mass, the greater your metabolic capacity to burn fat and the more weight you will lose in the future. Changes in size are more likely to occur if you exercise aerobically every day.

In the first few weeks of following *Choose to Lose* you will be replacing the fat you are losing with the muscle you are building from exercising. This muscle has weight. Actually, muscle is heavier than fat. In the long run, added muscle will help you burn fat. In the short run, because the muscle you are building may weigh as much as the fat you are losing, you may not see the loss registered on the scale. This is another reason not to measure your progress by stepping on the scale fifteen times a day.

*It takes 4 grams of water to store 1 gram of carbohydrate. When you follow a low-carbohydrate diet, you reduce your carbohydrate stores and thus the weight you lose is mostly water. Two large glasses of water and you've gained the weight back.

CHOOSE TO LOSE WORKS AND KEEPS ON WORKING — FOREVER

Choose to Lose is an eating plan for life. It is not a quick fix. Many people think of a diet as a dreadful, but short-term, endeavor. They decide to lose weight. They jump on Diet Island, where they starve, are deprived, eat powders, drink liquids, or balance exchanges. Everything they do is totally unnatural. And then, when they have lost weight (or given up), they jump off the island and return to their old eating habits. They gain back all the weight they lost because they made no permanent lifestyle changes. Sound familiar?

Choose to Lose is not an island. It is not temporary (or dreadful, either) because you are full and satisfied and can follow it forever.

But if your goal is to lose 40 pounds in the next three weeks so that you can fit into your bathing suit or the fancy silk dress that you haven't been able to zip for 20 years, then you may be disappointed. Following *Choose to Lose,* you will lose that 40 pounds, but it will take longer than three weeks. And the results will last. You will learn to love a low-fat diet so that you can maintain your weight loss for your lifetime, as well as improve your long-term health prospects.

Tools and Skills to Take Control

You now know the science behind *Choose to Lose.* Next you will learn the method that will help you apply *Choose to Lose* to your life. For instance you know that *fat* is the culprit that is making you fat, but do you know the sources of fat in your diet? You know that cheesecake and ice cream sundaes are full of fat, but how about whole milk and granola bars? Are steaks high-fat or high-protein? You want to reduce the fat in your diet but by how much? *Choose to Lose* gives you insight and the tools to take control.

Following *Choose to Lose* —

You discover where fat lurks in your diet and what changes you need to make to lose weight.

You determine a FAT BUDGET so that you know how much fat you can "afford" to eat and still reach and maintain your desirable and healthy weight.

You keep track of the fat you eat so that you know you are making enough changes to reach your goal.

You eat according to the base of the Food Guide Pyramid to ensure that you are eating a healthy diet.

Choose to Lose puts you in control.

A New Outlook on Food

Choose to Lose is not restrictive. It is an eating plan you can follow for a lifetime because it is palatable, practical, and flexible. You can choose to eat anything you want as long as it fits into your Fat Budget.

You're the boss. No one can tell you what you can or cannot eat. But knowing that you can eat high-fat favorites gives you the freedom not to eat them. You don't have to be frantic. Do you regard every high-fat goodie as a last-chance opportunity? When you see a piece of cheesecake, do you view it as the last cheesecake on earth? You know you shouldn't eat it, but you can't pass up the opportunity. You might even eat it because you want to get it out of the way so you won't eat it. Is this logical behavior? This is a gut reaction.

However, knowing that the cheesecake has 160 fat calories and that you can fit it into your Fat Budget (albeit you will have to eat low, low-fat for a few days to compensate), frees you. You can relax about it. Maybe you'll have it another day.

You won't believe it now, but having a Fat Budget and knowing the "cost" of foods will change you. You will go to a wedding and instead of wildly eating everything in sight because you'll never have the chance again, you will say, "Yuk, that greasy chicken wing isn't worth 120 fat calories" or "Yum, that slice of Black Forest cake is worth a major (but temporary) dent in my Fat Budget." Knowing you can eat the way others do relieves the pressure to gorge yourself whenever enticing food appears.

The goal of *Choose to Lose* is for you to move from a high-fat taste to a tasty, low-fat taste. Today you may say, "Not me. I'll never lose my fat tooth," but in a short time you may be repelled by food you once thought you couldn't live without.

Tastes Change

One Chooser to Lose told us he had given up fast food altogether, but for his birthday, he decided to treat himself to a Quarter Pounder. He bought the burger, but couldn't stomach eating it. He had made the lifestyle switch. This is not unusual. Scientific studies have shown it takes about 12 weeks to change from a high-fat to a low-fat taste. It may (and probably will) happen to you.

A Healthier You

Overweight people have all sorts of health problems, which they might have avoided had they followed *Choose to Lose*. By eating a diet low in fat and high in carbohydrates and fiber you can reduce your risk of diabetes, heart disease, stroke, kidney disease, cancer of the breast, prostate, and colon, bone and joint disorders, female sterility, pregnancy problems, and premature death. You will also improve your regularity.

People following *Choose to Lose* have reported major reductions in blood cholesterol (some as much as 100 points), significant reductions in blood pressure, and normalization of blood sugar. Some have told us they no longer suffer from swollen ankles. These short-term changes are just the tip of the iceberg. They predict vastly improved health in the future.

Focus on Health

Studies have shown that the people who are most successful at losing weight and keeping it off are those whose prime motivation is their health rather than their appearance. They know that by reducing their intake of fat and eating nutritious foods they are reducing their risks of many chronic diseases. Weight loss will come as a side effect. Think of *Choose to Lose* first as your ticket to good health.

Quality of Life

Changing the way you eat will give you a new lease on life. You will be introduced to a whole new variety of delicious, tasty low-fat foods. Freed from the bonds of limited calories, you will be able to enjoy a baked potato, a bowl of cereal, half a cantaloupe whenever your heart desires. Encouraged by the many benefits of home-cooked meals, you may find you love to cook and are a talented chef.

More Energy and Stamina. Low-fat/high-carbohydrate eating will keep your glycogen stores filled up. As a result you will have more energy and more stamina. No more dragging yourself around from hour to hour after no breakfast and lunch, and a high-fat dinner. With your new infusion of energy, you may have to make a new set of friends twenty years younger who are able to keep up with the new YOU.

If you are obese now, you will find that such simple activities as walking are a lot easier after losing weight. No more huffing and puffing. No more embarrassment at stuffing yourself into chairs or navigating through narrow passages. Sex will become a delight, not a defeat. The bottom line: you'll feel good about yourself.

Enough Talking — Let's Get Started

Read the next chapter to see how to make *Choose to Lose* work for you.

Remember:

1. Not all calories contribute equally to weight gain or loss.
2. Carbohydrates are not fattening because
   a. you have a limited capacity to store carbohydrates, so almost all the carbohydrates you eat are burned within a few hours.
   b. under normal conditions, carbohydrate is never converted to fat and thus never leads to weight gain.
3. Overconsumption of empty carbohydrate calories will not turn to fat but it will stall weight loss because the body will burn those calories in preference to fat in the fat stores.

4. Fat calories are fattening because
   a. all the fat you eat is immediately stored.
   b. the capacity to store fats is essentially unlimited.
   c. fat is the stored energy of last resort.

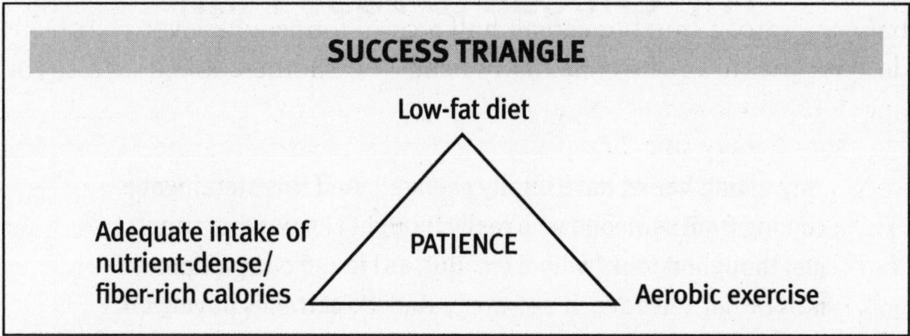

5. *Choose to Lose* is a simple, practical, and flexible plan based on three strategies for maximizing weight loss (Success Triangle):
   a. Eat low-fat because the less fat you add to the fat stores, the less you will have to remove.
   b. Eat lots of vegetables, fruits, low-fat whole grains, and low- and non-fat dairy to:
      1. maximize your metabolic rate
      2. ensure that you consume enough vitamins, minerals, and fiber for long-term good health
      3. keep you full, satisfied, and content
      4. increase your energy and stamina.
   c. Engage in regular aerobic exercise to:
      1. build and maintain muscle
      2. force the body to burn fat rather than protein from muscle
      3. maximize your BMR.
6. *Choose to Lose* puts you in control and teaches you to make food choices that will keep you thin for life.
7. *Choose to Lose* is a healthy way of eating and will reduce your risks of many chronic diseases.

2

The *Choose to Lose* Plan

> **"My eating habits have totally changed. And this statement is coming from someone who really thought I knew what to eat; I just thought it took 'willpower.' But, as I found out, 'willpower' is not what's needed, it's strategy. And it's certainly paying off. I can honestly say this program has changed my eating habits for life."**
>
> **— Pat Frohe, Tallahassee, Florida**

THE *CHOOSE TO LOSE* PLAN is simple. You need to know:

1. Your personal **Fat Budget** — the maximum number of fat calories you can eat each day and still reach and maintain your desirable and healthy weight. This is yours alone. It is based on the total number of calories needed to satisfy *your* basal metabolic rate (BMR) at *your* goal weight.
2. The number of **fat calories** in foods.
3. Approximately how many total calories you should be eating to fuel your BMR + physical activity.
4. Food Guide recommendations so you can meet them. (See Chapter 13, "Ensuring a Balanced Diet.")

YOUR FAT BUDGET: THE KEY TO SUCCESS

Your Fat Budget is a powerful tool. It gives you a framework for making choices. It is like your salary. If you went shopping and didn't know your salary, how could you make choices? How could you decide whether you could afford a $250 television set or a $40 pair of shoes? The same is true of food. One of the reasons Americans are overweight is that they have no idea how much fat is in the food they eat and they have no way of judging if they can afford it. Your Fat Budget puts all foods into perspective so that you can make choices. If you know that a 3-ounce bag of Fritos contains 270 fat calories and your Fat Budget is 315 for the whole day, you might decide you don't want to blow your day's fat intake on a tiny bag of chips. On the other hand, a 1-inch wedge of New York cheesecake (270 fat calories) on your birthday might be worth every luscious fat calorie.

Knowing your Fat Budget and the calories of fat in different foods, **you can choose to eat any combination of foods** — even your high-fat favorites — **as long as you stay within your Fat Budget.**

DETERMINING YOUR VERY OWN FAT BUDGET

Choose to Lose is based on a Fat Budget that is tailored to you. It is based on your sex, height, frame size, and goal weight. You probably know what your goal weight is without looking at the following tables. Skip to Table 2, page 26, if you know what you should weigh. Otherwise, if you want to determine your desirable weight or just want to see if the weight table agrees with your perception of perfection, consult Table 1, page 24.

Step 1: Determine Your Desirable Weight

Table 1 lists desirable weights according to sex, height, and frame size — small, medium, or large. (To determine your frame size, see box.)

How to Determine Your Frame Size

Place your left thumb and middle finger around your right wrist and squeeze your fingers together. If the thumb and finger overlap, you have a small frame. If they just touch, your frame size is medium. If they do not touch, you have a large frame. This method is crude, but adequate for our purposes. (If you know you have a smaller or larger frame than this method shows, use your actual frame size and ignore the results of the wrist test.)

Table 1. Desirable Weights* for Adults Age 25 and Over

HEIGHT WITHOUT SHOES		FRAME		
FEET	INCHES	SMALL	MEDIUM	LARGE
Women				
4	8	92–98	96–107	104–119
4	9	94–101	98–110	106–122
4	10	96–104	101–113	109–125
4	11	99–107	104–116	112–128
5	0	102–110	107–119	115–131
5	1	105–113	110–122	118–134
5	2	108–116	113–126	121–138
5	3	111–119	116–130	125–142
5	4	114–123	120–135	129–146
5	5	118–127	124–139	133–150
5	6	122–131	128–143	137–154
5	7	126–135	132–147	141–158
5	8	130–140	136–151	145–163
5	9	134–144	140–155	149–168
5	10	138–148	144–159	153–173
5	11	142–152	148–163	157–177
6	0	146–156	152–167	161–181
Men				
5	1	112–120	118–129	126–141
5	2	115–123	121–133	129–144
5	3	118–126	124–136	132–148
5	4	121–129	127–139	135–152
5	5	124–133	130–143	138–156
5	6	128–137	134–147	142–161
5	7	132–141	138–152	147–166
5	8	136–145	142–156	151–170
5	9	140–150	146–160	155–174
5	10	144–154	150–165	159–179
5	11	148–158	154–170	164–184
6	0	152–162	158–175	168–189
6	1	156–167	162–180	173–194
6	2	160–171	167–185	178–199
6	3	164–175	172–190	183–204
6	4	168–179	177–195	188–209

Courtesy of Metropolitan Life Insurance Company, New York, N.Y., 1959. For persons between 18 and 25 years of age, subtract 1 pound for each year under 25.
*Weight in pounds, without clothing.

Find your height along the left-hand column of Table 1 under the heading **Women** or **Men.** Look across to the weight range listed under your frame size and choose the weight within the range that is right for you. Fill in Step 1 of the worksheet on page 33. Remember. This is your goal weight, not your current weight.

Step 2: Determine Your Minimum Daily Total Caloric Intake

Table 2 (pages 26–27) lists minimum daily total caloric intakes for various goal weights. This is the number of calories needed to satisfy your basal metabolism at your goal weight. In other words, this is the minimum amount of calories (energy) you need each day to maintain your vital functions when you are *completely* at rest. This is a MINIMUM. We repeat. THIS IS A MINIMUM! THIS IS A FLOOR. Your body needs more than this MINIMUM total caloric intake to operate. Remember: BMR + physical activity. You walk, you climb steps, you may even run, bike, or swim. You need more calories than this minimum amount in order to power your physical activity. How many more depends on how much physical activity you do each day. For most people 300 to 500 total calories in addition to the minimum total caloric intake will be sufficient. But if you eat a few hundred total calories beyond even that amount as nutrient-dense/fiber-rich foods, don't worry. They will be burned off and not be stored as fat.

So if you are a woman who wants to weigh 140 pounds, you need to eat a *minimum* of 1638 calories a day. You should actually be eating about 2000 or 2100 total calories a day. That's right. You won't gain weight if you eat more than 2000 calories a day if the calories come from low-fat, nutrient-dense/fiber-rich foods. In fact, if your minimum total caloric intake is 1638 and you eat less than that, you will probably not lose weight. (See Chapter 1, "Fat's the One," to understand the science.)

You notice we said nutrient-dense/fiber-rich foods. That doesn't mean that if you have eaten only 1400 total calories by the end of the day and your minimum is 1638 that you should chug-a-lug three 12-ounce sodas (540 total calories) to exceed your minimum. Never! Those empty calories will do nothing for your health or your weight

Table 2. Minimum Total Caloric Intake and Fat Budget Based on Goal Weight and Sex

GOAL WEIGHT	MINIMUM DAILY TOTAL CALORIC INTAKE	FAT BUDGET
Women		
90	1053	211
95	1111	222
100	1170	234
105	1228	246
110	1287	257
115	1345	269
120	1404	281
125	1462	292
130	1521	304
135	1579	315
140	1638	328
145	1696	339
150	1755	351
155	1813	363
160	1872	374
165	1930	386
170	1989	398
175	2047	409
180	2106	421
185	2164	433
190	2223	445
195	2281	456
200	2340	468

Table 2. Minimum Total Caloric Intake and Fat Budget Based on Goal Weight and Sex (continued)

GOAL WEIGHT	MINIMUM DAILY TOTAL CALORIC INTAKE	FAT BUDGET
Men		
110	1430	286
115	1495	299
120	1560	312
125	1625	325
130	1690	338
135	1755	351
140	1820	364
145	1885	377
150	1950	390
155	2015	403
160	2080	416
165	2145	429
170	2210	442
175	2275	455
180	2340	468
185	2405	481
190	2470	494
195	2535	507
200	2600	520
205	2665	533
210	2730	546
215	2795	559
220	2860	572
225	2925	585
230	2990	598
235	3055	611
240	3120	624
245	3185	637
250	3250	650

loss. It means you could eat some fruit and a bowl of cereal with skim milk and make sure tomorrow you plan enough nutritious foods into your day to meet and exceed your minimum.

You may wonder why we ask you to determine total calories when *Choose to Lose* is based on limiting fat, not total calories. This is why: It is not enough just to limit fat. You need to eat a lot of nutritious calories. Knowing your minimum total caloric intake is important so that you don't eat *too few* total calories.

To determine your minimum total daily caloric intake, first find your goal weight along the left-hand column of Table 2 under the heading **Women** or **Men.** Look across to the column labeled **Minimum Daily Total Caloric Intake** to find your number. Fill in Step 2 of your worksheet.

Why You Must Eat MORE Than Your Minimum Total Caloric Intake (Assuming the Calories Come from Low-Fat Nutritious Foods)

1. To ensure that you get enough energy (calories) to fuel your basal metabolic rate + physical activity.
2. To keep up your basal metabolic rate. If your total caloric intake falls below the minimum amount you determined, your basal metabolic rate will decrease and you will lose fat more slowly.
3. To ensure that you consume enough vitamins, minerals, and fiber for long-term good health.
4. To ensure that you won't be hungry. Hunger leads to bingeing and eating foods that work against your healthy-eating/fat-loss goals.

Step 3: Determining Your Fat Budget

To determine your Fat Budget, find your goal weight again and look across to the column labeled **Fat Budget.** Fill in step 3 of your worksheet. Your Fat Budget is 20 percent of your minimum daily total caloric intake, or BMR. This may seem like a low percentage considering Americans eat on average 34 to 45 percent of their calories as fat, but you will not find it a hardship. You will be eating so much delectable

Fat Calories, Not Fat Grams

Choose to Lose uses fat calories rather than grams for three reasons.

- First, grams are a mystery to most people, while calories are familiar. You have lived with calories all your life. You probably even know approximately how many total calories you should be eating. If you see that a chocolate bar contains 99 calories of fat, familiarity helps you judge how fattening it is. What does the same amount of fat in grams (11 grams) mean to you?
- Second, since your Fat Budget is a percentage of your minimum daily total caloric intake, it makes more sense to express your Fat Budget in the same units as your total intake — that is, calories. To say that out of your daily intake of 2000 total calories no more than 315 calories should come from fat is easy to visualize. Compare this to "out of your daily intake of 2000 total calories you should eat no more than 35 grams as fat." Huh? Calories are units of energy. Grams are units of weight.

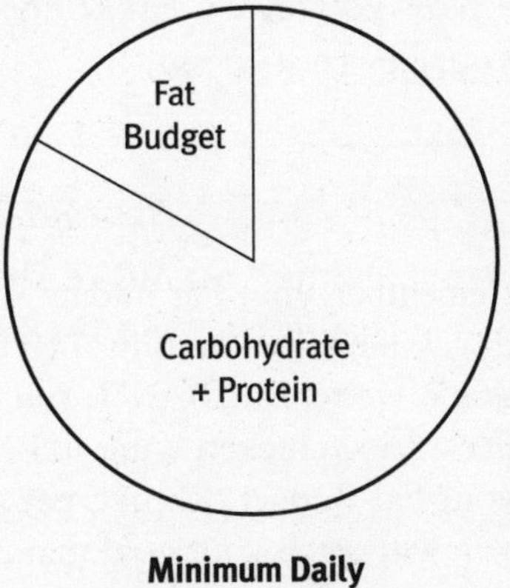

Minimum Daily Total Caloric Intake

- Third, expressing fat content as grams has less impact than expressing it as calories since each gram of fat equals 9 calories. "Only 5 grams of fat per cookie," you say as you cram five cookies in your mouth. But would you eat them with abandon if you knew each cookie contained 45 fat calories?

low-fat food and you'll feel so much better without all that gloppy grease churning around your insides, you will probably never even get close to reaching your Fat Budget each day.

If you weigh quite a bit more than your goal weight, you may feel that the Fat Budget for your goal weight is too restrictive. For example, if you weigh 250 pounds and want to weigh 140, the Fat Budget for 140 pounds is 328 fat calories. If a limit of 328 fat calories a day seems impossible to achieve, pick an attainable weight somewhere between 250 pounds and 140 pounds — perhaps 180 pounds. Determine a Fat Budget for the intermediate weight (421 fat calories) and adhere to it until you weigh 180 pounds. When you attain that weight, adopt the Fat Budget for your next goal, 140 pounds. (This will be your Fat Budget for the rest of your life.) Your weight loss may be somewhat slower this way, but if you feel more comfortable working in stages, you are more likely to follow *Choose to Lose* until you reach your final goal. After all, a diet plan works only if it fits your individual needs and you stick with it.

Now you have a Fat Budget. Engrave it in your mind. This number is your key to success.

Your Fat Budget Is a Ceiling

Remember, your Fat Budget is a ceiling, *not* a goal. If at the end of the day you have eaten only 150 fat calories and your Fat Budget is 315, don't, we repeat DON'T, run out to Kentucky Fried Chicken and order a fried chicken wing to boost up your fat intake so that it equals your Fat Budget. Rather, pat yourself on the back and smile because you will reach your goal that much sooner.

A CASE STUDY

Meet Ellen. Ellen is trying to lose weight. She also wants to learn to eat healthier. Her example should clarify each step of the *Choose to Lose* system for you.

Here's how Ellen figures out her Fat Budget. (See Ellen's worksheet, page 32.) Ellen is 5 feet 5 inches tall and weighs 185 pounds. She determines by the wrist test that she has a medium frame. To find her desirable weight, she locates her height (5′5″) in the left-hand column of Table 1 under the heading **Women.** Reading across to the **Frame** column labeled **Medium,** she finds that her weight range is 124–139. Ellen knows that she will look and feel great at 135 pounds.

Table 1

HEIGHT WITHOUT SHOES		FRAME		
FEET	INCHES	SMALL	MEDIUM	LARGE
Women				
5	4	114–123	120–135	129–146
5	5	118–127	**124–139**	133–150
5	6	122–131	128–143	137–154

Next, Ellen needs to determine her minimum daily caloric intake. She locates her goal weight (135) on Table 2 under the heading **Women.** Looking across to the column labeled **Minimum Daily Total Caloric Intake,** she finds her number — 1579. Ellen remembers that this number is a floor. She must eat more than 1579 nutrient-dense/fiber-rich total calories if she wants to lose weight.

Looking across to the column labeled **Fat Budget,** Ellen finds her daily Fat Budget — 315 fat calories.

Table 2

	GOAL WEIGHT	MINIMUM DAILY TOTAL CALORIC INTAKE	FAT BUDGET
Women			
	130	1521	304
	135	**1579**	**315**
	140	1638	328

See page 33 to fill out your own worksheet.

Worksheet to Determine Your Daily Fat Budget

Name ELLEN Date 3/27/99

STEP 1: DETERMINE YOUR GOAL WEIGHT

A. Sex: ☐ male ☒ female

B. Height: 5 feet 5 inches

C. Frame (wrist method): ☐ small ☒ medium ☐ large

D. Weight range (table 1): 124–139

E. Goal weight: 135

STEP 2: DETERMINE YOUR MINIMUM DAILY TOTAL CALORIC INTAKE

Minimum daily total caloric intake (use table 2): 1579

STEP 3: DETERMINE YOUR FAT BUDGET (use table 2)

20% of 1579 (your minimum daily caloric intake) = 315 (your Fat Budget)

Fat Budget = 315 Fat Calories

Worksheet to Determine Your Daily Fat Budget

Name______________________________ Date _________________

STEP 1: DETERMINE YOUR GOAL WEIGHT

A. Sex: ❑ male ❑ female

B. Height: _______ feet ______ inches

C. Frame (wrist method): ❑ small ❑ medium ❑ large

D. Weight range (table 1):______________

E. Goal weight:______________

STEP 2: DETERMINE YOUR MINIMUM DAILY TOTAL CALORIC INTAKE

Minimum daily total caloric intake (use table 2):_______________

STEP 3: DETERMINE YOUR FAT BUDGET (use table 2)

20% of ______________________________ = _______________
(your minimum daily caloric intake) (your Fat Budget)

Fat Budget = _______ Fat Calories

WHAT DOES YOUR FAT BUDGET MEAN IN TERMS OF FOOD?

The second part of the *Choose to Lose* plan is knowing the fat calories in foods. What does Ellen's budget of 315 fat calories mean in terms of foods? What about your Fat Budget? How does this number translate into food choices? Here are some foods and the fat calories they contain. As you read, keep your own Fat Budget in mind.

FOOD	FAT CALORIES
Egg (yolk)	50
Margarine (1 tablespoon)	100
Potato chips (18 chips)	90
Peanut butter (1 tablespoon)	75
Baked potato with skin	0
Wendy's potato with cheese	210
Italian dressing (4 tablespoons)	280
Cheddar cheese (1 oz)	85
Whole milk (1 cup)	75
2% reduced fat milk (3 cups)	135
Skim milk (1 cup)	0
Broiled lamb loin chop, untrimmed	144
Chicken breast baked without skin	13
Chicken breast, batter-dipped and fried, with skin	166
Stouffer's Chicken Breast in Barbecue Sauce	210
Plain white rice (1 cup), no fat added	0
Weight Watchers Santa Fe Style Rice & Beans	70
Budget Gourmet Swedish Meatballs	310
Swanson's Kids Fun Feast Chompin' Chicken Drumlets	230
Lunchables Fun Pack (bologna + American cheese + cookie)	300
Hardee's Big Country Breakfast, Sausage	594
Kit Kat candy bar	100
Ben & Jerry's butter pecan ice cream (1 cup)	440

Are any of these foods familiar? Can you understand now why you haven't been mistaken for a stick recently?

Using Your Fat Budget to Make Choices

Your Fat Budget will make you food-smart. You won't eat just because a food is there. Before you pop a handful of peanuts (about 35 kernels) into your mouth (knowing that 35 peanut kernels cost about 125 fat calories and your Fat Budget is 315 fat calories for the entire day), you can decide if that fleeting pleasure is worth more than one-third of your budget.

Having a Fat Budget and knowing the fat calories in foods helps you make better choices. Instead of gulping down a cup of cream of mushroom soup at 155 fat calories, you might choose to start dinner with six large boiled shrimp (6 fat calories) dipped in cocktail sauce (0 fat calories). You might decide that Post Wheat Chex at 10 fat calories a cup is a better choice than Kellogg's Cracklin' Oat Bran at 50 fat calories per 3/4 cup.

Although your goal is to move from a high-fat/low-complex carbohydrate diet to a low-fat/high-complex carbohydrate diet, you may still want to include some sinful favorites. This eating plan is for life, after all. If you really want to eat a slice of Aunt Tilly's chocolate mousse pie at a cost of 250 fat calories, you can. You will just have to balance the fat calories with low, low-fat choices for several days. With *Choose to Lose* you never have to feel guilty or deprived.

You will soon find that you prefer many foods that are low in fat. Knowing the "cost" of commercial foods, you will appreciate how easily and deliciously homemade low-fat recipes fit into your Fat Budget. For example, imagine a dinner starting with *Asparagus Soup** (0 fat calories), followed by *Cajun Chicken* (28 fat calories) on a bed of rice (0 fat calories), steamed broccoli (0 fat calories) with a *Dijon-Yogurt Sauce*, baked sweet potato (0 fat calories), and for dessert a dish of non-fat frozen yogurt (0 fat calories) covered with strawberries (0 fat calories). The total cost: 28 fat calories. Compare this to a meager dinner of Stouffer's Macaroni and Cheese (1 cup) at a cost of 150 fat calories.

---

*Italicized recipes come from either *Choose to Lose* or *Eater's Choice Low-Fat Cookbook* by Dr. Ron and Nancy Goor (Houghton Mifflin, 1999).

Percentage of Fat in Foods: A Misinterpretation of the Guidelines

You have probably read healthy-eating articles urging you to choose only foods with less than 30 percent of their calories coming from fat or seen foods listed according to the percentage of fat they contain so that you can choose those with low percentages. This is a misunderstanding of the health guidelines that recommend that you eat less than 30 percent* of your total daily caloric intake as fat and could lead to eating massive amounts of fat.

Let us explain. Percentage of fat in individual foods is irrelevant. For example, a malted milk has 1060 total calories. Of those total calories, 225, or 21 percent, are fat calories. While 21 percent is a relatively low percentage of fat, how do 225 fat calories fit into your Fat Budget? Drink two malteds and you consume a whopping 450 fat calories. Yet each drink is still only 21 percent fat. If a healthy diet could include any food as long as it had less than 30 percent of its calories from fat or even less than 21 percent of its calories from fat, you could drink ten of these malteds and accumulate 4500 fat calories. On the other hand, you wouldn't be allowed to eat a food if 8 of its 10 total calories came from fat because it would be 80 percent fat. Eight fat calories would hardly make a flea fat.

If you eat only foods with less than 30 percent fat, not only could you accumulate enough fat calories to obliterate your Fat Budget, your diet would be unnecessarily restricted. You could never cook with even a teaspoon of oil because all oils are 100 percent fat. Cheese, beef, and nuts would be totally verboten because they contain more than 30 percent of their calories from fat. Your food choices would be limited as would be your healthy-eating attention span.

The percentage you used to determine your Fat Budget is the only percentage that should ever concern you (you can tuck it away in the back of your brain now that you've determined your Fat Budget). The only information you need to know to choose a food is the number of fat calories it contains and how it fits into your Fat Budget.

*All the major health agencies recommend eating less than 30 percent of your total caloric intake as fat. We think 30 percent is too high. This percentage was chosen despite the fact that many health educators agree that 20–25 percent is a healthier range because they felt the public might balk at fat intakes this low.

DETERMINING THE FAT CALORIES IN FOODS

Now that you have your own Fat Budget, you are probably eager to find out how much fat you have been eating. How much does that handful of innocent sunflower seeds cost you? Is the Arby's chicken breast sandwich you had for lunch a low-fat choice? To help you learn the fat calories in foods so that you can make better choices, the *Choose to Lose* Food Tables list fat calories and total calories for hundreds of foods. You will want to become familiar with these tables. Not only can you use them to look up the fat calories of foods you ate or plan to eat, you can browse through them to find low-fat alternatives for high-fat foods.

THE FOOD TABLES

The Food Tables on pages 305 to 566 are an invaluable source of information — and surprises. They give you insight into why you did not wear a bikini last summer. Would you have thought that 15 potato chips have 95 fat calories? Four ounces of lean flank steak has 140 fat calories? Nine little cashews have 60 fat calories? Keep your Fat Budget in mind when you browse through the tables.

Organization

The Food Tables are organized alphabetically into major food groups, such as **BEVERAGES, DAIRY AND EGGS, FAST FOODS, FATS AND OILS**, and so on. You will find a Table of Contents on page 307 to help you locate these major food groupings. You might want to put your own identifying tabs at the beginning of each section so that you can turn to them quickly.

The food categories within each major group are also organized

Quick Find: Food Tables Index

If you have difficulty locating specific foods in the Food Tables, check the Food Tables Index (pages 567 to 579) following the Food Tables. In this index, foods are listed alphabetically, rather than by group.

alphabetically. For example, the first subheading under **DAIRY AND EGGS** is ***Butter.*** The next is ***Cheese.***

Within each food category, foods are listed alphabetically according to type. The first entry under ***Cheese*** is American; then Blue, then Camembert, and so on.

How Much Did You Actually Eat?

The amount that you ate and the portion size listed in the Food Tables may differ. You must adjust the numbers of total calories and fat calories listed in the tables to the amount you actually ate. For example, if one-half cup of Ben & Jerry's Chubby Hubby ice cream contains 210 calories of fat and you ate 1 cup (or 2 times as much as the listed amount), you consumed 210 × 2, or 420 fat calories. You would also double the total calories (350 total calories × 2 = 700 total calories). On the other hand, if you ate a quarter-cup of the ice cream or half a serving, you consumed 210 ÷ 2, or 105 fat calories, and 350 ÷ 2 = 175 total calories.

Comparison Value

Data for many foods such as cheeses and meats is given in 1-ounce amounts to help you compare fat contents so that you can make informed choices. However, keep in mind that although 1-ounce portions are listed for a food, 1 ounce may be much less than you eat and you must adjust your figures accordingly.

No One Eats a 1-Ounce Piece of Steak. For instance, at first glance the meats (beef, veal, lamb, pork, poultry, fish) all look like low-fat foods because the fat values are given for 1-ounce portions. But because you rarely eat 1 ounce of meat, you must multiply the calories and fat calories for a 1-ounce portion by the number of ounces you actually ate. If you ate 4 ounces of lean ground beef, you would multiply the fat content of lean ground beef (47 fat calories) by 4 to determine the number of fat calories you actually ate: 4 × 47 = 188 fat calories.

Find Your Favorites in the Food Tables

If you are dying to know the cost of that pecan Danish you bought at the supermarket to sustain you while you shopped or the 8-ounce bag of cheese twists you emptied while watching your favorite football team crush their perennial rival, turn to the Food Tables and take a look. You might also want to make a list of five to ten of your favorite foods. Next, using the Food Tables, determine the fat calories for the amount you usually eat. Then find low-fat alternatives by scanning through the same section of the Food Tables. For example, if your Sunday morning staple — 4 bacon slices — is beginning to taste greasy, look up bacon in the Food Tables under **SAUSAGES AND LUNCHEON MEATS** to find out why. Each thin (8 gram) strip contains 35 fat calories; 4 strips contain 140 fat calories. Look on a little further and find Canadian bacon in the same section. For a similar taste, more food, and less fat, 2 slices of Canadian bacon will cost you 18 fat calories.

READING LABELS

Food labels are another source of information on fat calories. They are the best source for commercial foods because they contain the most up-to-date information. In Chapter 6, "Putting *Choose to Lose* to Work for You," you will learn how to read food labels.

Coming Next . . .

Before you try to discover where fat lurks in every aspect of your own diet, read on to see how little food Ellen eats, what it costs her, and what insights she gains from keeping track. Don't be surprised if you get the eerie feeling you are reading about yourself.

Remember:

The *Choose to Lose* plan is simple. You need to know:

A. Your personal **Fat Budget,** which is the maximum number of fat calories you can eat each day to reach and maintain your desirable and healthy weight. Your Fat Budget provides a framework for making food choices.
   1. Use the step-by-step guide to *Choose to Lose* to:
      a. determine your desirable weight (Table 1).

b. determine your minimum daily total caloric intake (Table 2).
c. determine your Fat Budget (Table 2).
2. Your Fat Budget is the key to success because:
a. You can judge the "cost" of any food so that you can make wise food choices.
b. You can eat any combination of foods as long as they fit within your budget.
c. You can budget in high-fat splurges.

B. The fat calories in foods.
1. Use the Food Tables to determine fat calories.
2. Adjust the calories and fat calories for the actual amount you consume.
3. Don't forget that meat, cheese, and some other foods are listed in 1-ounce portions.

C. Your minimum total calorie intake so that you will be eating enough nutrient-dense/fiber-rich calories to meet the recommendations of the Food Guide Pyramid.

3
Self-Discovery

"I never realized how much fat I take in during one day, and now that I have been cutting back on the fat calories I have lost 12 lbs in 4 weeks and yet I'm not starving myself or depriving myself of all of those 'forbidden foods' on every other diet."

— Janice Douglass, Newark, Delaware

MOST ALL OF US THINK we eat *no* fat and thus are mystified that we are overweight. Ellen would also insist that she eats no more fat than her thin friends and can't understand why she is fat and they are not. Let's look at her intake for a day for the real story.

ELLEN'S FOOD RECORD: AN ILLUMINATING EXAMPLE

Ellen is going to keep track of everything she eats without making changes. She wants to assess the damage before she begins making repairs but first needs to have a picture of how much fat she currently eats and which foods are the high-fat culprits. Ellen is going to write down the times she ate as well as the exact amount of every morsel she lets slip between her lips. She will use the Food Tables on pages 305 to 566 and food labels to determine the fat calories of the foods she eats.

Breakfast: Don't Miss It

Ellen is supposed to be at work by 8:30 A.M., which means getting out of the house by 7:45. Ellen hates getting up in the morning, but she hates missing breakfast even more. She allows herself time to grab a muffin and a cup of coffee in the coffee shop in her building.

She records her muffin and coffee in her food record as she sits down to eat them. For the first time, Ellen questions her choice. It is a very large muffin and tastes a lot like pound cake. No wonder muffins are listed in the **SWEETS** section of the Food Tables. Perhaps she'll skip it this morning. A little voice inside her whispers, "This food record is for you alone. You are not keeping it to impress anyone with how little you eat. You need a baseline record so that you can make changes — and lose weight. *Eat the muffin!*" By being honest with herself, Ellen has already made a step toward being a thin person.

Ellen's first entry in her food record looks like this:

Ellen's Breakfast

TIME	FOOD	AMOUNT	TOTAL CALORIES	FAT CALORIES
8:15 A.M.	Blueberry muffin	1 large	400	**140**
	Coffee	1 cup	0	**0**
	Half-and-half	1 tbsp	20	**15**
			Accumulated Fat Calories:	**155**
			Ellen's Fat Budget:	315

Ellen was wise to eat breakfast. Eating three meals a day (low-fat, high-carbohydrate meals, natch) is *extremely* important for weight loss. You want to start burning calories as soon as you get up. In addition, when you skip breakfast, you are apt to satisfy your mid-morning cravings with a high-fat snack. However, the blueberry muffin was a poor choice for breakfast. It tastes a lot like pound cake because it is made with all the same high-fat ingredients — butter, lard, or vegetable oil, whole milk, sometimes cream. Next time, Ellen will have to decide if the blueberry muffin is worth almost half her Fat Budget.

Ellen uses half-and-half in her coffee. At 15 fat calories for each tablespoon, the fat adds up. How about 2% milk? 1% milk? Skim milk? Why not drink it black?

For better breakfast choices, see Chapter 6, "Putting *Choose to Lose* to Work for You."

Midmorning Snack

At about 10:30, Ellen's friend Lisa often stops by with a treat. By this time, Ellen is already famished. Lisa reaches into a small white bag and, with a flourish, pulls out a glazed doughnut and a cup of coffee. Normally, Ellen eagerly grabs the doughnut, considering it her splurge for the day. But today Ellen regards the doughnut with suspicion. She *knows* it's not going to look good on her food record. "How bad is it?" she wonders as she turns to the **SWEETS** section of the Food Tables. 130 fat calories! She utters a deep sigh as she records the number.

Ellen's Morning Snack

TIME	FOOD	AMOUNT	TOTAL CALORIES	FAT CALORIES
10:30 A.M.	Glazed doughnut	1 large	240	**130**
	Coffee	1 cup	0	**0**
	Half-and-half	1 tbsp	20	**15**
			Total Fat Calories:	**145**
			Accumulated Fat Calories:	**300**
			Ellen's Fat Budget:	315

Instead of the doughnut, she could have eaten a carton of nonfat flavored yogurt (0 fat calories), a bagel (9 fat calories) with jelly (0 fat calories), or an orange (0 fat calories) or all three.

Lunch: Sit Down, Choose Well, and Enjoy

The Fast-Food Option. Ellen looks forward to lunch. She knows she's always "good" at lunchtime. A salad should balance out the doughnut. Ellen has second thoughts about her restaurant choice — a fast-food restaurant. Next time she will have first thoughts. But it won't be easy. Friends or co-workers often choose restaurants that offer no food options for fat-conscious diners. Peer pressure is a poor excuse. Ellen has to stand up for her rights. The companionship of people who don't

Your Mother Was Right

Many adults have established the pattern of eating no breakfast and no lunch or a very light lunch. For some, the mistaken notion is that starvation is the way to lose weight. For others, breakfast and lunch may be the casualties of life in the fast lane. In either case, it is much healthier to eat three well-balanced meals than to starve yourself all day. Stuffing into one meal all the calories and nutrients needed for functioning at peak efficiency makes for very low efficiency peaks.

The chances are that if you eat no lunch, not only will you feel like a limp rag all afternoon, you will be so hungry and tired when you get home you will tear off the kitchen cabinets so that you can get to the snacks. And then, still not satisfied, you will lock your kids in the closet so that you can consume their dinners, too. Eating at least three meals a day is important for weight loss, too, because your body needs the fuel to maintain your basal metabolic rate (see Chapter 1, "Fat's the One").

Treat yourself like the special person you are. Take advantage of lunch time to unwind and enjoy a delicious, nutritious, low-fat meal.

respect your diet wishes is not worth a lethal gash in your Fat Budget. Ellen might even convince her friends that they can enjoy a restaurant with low-fat options.

But today Ellen feels she can handle a fast-food restaurant. Her selections from the salad bar are bound to look good on her record. Ellen tops her salad with two ladles of Italian Caesar dressing and a tablespoon of sunflower seeds. Ellen notices that the ladle handle is marked 1 oz. She takes a biscuit — "It is so little and dry, it must be fat-free" — and a pat of butter — "just a tiny square." She chooses a diet cola.

Ellen determines the fat calories for the tiny salad (319 fat calories) and biscuit (130 fat calories) she ate. She is dismayed. She ate 485 fat calories and is hungrier after eating than she was before she started.

Salad Bar: Diet Trap. Ellen's notion that the salad bar is a good diet choice is wishful thinking. Most of the salad bar options are made with high-fat ingredients — mayonnaise salad with a macaroni or two, oil

Ellen's Lunch

TIME	FOOD	AMOUNT	TOTAL CALORIES	FAT CALORIES
Noon	Mixed salad			
	lettuce	2 cups	20	**0**
	tomato	½	12	**0**
	cucumber	¼	10	**0**
	carrot	¼	8	**0**
	Italian Caesar dressing	4 tbsp*	308	**280**
	Sunflower seeds	1 tbsp	50	**39**
	Biscuit	1	280	**130**
	Butter	1 pat (tsp)	36	**36**
	Diet Coke	8 oz	25	**0**
			Total Fat Calories:	**485**
			Accumulated Fat Calories:	**785**
			Ellen's Fat Budget:	315

*Two ladles = 2 fl oz = 4 tablespoons

salad with a few slices of cold cuts or green beans, sour cream salad with walnuts and apple slices. Eating at a salad bar takes much care. Ellen's choice of salad vegetables — carrots, tomatoes, and cucumbers — could have been diet-wise. Salad vegetables are complex carbohydrates that are high in fiber, vitamins, and minerals and almost completely fat-free. Ellen then added her favorite "vegetable": salad dressing.

Salad Dressing. Like many people who eat salad only because it is a socially acceptable way of consuming salad dressing, Ellen poured a lot

Beware: The small salad dressing ladle at the salad bar holds 2 tablespoons of salad dressing. Two dunks and you have a quarter cup of dressing. Four dunks and you have a half cup of dressing.

— 2 ladles or 4 tablespoons of Italian Caesar dressing (70 fat calories per tablespoon or 280 calories for four) — on her salad. A low-, reduced-calorie, or nonfat dressing would have been a better choice, although creamy type reduced-calorie dressings at fast-food restaurants are still high in fat calories. (At Wendy's the reduced-fat Hidden Valley Ranch dressing has about 25 fat calories per tablespoon; 100 fat calories per 4 tablespoons.) Using two tablespoons of dressing instead of four cuts the fat calories in half, and one tablespoon cuts even more. Better yet, use the fork method.

Better than that, avoid salads altogether.

Salads have acquired a reputation as good diet food. But they are not. In general they contain mostly lettuce, which has little to recommend it nutritionally. To make the lettuce palatable, you pour on high-fat dressing. If you resist the high-fat dressing, you are eating lettuce with lemon juice. Why bother? Eat a baked potato or steamed broccoli instead and have a filling, nutritious treat.

The Fork Method

To reduce the amount of salad dressing you use, have the salad dressing served on the side. Then dip your fork into the dressing and spear a biteful of salad. You will get a taste of dressing but most of the fat will remain in the bowl, not in you.

Sunflower Seeds. Ellen sprinkled a tablespoonful of sunflower seeds over her salad. That casual gesture cost her 39 fat calories. Seeds and nuts are high-fat foods.

Biscuits. Biscuits don't *look* fattening. A biscuit made in your kitchen may have about 50 fat calories, depending on the size and how much butter or shortening you add. A fast-food biscuit ranges from 63 fat calories (Denny's) to 148 (Church's Fried Chicken). Add a pat or two of butter (36 fat calories per pat) and your "innocent" biscuit creates a 99-to-220 calorie dent in your Fat Budget.

Diet Drinks. Ellen chose a diet drink to save calories. This choice is a throwback to the old diet mentality. While neither regular nor diet soda has any redeeming qualities, neither contains fat and thus neither is fattening. In addition, diet drinks contain unhealthy sugar substitutes. Both are poor choices. Choose skim milk, 100 percent fruit juices, or water.

Fast-Food Restaurants: The Ultimate Poor Choice. Although fast-food restaurants may lure you because they are quick, relatively inexpensive, and everywhere, the price of convenience is more than your body can afford. Think of what all that fat will cost in terms of fat calories (and your long-term health) and stay away. Even if you come with good intentions, the temptation to splurge on a Jack-in-the-Box Ultimate Cheeseburger (711 fat calories), Burger King Double Whopper with cheese (570 fat calories), a clam dinner (468 fat calories) at Long John Silver's, an Arby's Roast Chicken Club (279 fat calories), a side dish of creamed spinach (220 fat calories) at Boston Market, or even a medium order of fries (170 fat calories) at Wendy's is hard to resist. As you can see from Ellen's food record, even the salad bar can be a disaster. Choose a sandwich shop, a real restaurant, or for your best bet, bring your lunch from home. For more on eating out, see Chapter 10, "Taking Control: Eating Out."

Water Is a Poor Substitute for Food

It is easy to spot dieters. They are the people walking around their worksites holding a super-size plastic container of water. Every diet course they have taken (and there have been many) requires that they drink at least eight glasses of water a day. Why? The first reason: restricted-calorie diets produce excess nitrogen wastes (toxins), which must be flushed from the kidneys. Drinking water washes out the kidneys and prevents kidney damage. Second reason: restricted-calorie diets make you hungry. Water helps fill you up.

The best reasons to drink water: you are thirsty and you enjoy it. But if you are doing it to lose weight, throw away your Big Sipper and start eating real, nutritious, low-fat food. Food is a lot more filling and satisfying and will really help you lose weight.

Afternoon Snack

Ellen's lunch has left her ravenous. The grumbles from her stomach are so loud, she can hardly pay attention to her work. She has drained her trusty Big Sipper of 48 ounces of water, but she is still hungry. Finally she succumbs and opens a Cup of Noodles she has squirreled away in her desk. "At least it's pasta and good for you." She smiles as she fills the styrofoam container with boiling water. After finishing every last noodle, Ellen checks the label. "130 fat calories! Why didn't I check the label first?" Ellen is stumped. How could a cup of noodles have so much fat? The answer: Ramen noodles are quick-fried in partially hydrogenated cottonseed oil.

Ellen's Afternoon Snack

TIME	FOOD	AMOUNT	TOTAL CALORIES	FAT CALORIES
3 P.M.	Cup of Noodles (chicken)	1 cup	300	**130**
			Total Fat Calories:	**130**
			Accumulated Fat Calories:	**915**
			Ellen's Fat Budget:	315

Always read the label *before* you eat the food. Repenting afterward doesn't take the fat back. Never make assumptions. Unadulterated pasta with a low-fat sauce is a good choice, but look what happened when it was turned into Cup of Noodles.

Dinner: Take Time to Eat Hearty

Ellen works without break until 6 P.M. By the time she leaves work she is ready to eat her handbag. On her way home, she runs into the grocery store to pick up a frozen dinner. She grabs Healthy Choice Country Breaded Chicken. She always chooses Healthy Choice, Weight Watchers, or some other lean or light frozen dinner. Ellen doesn't actually like the taste of frozen dinners, but she generally cooks only on weekends if she cooks at all.

Ellen doesn't even take off her coat. She immediately goes to the kitchen and sticks her meal in the microwave. While her chicken is being zapped, she cuts a head of iceberg lettuce into fourths and sticks one on her plate. She covers it with 4 tablespoons of fat-free dressing, congratulating herself that she didn't repeat her lunchtime dressing fiasco. Ellen does not linger over dinner because there is little to linger over. As Ellen puts her dinner dishes in the sink, she is once again struck by how quickly she is finished with dinner. Is it because she is eating almost no food? Ellen checks the Healthy Choice label. At 80 fat calories, the Country Breaded Chicken is not much of a weight loss buy.

Ellen is proud of her dessert choice. She had chosen Entenmann's fat-free oatmeal cookies. No fat calories at last! She opens the box and begins to munch. She isn't paying much attention because her favorite sit-com is on. During one of the many breaks, she looks down to find she has emptied half the box. She is shocked. "Well, at least I've added no fat," she rationalizes. She looks at the label. Serving size = 2 cookies; total calories per serving, 100; number of servings, 11. In her head she multiplies 100 times 11. 1100! And she had half, so half of 1100 is 550 total calories. "That can't be good." Right.

Ellen is like millions of working people. They labor all day, spend

Ellen's Dinner

TIME	FOOD	AMOUNT	TOTAL CALORIES	FAT CALORIES
7 P.M.	Breaded Country Chicken (Healthy Choice)	1 meal	280	**80**
	Iceberg lettuce	¼ head	20	**0**
	Fat-free dressing (ranch)	4 tbsp	60	**0**
	Oatmeal Raisin Cookies (fat-free)	11	550	**0**
			Total Fat Calories:	**80**
			Accumulated Fat Calories:	**995**
			Ellen's Fat Budget:	315

hours in the car getting to and from work, and when they get home, they don't feel like cooking. They feel pressured for time and want food that takes no effort to prepare and seconds to cook. This mindset has created a multibillion dollar frozen-food industry. In a dream world, one shoves a box into a microwave oven and produces delicious, healthy meals. This is the real world. Meals in a box do not taste great and they are often shot up with fat and sodium as well as artificial colors and flavors. Whether or not they are overladen with fat, they are usually underladen with food and fiber. Read the section on frozen dinners in Chapter 4, "Where's the Fat?" and check out the fat calories of frozen dinners in the Food Tables.

If you want to lose weight or are just interested in good health and sensory pleasure, you need to cook your own food using real ingredients. Ellen is cheating herself by eating plastic food. Cooking does take time, but it doesn't have to take a lot of time. Try the recipes in *Choose to Lose* or *Eater's Choice Low-Fat Cookbook* for scrumptious proof. See Chapter 9, "Cooking Low-Fat and Delicious" for ideas on cooking low-fat.

"Healthy" Frozen Dinners: Watch the Fat. The Healthy Choice Breaded Country Chicken has 80 fat calories. With very little effort, Ellen could make *Grainy Mustard Chicken* (page 274), which has 23 fat calories, or *Shrimp or Scallop Curry* (page 278) at 35 fat calories. Both *Choose to Lose* dishes take little time to prepare and taste superb.

The Missing Vegetable. Ellen's diet is vegetable-bare. Vegetables can be steamed in minutes, add color and texture to your meal, contain fiber, vitamins, and minerals, and taste wonderful. They fill you up, improve your bowel function, and help reduce your risk of breast and colon cancer. What's more, vegetables can be eaten with abandon because they contain little or no fat. Even if you insist on eating boxed meals, enhance them with a baked plain potato or sweet potato and a steamed vegetable or two.

Fat-Free Is Not Free. Eating fat-free cookies will *not* make you lose weight as many people hope. In fact, eating all those empty calories will slow down or stop your weight loss. Although the oatmeal fat-free cookies have no fat, they are high in empty calories. Ellen's body must now burn those 550 calories (because it can't store them) in preference to burning stored fat. In Ellen's case, so much new fat is coming in, there is no chance she will be burning any stored fat. But in the future when she is eating a nutritious, low-fat diet, eating all these empty calories will stall her weight loss.

High-Fat Snacks: Dieter's Downfall

The evening stretches before Ellen. She has made no plans. She settles comfortably into her favorite chair, turns on the television, and eyeballs the peanuts in the bowl beside her. "Not too many," she thinks. "But I'd better keep track of what I eat. I'll measure the amount of peanuts in the bowl before I eat them and then I'll measure the amount left when I go to bed." Good idea, Ellen. The bowl originally held one-half cup of peanuts. Ellen finds nothing in the bowl when she goes to bed. Where could all the nuts have gone? Ellen feels a bit foolish that she has consumed the whole bowl of peanuts. She feels sick when she reads the label on the peanut jar and determines that her mindless snack cost her 320 fat calories. She adds the item to her food record.

Ellen's Evening Snack

TIME	FOOD	AMOUNT	TOTAL CALORIES	FAT CALORIES
9–11 P.M.	Roasted salted peanuts	½ cup	420	**320**
			Accumulated Fat Calories:	**1315**
			Ellen's Fat Budget:	315

Snack foods are the downfall of weight watchers. Snacks are almost always filled with fat, and they are addictive. It is impossible to eat one peanut. One peanut leads to another and another, until finally the whole bowl of peanuts has disappeared. The snacker eats without thought. When the bottom of the bowl appears, she wonders where all the peanuts, potato chips, corn chips, or cheese twists went.

A tip for snackers: put out no more than you plan to eat. Another tip: don't put out high-fat snacks at all.

The Final Tally

Ellen ate 995 fat calories from breakfast through dinner. By the time she finished her evening snack and went to bed she had eaten 1315 calories of fat. Is it any wonder she is fat? Remember, Ellen's Fat Budget is 315. She's *only* 1000 fat calories over her budget. You may think that her fat calorie intake is unusually high, but if you look back at Ellen's food record, you will see nothing out of the ordinary. In fact, you may have noticed that she had no rich desserts. Fat calories have a sneaky habit of adding up.

Before You Know It. Think of your $1000 monthly credit card bill. "What?" you cry in despair. "There's a mistake. I didn't buy anything big." You take out your calculator. A pair of $45 shoes, a gasoline charge of $9, a TV repair bill for $70, and on and on. Nothing over $100. But it all adds up to a whopping $1000. Fat calories work the same way. A 45-fat-calorie glass of 2% milk, a 9-fat-calorie cracker, a 70-fat-calorie slice of bologna. They all add up until you've bankrupted your budget.

A Food Record = An Education

Reviewing her day's food intake has been invaluable for Ellen — and quite a surprise. She no longer wonders why she is overweight or why she is always hungry. She has discovered the sources of fat in her diet. She won't regard biscuits with such a friendly eye. She'll think before dumping 4 tablespoons of high-fat dressing on her salad. She'll read labels before she gobbles down food she assumes is low in fat. Ellen now knows where to cut out fat. She knows which foods she still wants to eat and how much it will cost her — fat-wise.

Ellen is amazed to see how little food she eats in terms of bulk. Fruit, vegetables, and whole grains are noticeably missing from her diet. Looking over her food record, Ellen can already think of many places where she could have enjoyed nutritious food instead of her high-fat choices. (Read page 134 to see how Ellen turned her meal plan around.)

Ellen's Baseline Food Record

TIME	FOOD	AMOUNT	TOTAL CALORIES	FAT CALORIES
8:15 A.M.	Blueberry muffin	1 large	400	**140**
	Coffee	1 cup	0	**0**
	Half-and-half	1 tbsp	20	**15**
10:30 A.M.	Glazed doughnut	1 large	240	**130**
	Coffee	1 cup	0	**0**
	Half-and-half	1 tbsp	20	**15**
Noon	Mixed salad	2 cups	30	**0**
	Caesar dressing	4 tbsps	300	**280**
	Sunflower seeds	1 tbsp	50	**39**
	Biscuit	1	280	**130**
	Butter	1 pat (tsp)	36	**36**
3 P.M.	Cup Noodles	1 cup	300	**130**
7 P.M.	Breaded Country Chicken (Healthy Choice)	1 meal	280	**80**
	Iceberg lettuce	¼ head	20	**0**
	Fat-free dressing (ranch)	4 tbsp	60	**0**
	Oatmeal Raisin Cookies (fat-free)	11	550	**0**
9–11 P.M.	Oil-roasted salted peanuts	½ cup	420	**320**
		Total Calories:	3006	
			Total Fat Calories:	**1315**
			Ellen's Fat Budget:	315

YOUR TURN

Here's Looking at You, Kid

After seeing how Ellen's food intake adds up, are you curious to know how many fat calories you ate yesterday? Last weekend? Do you still think you eat no fat?

The Past Helps Determine the Future

Perhaps you don't even want to think about the past. You want to begin afresh and discard your old eating habits. And you want to get started NOW. Fine. But do you really know your old habits? Before you opened this book, you didn't know your Fat Budget or the number of fat calories in foods. You just ate. You chose certain foods. Some foods you ate without thought. You probably knew that some foods were fattening, but you didn't know how fattening. You might have even avoided some low-fat foods because you thought they would make you gain weight.

To make changes that will last, you need to know what to change. It is helpful to evaluate the foods you are eating to discover the sources of fat in your diet. For invaluable insight, keep a base-line record for three days — two weekdays and one weekend day. Write down *everything* you eat — honestly.

Even if you want to start making changes NOW and are too impatient to keep a baseline record, you need to start keeping track of everything you eat.

> You will find a step-by-step guide to keeping food records in the Appendix, "The Nitty-Gritty of How to Keep a Food Record," page 588. The inexpensive *Choose to Lose* Passbook and refills (see order form on the last page of this book) make keeping track of your food intake easy and convenient.

People Who Keep Track Lose Weight

Food records are the ticket to weight loss success. Keeping food records will give you insight and understanding about your eating habits and will empower you to take control. You may think you eat a low-fat diet (Ellen certainly thought she did), but to lose weight you must know *exactly* what you are eating. You learn which foods are making you fat so that you can make changes. You discover the high-fat foods you can't live without, the ones you can eat less of or less often, and the ones you can eliminate altogether. You know how many fat calories you have eaten so you can save for splurges. You learn if your diet is balanced, if it is filled with enough fruits, vegetables, whole grains, and nonfat dairy for good health. How do you know if you don't write everything down?

Recording what you eat is your key to success. Significant changes can be made only if you know what to change.

You Can't Fool Your Body. Your records must be accurate and honest. Compulsive dieters have been known to be expert at fooling themselves. As the Nobel laureate physicist Richard Feynman so wisely advised when referring to the pitfalls of scientific investigation, "You must not fool yourself, and you are the easiest person to fool."

Honesty Is the Only Policy

To determine why they were overweight, participants in a diet study wrote down everything they ate. Their records showed that their diets were perfect — very low in fat — yet they weren't losing weight. The group was then put in a metabolic ward in a hospital where they were fed only the foods they had recorded in their food records. They all lost weight. Conclusion: You can't fool your body. The only way to know you are eating below your Fat Budget and eating enough total calories is to keep honest and accurate food records.

Keeping Track Is Only a Temporary Drag. After you get used to keeping food records, it won't even take much time. You don't eat that many foods at a meal. Just record your food as you eat. Keep a running subtotal so that you know how each food you eat impacts your Fat Budget. Don't save up all your entries until the end of the day.

You won't have to keep records forever. Soon you will learn the combinations of foods you like to eat that fit within your budget and you won't have to write everything down. Of course, you will continue to refer to the Food Tables or to food labels to see how a new food fits into your Fat Budget.

When You Start Keeping Food Records:

- Record *everything* you eat and drink.
- Eat normally. *Do not change* your diet for the record — it's for you! Do not refrain from eating something just because you have to record it.
- Record food *when* you eat it. Otherwise, later you will forget what you don't want to remember.
- Record the time you eat each food item.
- *Measure* (if necessary, estimate) amounts as accurately as possible.
- *List components* of sandwiches, salads, stews, and so on separately.
- *Ask* questions at restaurants. (See Chapter 10, "Taking Control: Eating Out.")
- Read the Appendix, "The Nitty-Gritty of How to Keep a Food Record," before you begin.

As you begin to record what you eat, you are going to become keenly aware of the fat in food. Read the next chapter to learn "Where's the fat?" You're in for a lot of surprises.

Remember:

1. Eat breakfast and lunch! If you don't, you will be exhausted all day and binge all night. On the positive side, eating lots of delicious, nutritious food will keep you full, satisfied, energetic, and healthy.
2. Avoid fast-food restaurants. They offer too much fat and temptation.
3. Avoid salad bars and salads. They are high-fat outlets.
4. Eat a full, nutritious low-fat dinner. Besides being enjoyable and keeping you from being hungry, nutritious food supplies vitamins, minerals, and fiber you need for good health.
5. Avoid high-fat snacks.
6. Keep a baseline food record to help you discover the sources of fat in your diet.
7. Keep thorough, accurate food records. Recording what you eat is your key to success. Significant changes can be made only if you know what to change.

4

Where's the Fat?

> "Finally a new way of looking at food and understanding the 'hidden fats' that many people don't realize they are ingesting."
>
> — Connie Giliberto, APO

YOU MUST HAVE NOTICED that you are not the only overweight kid on the block. In fact, if you choose your friends wisely, you can feel virtually skinny by comparison. In America, being overweight is an epidemic. More than 54 percent of the population — 97 million people — is overweight; of those, 39 million are obese. Just look around you. The United States is now the fattest nation in the world. We're number one!

Don't Blame Mutant Genes

Why are we so fat? It is not because half of all Americans have psychological problems. It is not because our genes have mutated in the last twenty years so that we get fat just by sniffing cheeseburgers. It is not because we all have slow metabolisms. It is not even that we have suddenly become gluttons. It is because our food system is so ultra high-fat and we can afford to revel in it.

Years ago people ate simpler food — meat (probably not too often), vegetables, fruits, and grains. Some grew their own food and others bought at small grocery stores. There were no processed foods.

The Wealthy Could Afford to Get Fat and Chronically Ill

The people who did get fat were generally the wealthy because they could afford to eat high-fat foods more often. They could afford beef, butter, and cream. But the ordinary person ate a high-fiber, low-fat diet: a lot of grain, some vegetables, and a little meat.

People were also more active. Their jobs were more physical (more farmers and laborers) and people had to walk to get anywhere. They didn't choose to walk because they loved exercise; cars just weren't available.

Eating Out Was Rare

Eating out at restaurants was limited to the wealthy. For most people, eating out was a rare treat — birthdays, anniversaries, or graduations. Even in the 1950s and 60s, middle-class people enjoyed dining out only on very special occasions. It was much more difficult to become fat in earlier times because the food system and culture protected us.

Too Busy to Be Sensible

But times have changed. What is different is our lifestyle and the food system created to feed our lifestyle. We are busy, busy, busy. We are too busy to cook, too busy to buy fruits and vegetables, too busy to walk. In the last 20 years, restaurants and fast-food chains have proliferated because we can now afford to eat out or carry-in every night and we do. Dining out guarantees a high-fat, low-fiber diet. Because we eat so few fruits and vegetables, we never feel full. We choose from hundreds of different types of fat-laden chips and cookies because we are con-

Upending the Food Guide Pyramid

Americans (in conjunction with the food industry) have turned the Food Guide Pyramid, the Guide for Healthy Eating (see page 109) on its head. Instead of eating most of our foods from the base, which consists of filling, nonfattening fruits, vegetables, and whole grains, we are stuffing ourselves with the restricted foods from the top — pizza, butter, burgers, chips, ice cream — high-fat offerings. If Americans followed the Food Guide Pyramid as it was designed, they would be among the healthiest and thinnest people in the world.

stantly hungry. Everywhere we turn there is some tempting, convenient, high-fat offering.

So don't blame your metabolism or your heredity. Don't think of yourself as unique, the only person who has succumbed to the American Diet. You are like 50 percent of the population who have trouble resisting the high-fat cornucopia. But just because it is also happening to millions of other Americans doesn't make it right. In fact, it is downright unhealthy.

Not only does eating an unhealthy diet contribute to big, unsightly, mushy bodies, make just existing and moving about an uncomfortable endeavor, it also increases your risk of developing diabetes, heart disease, stroke, and/or cancers. In fact, eating just one meal loaded with fat increases the risk of heart attack or stroke. A study of healthy men and women with normal cholesterol levels fed high-fat and nonfat meals showed the harmful effect of a high-fat meal on artery function. The high-fat meal (450 calories of fat, 126 calories of saturated fat) used in the study consisted of an Egg McMuffin, Sausage McMuffin, and two McDonald's hash brown patties. The oxidized triglyceride-rich lipoproteins formed as a result of eating so much fat prevented the arteries from relaxing or opening up normally following the meal. The effect lasted up to four hours and was not seen in people fed a low-fat meal. The effect on the artery wall is similar to that caused by high blood cholesterol or by smoking and is considered to be a precursor to atherosclerosis.

It doesn't have to be this way. You can face down the American high-fat food system. You can put the Food Guide Pyramid back on its base. You don't have to starve. You don't have to give up your favorite treats. You just need to know what is making you fat and what you can do to get thin and healthy.

This chapter is devoted to showing you "Where's the fat?" so that you can control it.

WHERE'S THE FAT?

Fat is everywhere. Some is visible, like the strip of fat on a sirloin steak. But much is hidden, like the fat marbled throughout the steak and the fat added to processed frozen dinners. We Americans eat entirely too much of it. And there you are in the midst of it. To root it out of your diet you must know where it exists. You must dispel old notions. Do you consider beef a high-protein food or a high-fat food? Do you view whole milk and cheese as calcium-rich and wholesome rather than fat-laden? Do you consider skinny, little dried-out crackers good diet snacks or high-fat foods? The following discussion should start you down the road to a new, lean body.

> **Don't Commit Hara-Kiri!**
>
> NOTE: You may be disheartened to discover that fat lurks in so many of the foods you love, but don't lose hope. You'll learn how often and in what amounts you can comfortably fit them into your Fat Budget. True, you won't be eating your favorite high-fat foods in such large quantities or as often as before, but is this so bad? You won't just inhale that piece of apple pie, you'll *really* taste it. You will also learn to like foods you once scoffed at — vegetables and fruit, for example. Just relax. You're going to love being thin and eating healthfully. And in just a few pages, you'll find a whole chapter devoted to great weight loss food choices you once thought were forbidden fruit.

RED MEATS

Beef: A Dieter's Undoing

(pages 432–434, Food Tables)

If you think of beef as a high-protein food that builds bulging biceps and triceps, revise your thinking. Beef is a high-fat food that creates bulging hips, tummies, and thighs. Removing the layer of fat that hugs the edge of a piece of beef will not eliminate all the fat. Invisible fat is marbled throughout. The better the quality of beef (the higher the grade), the more fat it contains. Even meats that appear to be lean contain a lot of fat. For instance, braised flank steak trimmed of fat has 35 fat calories *per ounce*. That's 140 fat calories per 4 ounces. The leanest beef is broiled top round trimmed of fat, which has 16 fat calories per ounce.

Don't be taken in by the new food labels on meat. The label that boasts "80% Lean, 20% Fat" is only describing how much of the food is muscle — 80 percent — and how much is fat — 20 percent. At 50 fat calories an ounce, this ground beef is far from lean.

Take a look at the **MEATS** category of the Food Tables for an eye-opener. The entries are listed in 1-ounce portion size for easy comparison. Of course, no one eats just 1 ounce of meat. Be sure to multiply the fat calories by the number of ounces you eat. For instance, braised short ribs have 107 calories of fat per ounce. If you eat 4 ounces of short ribs, multiply 4 x 107 for a total of 428 fat calories. How does 428 fat calories fit into your Fat Budget?

Visions of Hamburger

The image of a big, juicy hamburger often ravages the subconscious of dieters. Until that need is satisfied, dieters can think of nothing else, often stuffing themselves with every food in sight to satisfy that craving. Good news: you *can* fit the hamburger into your Fat Budget. But it won't be often. (For the leanest* hamburger meat you can buy, ask the butcher to trim the fat off a round steak and grind it.)

If You'd Trade Your Right Arm for a Hamburger

If nothing but a large, 100 percent regular ground beef hamburger will satisfy you, save fat calories for your splurge by keeping your fat intake extra-low for a few days. Remember: saving for a splurge is a better policy than paying back for it later. When you splurge without prepayment too often, you're off *Choose to Lose*. (See Splurging, page 181.)

Dieter's Plate: Guaranteed to Put on Weight

Now that you know about the fat in beef, you will scorn the diet plate offered in thousands of American restaurants. The "Diet Special" fea-

*Don't get taken in when a meat package boasts "Lean!" See page 151 in Chapter 7, "Decoding Food Labels."

tures a hamburger (without the bun), whole-milk cottage cheese, and a canned peach half. The star of the dieter's special contains 58 fat calories per ounce or about 348 fat calories per 6-ounce patty. (Aren't you glad they spared you the 18 fat calories for the bun?)

Be wary of any restaurant's diet claims. When a restaurant that specializes in hamburgers claims it has a heart-healthy menu, be suspicious. Perhaps the menu is heart-healthy, but you won't be eating the menu.

Veal: Expensive and Fattening (pages 438–439, Food Tables)

Veal may be pale in color like chicken and turkey breast, but that is where the similarity ends. Veal is not diet food. Braised veal breast contains 54 fat calories per ounce or 216 fat calories for 4 ounces. Veal cutlet — before it becomes breaded and fried veal cutlet or even sautéed veal cutlet — has 27 fat calories per ounce or 108 calories per 4 ounces. Try turkey cutlets instead (2 fat calories per ounce). Pound them thin and your guests will think they are eating veal scaloppine. Try *Turkey with Capers* (page 276), and, from *Eater's Choice Low-Fat Cookbook, Turkey Scaloppine Limone.*

	FAT CALORIES	
	1 OUNCE	4 OUNCES
Turkey cutlet, braised	2	8
Veal cutlet, braised	27	108
Veal breast, braised	54	216

Lamb: Not Just the Wool Keeps It Warm (pages 435–436, Food Tables)

Lamb is also a poor choice for a dieter. It should come as no surprise that a 4-ounce rib lamb chop has 360 calories of fat. You can see it. Ground lamb is no substitute for ground beef. Four ounces of ground lamb has 320 fat calories. The best lamb choice is roasted leg of lamb trimmed of fat (20 fat calories per ounce, or 80 fat calories for a 4-ounce serving) or a broiled loin lamb chop trimmed of fat (19 fat calories per ounce, or 76 fat calories for a 4-ounce chop).

	FAT CALORIES 1 OUNCE	4 OUNCES
Loin lamb chop (trimmed)	19	76
Leg of lamb (trimmed)	20	80
Ground lamb, broiled	80	320
Rib lamb chop	90	360

But remember, if a lamb chop would make you incredibly happy, just make room for it in your budget — but rarely.

Pork: You Are What You Eat (pages 436–437, Food Tables)

It may be called the other white meat by the people who sell it, but except for roasted lean pork tenderloin at 12 fat calories per ounce (48 fat calories per 4 ounces), pork products are high in fat. Fat calories for pork products average around 65 per ounce (260 per 4 ounces). Spareribs, raw, contain 77 fat calories per ounce (308 per 4 ounces) and pan-fried loin blade has 94 fat calories per ounce (376 for 4 ounces). Your favorite roasted ham is 53 fat calories per ounce (212 per 4 ounces). It is not surprising that one skinny (ha!) strip of bacon has 35 fat calories. But would you kill for a B.L.T. sandwich? Make it with one strip of bacon. You'll get the taste without all the fat.

	FAT CALORIES 1 OUNCE	4 OUNCES
Pork tenderloin (lean, roasted)	12	48
Ham	53	212
Other pork products (average)	65	260
Spareribs	77	308
Pork, loin, blade, pan-fried	94	376
	per strip	4 strips
Bacon	28	112

Fitting It In

Although beef, veal, lamb, and pork are all high-fat foods, you can make room for them in your budget. If you want to eat an occasional (a *very* occasional) steak or chop, reduce the fat calories by trimming off the visible fat before you broil or grill it. Marinate less fatty cuts of meat to make them tender. Instead of choosing a thick steak or roast as the centerpiece of your meal, choose dishes in which small amounts of meat are added to vegetables, rice, or pasta.

Be sure to check the **MEATS** (beef, lamb, pork, veal, and game), **FAST FOODS, FROZEN, MICROWAVE, AND REFRIGERATED FOODS**, and **SAUSAGES AND LUNCHEON MEATS** sections of the Food Tables to familiarize yourself with the fat contents of meat products not mentioned in this chapter.

POULTRY

Chicken: Basically Great . . . (pages 443–447, Food Tables)

White-meat chicken without skin, properly prepared, trounces beef, veal, pork, and lamb in the low-fat marathon. A roasted chicken breast without skin has only 13 fat calories. Chicken has the potential to be part of an infinite number of low-fat, delicious recipes. Try *Sesame Chicken Brochettes* (page 275) or *Chicken with Rice, Tomatoes, and Artichokes* (page 275) and, in *Eater's Choice Low-Fat Cookbook, Kung Pao Chicken with Broccoli* and *Pineapple Chicken* for delectable proof.

Dark meat is fattier than light meat. A roasted drumstick without skin has 22 fat calories. But then one drumstick isn't much of a meal. A thigh without skin has 51 fat calories. A roasted thigh isn't all that much to eat, either. However, if you moderate the amount, dark-meat chicken can fit into your Fat Budget.

. . . But Easily Ruined

White-meat chicken cooked without skin is a low-fat food. Chicken *with* the skin has a fat content approaching or even surpassing that of many cuts of beef. A roasted chicken breast without skin has 13 fat calories. The same breast with skin has 69 fat calories. Take that breast with skin, dip it in flour and fry it, and the fat calories rise to 78. Batter-dip and fry it and the fat calories climb to 166. Eat it extra tasty crispy at a KFC fast-food restaurant and the breast becomes, at 250 fat calories, more than 19 times as fattening as the original bare skinless breast.

PREPARATION OF CHICKEN BREAST	FAT CALORIES
Roasted without skin	13
Roasted with skin, then skin removed	28
Roasted *with* skin	69
Fried *with* skin, flour-coated	78
Fried *with* skin, batter-dipped	166
KFC Extra Tasty Crispy	250

Go to Boston Market (formerly Boston Chicken) to view the myriad ways a skinny little chicken can become high-fat eats. You can choose half a chicken with skin for 300 fat calories, chicken Caesar salad for 410 fat calories, a chicken salad sandwich for 270 fat calories, or a chicken sandwich with cheese and sauce for 300 fat calories. If you were to ask the patrons of Boston Market if chicken is a high-fat food, we bet they would all deny it.

Keeping Chicken Low-Fat

Choose to Lose tips: Always remove the skin from chicken *before* cooking or the fat from the skin will be absorbed by the meat. If you remove the skin *after* you cook the chicken breast, your chicken breast will have accumulated 15 extra fat calories for a total of 28 fat calories. To keep the fat at a mere 13 calories a breast, flour and bake skinless chicken breasts in a shallow baking pan without fat or cook on a cast-iron griddle or barbecue grill without fat.

Redi-Serve Fat

The food industry is always up-to-the-minute in creating foods that will save us time, but at what expense? A new category of ready-made, prepared meats has cropped up in the market. But pick and choose carefully. For example, you can buy Redi-Serve breaded and cooked chicken patties (white meat) at a mere cost of 150 fat calories per patty. These patties have more fat than the veal patties made by Redi-Serve.

Tyson makes roasted chicken that you just heat up. At first glance it looks pretty low in fat — 25 fat calories for a chicken breast. That's only 12 more fat calories than if you roasted it yourself. Look again. With the skin, the chicken contains 120 fat calories a breast. Make sure you read the label carefully.

Turkey: Dieter's Pick

(pages 449–450, Food Tables)

One ounce of unadulterated turkey breast without skin has a mere 2 fat calories. Five ounces has only 10 fat calories. That means you can eat a lot of white-meat turkey without remorse. The white meat of deep-basted Butterball turkeys is not quite as fat-free. That's because it is basted with partially hydrogenated soybean oil. Each ounce contains 10 fat calories. This is not a lot of fat, but it can accumulate particularly with added gravy or mayonnaise. If you add the skin, Butterball turkey white meat is 17.5 fat calories per ounce (70 fat calories for 4 ounces!). Be sure to keep track of the fat-containing sauces and gravies you add to your turkey.

As you would expect, dark meat without skin is higher in fat than white-meat turkey without skin. Unadulterated dark-meat turkey contains 6 fat calories per ounce. That's 30 fat calories for five ounces. A whole leg roasted without skin is 70 fat calories. If you moderate your use of gravy, dark-meat turkey doesn't have to overload your Fat Budget. Read labels. Dark meat from deep-basted Butterball turkeys contains 26 fat calories per ounce. Even without dressings and gravies this can add up quickly to a lot of fat calories.

TURKEY	FAT CALORIES 1 OUNCE	4 OUNCES
White-meat turkey, roasted		
without skin	2	8
with skin	19	76
(Butterball) without skin	10	40
Dark-meat turkey		
without skin	6	24
with skin	29	116
(Butterball) without skin	26	104
Butterball young turkey, smoked	27	108
Ground turkey		
(93% fat free)	18	72
(frozen)	30	120

Ground Turkey: Deceptive Advertising

Ground turkey seems to be the answer to a weight-watching hamburger-lover's dreams. Because turkey is so low in fat, we assume that ground turkey is also low. Not necessarily. Even when the turkey package cries, "100% pure turkey," the ground turkey you buy may be loaded with turkey skin and fat. To ensure that your ground turkey is low in fat, grind your own raw turkey breast or cutlets in a food processor or meat grinder or have your butcher do it.

Prepared Fat

As with chicken, turkey is prepared into convenient high-fat ready-to-cook packages. Turkey Nibblers, a 1¼-inch × 1-inch niblet-shaped turkey patty (white meat — ha!), contains 20 fat calories. That's one bite. Five contain 100 fat calories. That's five bites. Still hungry? Eight contain 160 fat calories.

Duck: Super Splurge

(page 447, Food Tables)

Have you ever seen a duck frolicking in icy water in 15° weather? The duck is padded with fat to keep it warm and afloat. Eating duck is a sure way to increase your padding and sink your Fat Budget. Half a duck with skin contains 975 fat calories — without a sauce. Has duck à l'orange lost some of its appeal? Duck without skin has 29 fat calories per ounce. Perhaps your Fat Budget can handle a pancake or two of Peking Duck (without the skin).

Be sure to check the **POULTRY, FAST FOODS, FROZEN, MICROWAVE, AND REFRIGERATED FOODS**, and **SAUSAGES AND LUNCHEON MEATS** sections of the Food Tables to familiarize yourself with the fat in poultry and poultry products not mentioned in this chapter.

SAUSAGES AND LUNCHEON MEATS

(pages 480–487, Food Tables)

Beware of sausages and luncheon meats. They are crammed with fat (in addition to being filled with dangerous additives and excessive amounts of salt). A 3-ounce link of grilled bratwurst has 230 calories of fat; a 2.7-ounce smoked link sausage has 200. A 2-ounce all-beef frankfurter contains 150 fat calories. Even a 2-ounce chicken frank may contain 90 fat calories. The five little pepperoni slices atop your piece of pizza take a 50-fat-calorie bite out of your Fat Budget.

The Cold Facts about Cold Cuts

A thin slice of packaged bologna has 75 calories of fat. Slide two slices of bologna (150 fat calories) between two slices of bread (15 fat calories) slathered with a tablespoon of mayonnaise (100 fat calories) and you have created a sandwich worth about 265 fat calories. Doesn't sound like diet food.

Check out the figures in the box below. Don't get carried away. *Remember these numbers are for only one slice.*

You will also find a slew of low-fat cold cuts in the luncheon meats section. They range from about 2 fat calories (smoked turkey) to 6 fat calories (cooked ham) for a 3/4-ounce slice. This isn't much fat and it isn't much meat. Remember to multiply the fat calories by the number of slices you eat.

If you use the cold cuts that have less fat (read the labels) and make your sandwiches less thick, you can easily fit cold cuts into your budget. However, you might want to eat cold cuts less often. The cancer-causing nitrites and blood pressure–raising sodium won't make you fat, but they may shorten your life.

LUNCHEON MEATS	AMOUNT	FAT CALORIES
Chicken bologna	1 slice (32g)	50
Salami	1 slice (28g)	65
Beef bologna	1 slice (28g)	75
Pork and beef bologna	1 slice (28g)	70
Liverwurst	1 slice (28g)	75
Pork and turkey bologna	1 slice (38g)	100

Sliced Turkey Breast: A Deli Delight

The only luncheon meat you can eat with abandon is fresh-roasted turkey breast that you roast yourself (2 fat calories per ounce). You can also find turkey breast at many grocery-store deli counters for about 8 fat calories per ounce. However, if you are buying packaged turkey breast, read the label. Some turkey cold cuts have so much added fat, they approach or even exceed some beef or pork cold cuts. For example, one thin 28-gram slice of turkey bologna has 35 fat calories; one 28-gram slice of turkey cotto salami has 30 fat calories.

Wurst Is Worst

For those of you whose default dinner selection is hot dogs, take a look at the figures in the next box. Will you still rely on hot dogs to feed picky eaters (definition of picky eaters: those who are not hungry because they have been snacking all day) when you know you are stuffing 150 fat calories into their mouths? When you didn't plan for dinner, will you still boil up 150 calories of fat and nitrites? We hope not.

Don't let the words turkey and chicken fool you into buying poultry dogs. They are generally less fat-filled than beef hot dogs, but 70 to 100 fat calories is a hefty amount of fat for so little food.

"Light" or "lite" franks may also be higher in fat than the terms would lead you to believe. Oscar Mayer light beef franks (1.8 oz) and Ball Park SmartCreations Lite beef franks (1.8 oz) each contain 70 fat calories.

Really low-fat franks such as Healthy Choice Low Fat Franks or Hebrew National 97% fat-free at 15 fat calories for a 1.8 ounce frank or even the multitude of fat-free franks now on the market are best saved for rare occasions. Although a hefty amount of fat has been removed, the excessive sodium, unhealthy preservatives, and high-fat taste remain. If you continue to subject yourself to a high-fat taste, albeit from a low-fat food, you will never lose your craving for high-fat foods.

And then, there are sausages. Don't even think of knockwurst, kielbasa, or pork sausage. Even turkey sausage will make your Fat Budget shudder.

FRANKS AND SAUSAGES	AMOUNT	FAT CALORIES
Turkey polska kielbasa	1 inch (28g)	27
Polska kielbasa	1 inch (28g)	75
Frankfurter		
turkey	1 frank (57g)	80
chicken	1 frank (56g)	90
beef	1 frank (57g)	150
Cheese dog	1 frank (45g)	120
Beef knockwurst	1 link (85g)	210
Smoked link pork sausage	1 link (85g)	230

For more illuminating sausage and cold cut values, take a look at the **SAUSAGES AND LUNCHEON MEATS** section of the Food Tables.

SEAFOOD

(pages 364–374, Food Tables)

Just plain, unadulterated fish is an excellent choice for a dieter. Two calories of fat per ounce for raw cod, dolphin fish, haddock, lobster, pollock, and scallops; 3 for grouper, pike, snapper, sole, and sunfish; and 4 for flounder, monkfish, perch, rockfish, and shrimp hardly affects your Fat Budget.

The message for fatty fish is mixed. On the one hand, the fat they contain is omega-3 polyunsaturated fat, a potent lowerer of triglyceride levels and reducer of blood clots. On the other hand, fat is fattening. Here are some numbers of fat calories per ounce: sable fish 39, Pacific herring 35; Atlantic mackerel 35; Chinook salmon 27; Atlantic herring 23; sockeye salmon 22; butterfish 20; Pacific mackerel 20; orange roughy 20. Check the food tables for the fat calories of the fish you wish to eat.

But, wait. Orange roughy! Salmon! You don't have to eliminate any of these wonderful fish from your diet — just keep track of their fat calories and fit them in.

Of course, no matter how lean a fish is, when you bread and deep-

FISH	FAT CALORIES 1 OUNCE	FAT CALORIES 4 OUNCES
Cod, dolphin fish, haddock, lobster, pollock, scallops	2	8
Grouper, pike, snapper, sole, sunfish	3	12
Flounder, monkfish, ocean perch, rockfish, shrimp	4	16
Catfish	11	44
Butterfish, orange roughy, Pacific mackerel	20	80
Sockeye salmon	22	88
Atlantic herring	23	92
Chinook salmon	27	108
Atlantic mackerel, Pacific herring	35	140
Sablefish	39	156

fry it, or drown it in a cream sauce or butter, it is no longer a good diet choice. Bake, broil, poach, or grill fish with little or no fat or cover it with low-fat sauces to create delectable, guilt-free dishes.

Choose to Lose tip: Choose canned tuna packed in water. A whole can of undrained tuna packed in water has 30 calories of fat or 4.5 fat calories per ounce. A whole can of undrained tuna packed in oil (it's not fish oil; it's soy oil) has 297 calories of fat or 45 fat calories per ounce. Oil-packed tuna has ten times as much fat as water-packed.

You will find the fat calories of fish listed in the **FISH AND SHELLFISH, FAST FOODS,** and **FROZEN, MICROWAVE, AND REFRIGERATED FOODS** sections of the Food Tables.

MEALS IN A BOX

Gain Time and Weight

There has been an explosion in the food industry, which is capitalizing on our changing society. Fast-food restaurants, microwave dinners, and a seemingly endless proliferation of convenience foods have been developed for people who don't take time for food preparation. However, what you gain in convenience you often sacrifice in taste, health, and your weight.

The price is also steep in terms of dollars. For that little box called dinner, you are paying at least five times what it would cost if you were to purchase the ingredients fresh and make the dish yourself.

Be Label-Wise: A Survival Skill

Learning how to read a label is mandatory if you want to eat convenience foods. The food label tells it all. The package may advertise a low-fat entrée such as fish, turkey, or chicken, but you need to know more. The package may claim "25% less fat," but you need to know more.

You need to know exactly how much fat the food processors have added to that dinner to make it palatable. And it is often a lot. Be sure to check out Chapter 7, "Decoding Food Labels," for more information on deciphering food labels.

Ruining a Good Thing

We'd be wealthy if we had a dollar for every time we've been told, "I don't understand why I'm fat. I only eat fish and chicken." As you will see, preparation makes all the difference.

Most fish are naturally very low in fat. Here's what happens when food companies process fish into a frozen meal. Gorton turns about 2 ounces of pollock fillets (2 fat calories an ounce) into Crunchy Golden Fish Fillets (130 fat calories). Mrs. Paul transforms 2 ounces of haddock (2 fat calories an ounce) into Premium Fillets (100 fat calories). Sea Pak processes four large shrimp (4 fat calories each) into Jumbo Butterfly Shrimps (80 calories of fat). That's some kind of magic.

	FAT CALORIES	
FISH	**UNADULTERATED**	**PROCESSED**
Shrimp (1 oz)	4	80
Pollock (2 oz)	4	130
Haddock (2 oz)	4	100

Abracadabra! Watch the food processors transform a low-fat chicken breast (13 fat calories) into a super high-fat food:

CHICKEN	FAT CALORIES
Chicken breast without skin	**13**
Stouffer's Chicken Breast in Barbecue Sauce	210
Banquet Chicken Nugget Meal	210
Stouffer's Escalloped Chicken & Noodles	240
Swanson Fried Chicken, White Portions	280
Marie Callender's Chicken & Broccoli Pie	880

Even a turkey would be horrified to see how its lean flesh (2 fat calories per ounce for white meat; 6 fat calories per ounce for dark meat) can be turned into high-fat frozen turkey dishes.

TURKEY	FAT CALORIES
Turkey breast without skin (1 oz)	**2**
Stouffer's Turkey Tetrazzini	150
Swanson Turkey Pot Pie	190
Stouffer's Sliced Turkey Breast (Hearty Portions)	240
Marie Callender's Turkey Pot Pie	320

Kids' Killer Cuisine

If you think these numbers are high, just look at the amount of fat you can inflict on your children: The makers of Healthy Choice bring you Cosmic Chicken Nuggets at 230 fat calories a meal. Swanson tempts your kid with Chompin' Chicken Drumlets at 220 fat calories. Is it a surprise that obesity in children has reached epic proportions? Each dinner comes along with a subliminal message, too: Eating is fun only if you are eating fried chicken, hot dogs, or pizza. It is surprising that each package doesn't include a lifetime guarantee for future (or even present) obesity.

KIDS' FROZEN DINNER		FAT CALORIES
Kid Cuisine		
Circus Show Corn Dog	1 meal (249g)	140
Game Time Taco Roll-Ups	1 meal (208g)	160
Cosmic Chicken Nuggets	1 meal (257g)	230
Kids' Fun Feast (Swanson)		
Chompin' Chicken Drumlets	1 pkg (255g)	220
Frenzied Fish Sticks	1 pkg (198g)	130
Munchin' Mini Tacos	1 pkg (204g)	140

Resist Irresistibles

The food industry is getting smarter and smarter. They know that most parents can't resist the demands of their children. Not only do food manufacturers create health-risky frozen dinners for children, they have developed a whole line of irresistible lunch packs to start your children on the road to poor nutrition. Oscar Mayer offers a Bologna & American Cheese Lunchable for a 300–fat calorie treat. Their *Lean* Ham & Cheddar Lunchable has 200 fat calories and their *Lean* Turkey Breast & Cheddar Lunchable has 180 fat calories.

A Plea

Please, if you care about your children, don't buy them high-fat gaily packaged poison. Just say "no." Your children are developing eating preferences that will last them a lifetime. Teach them about the wonderful tastes of healthy food. Take the time to fix them real, nutritious meals. For more about children, see Chapter 14, "*Choose to Lose* for Children."

Lunch Express

Not only does the frozen-food industry expect you to be too busy to prepare dinner — it's banking on the fact that you haven't the time or the inclination to fix lunch either. Now available are a large variety of sandwiches that you can take to work and zap in your office microwave. But if you jump onto the lunch express, you will find that the express they refer to is not the five minutes you save by not preparing your lunch at home but the speed at which these high-fat lunches will nestle into your fat stores. Check out the chart below if you need inspiration for bringing your own bag lunch to work.

LUNCH IN A BOX	FAT CALORIES
Hot Pockets Beef & Cheddar	150
Croissant Pockets Pepperoni Pizza	140
Safeway Stuffed Sandwiches Ham & Cheese	140
Veggie Pockets Bar-B-Q Style	80

Eater Beware: The food industry is adding a tremendous amount of gratuitous fat to its frozen dinners — and to you, if you eat them.

Diet (?) Frozen Dinners

Don't assume that "diet" frozen dinners are low in fat. Although hype such as "Lite!" and "Low-fat!" may be used only if a product meets low-fat criteria — less than 27 fat calories per serving — names implying healthy or a good diet choice such as Lean Cuisine, Weight Watchers, and Healthy Choice may still be used even if the product is neither. Read the label. The dishes below would take a chunk out of anyone's Fat Budget and provide a modest amount of food. Why not make your own "diet" dinners?

DINNER IN A BOX	FAT CALORIES
Healthy Choice Country Breaded Chicken	80
Lean Cuisine Chicken Pie	80
Weight Watchers Santa Fe Style Rice & Beans	70
Budget Gourmet Low-Fat Chicken Oriental	70

Reduced Fat Means Reduced Food

If you read labels carefully, you can find a variety of frozen dinners that are truly low in fat. However, be advised that low-fat dinners may be low-fat because they are almost meat-free (many contain about ½ to 1 ounce of meat) and thus eating only one dinner leaves you hungry. (See page 144 to find out how to determine the amount of meat in a frozen dinner.) Eating modest amounts of meat is fine (and even healthy) but not if you need to consume another low-fat dinner to fill you up (45 + 45 = 90 fat calories) or you binge on high-fat snacks because the frozen dinner has left your stomach feeling empty and your psyche feeling unsatisfied.

A Box Does Not a Dinner Make

If you decide to buy commercial dinners, be sure to enhance them with fresh vegetables. Steam broccoli, carrots, cauliflower, or zucchini in a vegetable steamer. Bake a plain or sweet potato. Boil a pot of rice. Don't cheat yourself out of the vitamins, minerals, and fiber that you need to maintain your good health. Remember, this is an eating diet. You need lots of complex carbohydrates to make your fat vaporize.

PREPARED PASTA, NOT SO FASTA

(pages 426–432, Food Tables)

You probably jumped for joy when you found that pasta could and should be included in your eating plan. But, whoa, not just *any* pasta.

And not just any amount of pasta. Pasta is highly processed so it won't yell "stop" when you have eaten enough. *You* have to yell "stop" after 1½ or 2 cups so that you don't eat too much.

And, of course, pasta with high-fat sauces have to be eaten rarely and in very limited, budgeted amounts. Fettuccine with a cream sauce can blow away your Fat Budget for several days. Pasta with pesto may be more than a 500-fat-calorie splurge. Even a marinara sauce eaten at a restaurant can take a heavy toll on your Fat Budget.

If you are using prepared refrigerated pasta and pasta sauces, be sure to note the serving size when you figure the fat cost. You might think 140 fat calories for Alfredo sauce and 290 fat calories for pesto sauce with basil is a hefty amount of fat, but that ain't the half of it. The serving size for Contadina and DiGiorno sauces is ¼ cup. Who eats ¼ cup of a pasta sauce? Who eats ½ cup? A skimpy half cup of pesto sauce will cost you 580 fat calories. And that is without the pasta. Add

PASTA	AMOUNT	FAT CALORIES
Lipton Creamy Garlic Pasta & Sauce	2 cups	240
Cup of Noodles, chicken	2 cups	260
Kraft Macaroni & Cheese	2 cups	340
Lasagna (Italian Restaurant)	2 cups	477
Contadina Cheese & Herb Tortelloni with reduced-fat pesto sauce	2 cups	500

80 fat calories if you pour the pesto sauce over 1 cup (the serving size) of cheese & herb tortelloni. Only one cup? Add 160 fat calories for 2 cups of pasta for a grand total of 740 fat calories. Now you understand why people who complain, "I just eat pasta, but I can't seem to lose weight" never lose weight.

For no strain on your Fat Budget, make your own *Pasta with Spinach-Tomato Sauce* (page 289) at 14 fat calories a serving or *Asparagus Pasta* (page 288) at 26 fat calories a serving.

FAST FOODS

Fast Foods = Fat Foods (pages 319–358, Food Tables)

A person who wants to stay thin and healthy should avoid fast-food restaurants like the plague. Fast-food restaurants are a plague. Hundreds of thousands of Americans die from fat-related, or should we say fast-food-related, diseases (heart disease and breast, colon, and prostate cancers). What is happening in Japan proves a vivid illustration of the effect of fast food on health. In the past few years since fast food has come to Japan (a country in which both fast-food restaurants and heart disease were virtually nonexistent), heart disease has increased fivefold and obesity has become a problem. Here are some figures that help explain why:

FAST FOOD	FAT CALORIES
Roy Rogers Big Country Breakfast with Sausage	540
KFC Hot & Spicy Chicken: Breast & Thigh	550
Burger King Double Whopper with Cheese	570
Hardee's Big Country breakfast sausage	594

Baked potato, chicken, pasta salad, fish — what happens to all those good, low-fat foods when a fast-food restaurant gets hold of them?

FAST FOOD	FAT CALORIES
Hardee's **Bagel** with sausage, egg & cheese	**207**
Popeye's **Red Beans and Rice**	**153**
Jack-in-the-Box **Onion** Rings	**207**
Wendy's **Garden Veggie Pita**	**150**
Carl's Jr. **Zucchini**	**207**
Arby's **Turkey** Sub Roll Sandwich	**243**
Arby's **Baked Potato,** deluxe	**324**

Slim Pickin's

There are no "no-no's" in *Choose to Lose.* You can eat even the highest-fat fast foods if you can fit them into your budget. But is McDonald's large fries (200 fat calories) or Roy Rogers cole slaw (225 fat calories) worth all those fat calories? Fast-food restaurants offer you such poor choices, you will do yourself a favor by staying away. Even if you enter with noble intentions, a whiff of a burger or fries might weaken your resolve and cause a meltdown of your Fat Budget.

However, if you have been knocked on the head, tied up, and dragged to a fast-food restaurant, here are a few choices that won't totally destroy your Fat Budget.

FAST FOOD	FAT CALORIES
Roy Rogers or Wendy's Baked Potato, plain	0
KFC Corn on the Cob	15
Long John Silver's Flavorbaked Chicken	30
Chick-Fil-A Chargrilled Chicken Sandwich	30
Burger King Chunky Chicken Salad plain	36
KFC Tender Roast Chicken Breast Without Skin	39
Hardee's Roast Beef Sub	45
Arby's Light Roast Chicken Deluxe Sandwich	54
Carl's Jr. Barbecue Chicken Sandwich	54

The Rap on Wraps

A new offering at Taco Bell and Long John Silver's is the wrap. If you haven't encountered them, wraps are made by filling a tortilla with a variety of stuffings — chicken, shrimp, vegetables — and rolling it up. The ingredients may be healthy, but the glue that holds them together is generally fat — fat in the form of cheese, mayonnaise, or oil. Some wrap restaurants (there are restaurants dedicated solely to serving wraps) give you the option of making your own wrap, which means you can create them without adding fat. Otherwise you need to ask a lot of questions and be very careful before you choose a wrap.

Fast-food wraps are worth avoiding. At Long John Silver's the chicken classic wrap has 320 fat calories. The fajita wraps at Taco Bell range from 170 fat calories for a Veggie Fajita Wrap to 230 fat calories for a Chicken Fajita Wrap Supreme.

PIZZA

(pages 319–358, Food Tables)

Eating pizza is a national pastime. You don't even have to go to a restaurant to eat it. Call up your local pizzeria and within minutes, a steaming-hot pizza arrives at your door. It's so convenient. What a quick and easy way to add hundreds of fat calories to your fat bank. A 6-inch Domino's deep-dish pizza with Italian sausage has 309 fat calories; 2 slices of a 12-inch medium Pizza Hut Meat Lover's pan pizza contain 324 fat calories (but who eats 2 slices?). Jody Goodman-Block, a registered dietitian in New York, estimates that a slice of pizza in New York City may contain as much as 245 fat calories. Is it any wonder that so many Americans are obese?

It is possible, however, to survive a pizza party with your Fat Budget intact. Order pizza without the cheese. Have the pizza chef add mushrooms, onions, green peppers — whatever vegetables are available. Insist that no oil be drizzled over the pizza. The result will taste delicious (not like cheese pizza, but still good) and will have far fewer fat calories. But since you really have no idea how much fat the cheeseless pizza may contain, don't go out of your way to eat it.

RESTAURANT FOOD

Fat is what Americans love. Since restaurants want business, they create fat-filled dishes to attract patrons. It doesn't matter if the restaurant is vegetarian or a steakhouse — fat is an essential ingredient. Not a little fat, either. The portions are mammoth. And that is why we are mammoth.

Here are some staggering (and not atypical) numbers: fettuccine with creamed spinach, 730 fat calories; two chile rellenos with beans and rice, 864 fat calories; one serving of risotto, 990. You can make an effort to reduce fat in your order (see At Restaurants, page 184), but if you really want to lose weight and improve your health, make your restaurant visits a rare occurrence.

Chinese (pages 456–457, Food Tables)

We have grown up with the misconception that Chinese restaurants are a dieter's haven. All those vegetables and rice — what could be healthier? Lots. Although homemade Chinese dishes have the potential to be relatively low in fat, most Chinese restaurant food is astronomically high. Not only do many Chinese restaurants fry and deep-fry ingredients in large amounts of oil (generally corn or cottonseed oil), they often blanch them in oil (pass meat or vegetables through hot or warm oil before stir-frying) first. In addition, oil is sometimes added after the dish is cooked to make it shiny. Check out the fat calories in the next box. Does Chinese restaurant food still seem so enticing?

It may be possible to eat like a low-fat king at Chinese restaurants if you specifically order the meat and vegetables steamed rather than sautéed in oil and then mixed with the sauce. The food is equally as delectable as when cooked with tons of oil. The only drawback is that you don't really know if your low-fat request has been taken seriously or if the management is trying to please you by saying the dishes are oil-free. This problem always occurs when you eat out, because you are not in the kitchen yourself controlling the ingredients.

For a larger listing of Chinese dishes, look in the **RESTAURANT FOODS** section of the Food Tables.

Make your own delectable low-fat Chinese dishes at home. For a

CHINESE RESTAURANT DISH	AMOUNT	FAT CALORIES
Egg roll	1	103
Hunan shrimp (not deep-fried)	1 whole dish	755
Beef with vegetables	1 whole dish	1068
Chicken with cashews	1 whole dish	1075
Kung pao chicken	1 whole dish	1134
Orange beef	1 whole dish	1216
Chinese barbecued spareribs	1 whole dish	1232
Sweet and sour pork	1 whole dish	1509

real treat try *Spicy Chicken with Diced Carrots and Green Peppers* (in *Eater's Choice Low-Fat Cookbook*) or *Hot and Garlicky Eggplant* (page 282). Your Fat Budget won't even feel it.

Italian (page 466, Food Tables)

Unlike Chinese food, Italian food (in America*) has never been mistaken for low-fat fare. The basic ingredients of many dishes are cheese, oil, or meat. Fettucine Alfredo, for example, is a combination of cheese, butter, cream, and pasta. Lasagna is a combination of cheese, meat sauce, and pasta. (It is easy to see why pasta developed such a bad reputation considering the company it often keeps.) You know these dishes are high in fat, but would you believe 873 fat calories in a serving of fettucine Alfredo? 558 fat calories in a serving of eggplant parmigiana? The best choice at an Italian restaurant is the spaghetti with tomato sauce at 153 fat calories or the linguine with red clam sauce at 207 fat calories a serving. Neither are great buys. Is it becoming clearer why Americans who eat every meal out have trouble staying lean?

Mexican (pages 466–467, Food Tables)

Making choices at a Mexican restaurant is like choosing to die by firing squad or by electric chair. All of the dishes are riddled with fat —

*In Italy (except in restaurants that cater to American tourists), the portions are small. Pasta is served as a first course, *never* as a main course, and it is about one-fourth the size of a pasta dish served in an Italian restaurant in the United States.

ITALIAN RESTAURANT DISH	AMOUNT	FAT CALORIES
Spaghetti with tomato sauce	3.5 cups	153
Linguine with red clam sauce	3 cups	207
Spaghetti with meat sauce	3 cups	225
Linguine with white clam sauce	3 cups	261
Spaghetti with meatballs	3.5 cups	351
Spaghetti with sausage	2.5 cups	351
Veal parmigiana	1.5 cups	396
Lasagna	2 cups	477
Eggplant parmigiana	2.5 cups	558
Fettuccine Alfredo	2.5 cups	873

Sources: *Nutrition Action Healthletter;* Center for Science in the Public Interest; Lancaster Laboratories.

cheese nachos: 500 fat calories per serving; beef burrito with beans, rice, sour cream, and guacamole: 711 fat calories; two chicken enchiladas with beans and rice: 513 fat calories; taco salad with sour cream and guacamole: 639 fat calories per serving. What choices!

Check out the fat calories in Mexican restaurant food in the **RESTAURANT FOODS** section of the Food Tables and stay away.

Chain Restaurants (pages 455–456, Food Tables)

The chain restaurants — such as Applebee's, Bennigan's, Chili's, T.G.I. Friday's, and Hard Rock Cafe — have personality. They are loud and brassy, with young, peppy waiters and waitresses. If you could just enjoy the ambiance, they would be great places to go, but because the food choices are so high-fat, you put your Fat Budget in peril if you choose to eat there.

One of the reasons Americans are so much fatter is that we eat out so often at restaurants like these. Look at some of these statistics for chicken choices: chicken sandwich, 450 fat calories; grilled chicken, 270 fat calories; chicken fajitas with the works, 567 fat calories; Oriental chicken salad with dressing — 4 cups (!), 441 fat calories. How about an order of cole slaw (1 cup = 126 fat calories), onion rings (576 fat calories), or a loaded baked potato (279 fat calories)?

Chicken Out

People know that chicken is a low-fat food, so when they see a name like Chicken Out or Boston Market, they feel they have arrived at a low-fat island in a sea of high-fat sharks. But, alas, this is no haven from fat. Look at the fat calories of these chicken dishes: chicken pot pie, 410 fat calories; chicken Caesar salad, 410 fat calories; chicken salad sandwich, 270 fat calories; 1/4 dark-meat chicken with skin, 190 fat calories. Try some of the side dishes if you want to see your Fat Budget cringe: old-fashioned potato salad, 270 fat calories for 3/4 cup; chicken tortilla soup, 100 fat calories a cup; mashed potatoes, 80 fat calories for 2/3 cup.

Check out the fat calories in chicken restaurant food in the **RESTAURANT FOODS** section of the Food Tables and make delicious low-fat chicken dishes at home.

Fish Restaurants

You might also assume that a fish restaurant would be a healthy, low-fat choice because fish can be so low in fat and healthy. Again, don't be an assumption jumper. You can usually order your fish broiled with no fat, but the typical fare is breaded and deep-fried. The fried seafood combo contains 450 fat calories; the fried clams 423. But even items such as broiled salmon may contain 189 fat calories.

For the lowest-fat and healthiest fish dinner, have your fish broiled, baked, or blackened with no fat added. If you want to add a teaspoon of fat (34 fat calories), have the waiter bring you a pat of butter or margarine to spread over the fish.

Order a shrimp cocktail (a financial but not a fat splurge) and dip fearlessly into the cocktail sauce for an almost fat-free treat.

Watch out for the rum buns (60–80 fat calories each) and hushpuppies (2 for 54 fat calories), and be sure to avoid the extras such as cole slaw (81 fat calories per half cup) and tartar sauce (72 fat calories per tablespoon).

Check out the fat calories in fish restaurant food in the **RESTAU-**

RANT FOODS section of the Food Tables and see how much more and how much healthier you will eat if you prepare it at home.

Coffee Bars (pages 457–463, Food Tables)

The coffee bar is a popular new social center — a place to hang out, meet friends, sip, and chat. Coffee bars would seem to be safe low-fat havens because black coffee has absolutely no fat. However, coffee bars offer high-fat booby-traps that can blow your Fat Budget to smithereens. If you choose any of the coffees with nonfat milk and *no* whipping cream (be sure to ask), it will cost you between 0 and 10 fat calories. If you choose other fancy coffee drinks, hold steady for the blast. At Starbucks, an Espresso with whipping cream will cost you 180 fat calories. A short (8 ounces) Cafe Mocha (whipping cream is included) with whole milk costs 150 fat calories; the same with 2% milk costs 120. Even made with skim milk, it's 100 fat calories. Drink a venti (20 ounces) Cafe Mocha and you'll be spending 250 fat calories. The ultimate fat-extravaganza is a venti Cocoa with whipped cream at 270 fat calories. Check out the fat calories in Coffee Bar Coffees in the **RESTAURANT FOODS** section of the Food Tables.

The scones and muffins are also killers. A Very Blueberry Scone contains 210 fat calories, a maple oat nut scone contains 250. The lower cholesterol scones are no buy, either. A blueberry lower cholesterol scone contains 100 fat calories. Even a reduced fat apricot scone contains 50 fat calories.

Double those sky-high numbers and you get the muffin figures: a cranberry muffin weighs in at 5 ounces and contains 300 fat calories; a chocolate muffin contains 280 fat calories.

Then, there are the sweets to avoid: cupcakes (carrot cupcake: 210 fat calories), cakes (classic carrot cake: 370 fat calories), cookies (rolled oats 'n raisin cookie: 216 fat calories), croissants (butter: 210 fat calories), brownies (walnut brownie: 240 fat calories).

You do have one good choice: Starbucks says their bagels (except the ones with seeds at 32 fat calories) have 0 fat.

Check out the fat calories in Beverages and in Coffee Bar Extras in the **RESTAURANT FOODS** section of the Food Tables.

DAIRY PRODUCTS

Milk — A High-Fat Food (page 316, Food Tables)

It seems un-American to regard whole milk as anything but wholesome and pure. However, whole milk should be viewed as a high-fat and thus a fattening and unhealthy food. One glass of whole milk (8 ounces) contains 75 calories of fat. If your Fat Budget is 280, a glass of whole milk at each meal and one at bedtime would shoot your entire day's fat quota. That is not to say that all milk products are high in fat. We are fortunate to live in an age in which we can buy low-fat and nonfat milk products that are truly delicious.

Shift to Skim Milk. If you currently drink whole milk and find the thought of drinking skim milk disgusting, first switch to 2 percent milk. You probably think 2 percent reduced fat milk is truly low-fat. Two percent — it's barely above zero. But, a glass of 2 percent milk contains 45 calories of fat. How can a drink that is 2 percent fat contain so many fat calories? The 2 percent refers to the percentage of the weight of the milk that is fat, not the percentage of the total calories* that are fat. If the milk producers called this milk 38 percent fat milk (45 fat calories ÷ 120 total calories = 38 percent), would you rush to buy it?

At 45 fat calories a glass, 2 percent milk is a better choice than whole milk at 75 fat calories a glass, but it's still pretty high. After you get used to 2 percent milk, try 1 percent milk for a while. You'll find it tastes pretty similar, and you'll be down to 20 calories of fat per glass. **Skim milk has no fat and more calcium than whole milk** — so go for it!

Choose to Lose tip: One of our Choosers to Lose told us how she moved her husband from 2 percent to skim milk without his being any the wiser. When the half gallon of 2 percent milk became half full, she filled it up with 1 percent milk. The taste was almost the same. Her husband didn't notice any difference. When he had finished half of the combination milk, she again filled the carton up with 1 percent milk.

---

*Also irrelevant. See page 36.

No negative response. Eventually he was drinking pure 1 percent milk without complaint (or knowledge). And then she repeated the dilution method using skim milk. Try it yourself.

MILK	AMOUNT	FAT CALORIES
Nonfat skim milk	8 fl oz	0
1% milk	8 fl oz	20
2% milk	8 fl oz	45
Whole milk	8 fl oz	75

Butter: A Spread That Increases Your Spread (page 312, Food Tables)

Butter is 100 percent fat. A teaspoon contains 33 calories of fat; a tablespoon, 100. Each time you lift your knife to slather butter over a roll or piece of toast you are preparing to make a large dent in your Fat Budget. Keep track. Measure a specific amount — a half-teaspoon (17 fat calories) or a teaspoon (33 fat calories) — and then spread. Try eating the bread with jelly or apple butter or even plain.

Butter adds fat calories even when you can't see it — when you eat baked goods, cream sauces, casseroles, and vegetables. Avoid foods containing unknown amounts of butter. Whatever the amount, it's more than you need to get thin and stay thin.

You may have heard that margarine is a better choice than butter because margarine has so much less saturated fat (18 sat-fat calories per tablespoon) than butter (65 sat-fat calories per tablespoon). Saturated fat contributes to raising your blood cholesterol and thus your risk for heart disease.* However, **both margarine and butter are 100 percent fat and both have about 100 calories of fat per tablespoon.** In addition, it is now known that the trans fats (see page 92) in margarines make them behave like saturated fats in the body. So, although margarine has a heart-healthy reputation, it is just as fattening as butter, almost as heart-risky, and should be limited.

*See *Eater's Choice: A Food Lover's Guide to Lower Cholesterol* by Dr. Ron and Nancy Goor, 5th ed. (Houghton Mifflin, 1999), page 39, for a discussion of saturated fat.

Cooking Sprays

Pump cooking sprays such as I Can't Believe It's Not Butter! are not fat-free. One tablespoon contains 52 fat calories. Of course, one spray is not equal to one tablespoon. To find out exactly how much you use, spray the cooking pan with the amount you normally use, use a rubber spatula to gather what you have sprayed, and measure the amount with your measuring spoons. *Do not use cooking sprays with abandon.*

Cream in Your Coffee

(pages 314–315, Food Tables)

It hardly seems necessary to advise dieters that cream is full of fat. But do you know exactly how fat-laden it really is? It is necessary to keep a watchful eye on cream, for it sneaks into meals and wreaks havoc on diet goals. Cream in coffee can devastate your Fat Budget. One tablespoon of light table cream has 26 calories of fat. If you drink four cups of coffee a day, that adds up to 104 fat calories and you haven't eaten anything of substance. Half-and-half is better, but not great. One tablespoon of half-and-half has 15 fat calories. Half-and-half for your four cups of coffee quickly adds up to 60 fat calories.

However, if you want to use cream or half-and-half, why not drink less coffee? Or, better still, use whole milk at 5 fat calories per tablespoon, 2 percent at 3, or skim milk at 0. Best yet, drink your coffee black.

Warning: Some nondairy creamers may be almost as fattening as cream. One tablespoon of a powdered nondairy creamer, such as Coffee-mate, has 15 fat calories; a liquid nondairy creamer, such as Coffee-mate hazelnut, has 20 fat calories per tablespoon. Four cups of coffee and you have accumulated 80 fat calories of nondairy creamer. Is it worth it?

And then, if you add the 20 fat calories of the nondairy creamer to the 20 fat calories of your International French Vanilla Cafe instant coffee, each cup of coffee will cost you 40 fat calories; 4 cups will cost 160 fat calories. How does that fit into your Fat Budget?

Cream in Soups and Sauces: Riches to Avoid

Cream in soups can ruin your whole day. A bowl of cream of mushroom soup has about 155 calories of fat; New England clam chowder about 80, vichyssoise 210. These may be low estimates. When eating out, be sure to ask whether a soup contains cream; if it does, then choose another appetizer or figure out how to budget it in. If you decide to order it and after the first taste find it isn't divine, stop eating. Why spend all those fat calories for nothing?

And cream sauces! The half-cup of curry cream sauce the chef ladles over your chicken costs you 250 fat calories, the cheesy cream sauce 170. If you want cream sauces to be included in your Fat Budget, insist that they be served on the side. Then you can determine how many spoonfuls (at 30 fat calories each) you want. But remember. Don't fool yourself. Spoonfuls quickly become cups.

Whipped Cream: For Your Eyes Only

Whipped cream looks and tastes so light and airy it is hard to believe its fat content is so high. But it is. The two-tablespoon dollop that perks up your pumpkin pie has 52 fat calories. The little cloud that nestles atop your ice cream sundae may have more than 350 calories of fat. Whipped cream is not a diet food. Whipped cream splurges should be saved for very, very, *very* special occasions.

Cooking with "Cream"

It is possible to make delicious dishes by substituting nonfat yogurt, buttermilk, condensed skim milk, or skim milk in recipes that call for cream, sour cream, or whole milk. The divinely delicious *Deep-Dish Pear Pie* (in *Eater's Choice Low-Fat Cookbook*) made with nonfat yogurt has 368 fewer fat calories than *Deep-Dish Pear Pie* made with one cup of sour cream. *Buttermilk Waffles* (see page 296) are equally as scrumptious made with nonfat buttermilk (0 fat calories) and skim milk (0 fat calories) as with whole milk (75 fat calories per cup). *Red Potato Salad* (page 291) made with nonfat yogurt instead of sour cream saves 184 fat calories.

For a more detailed discussion of cooking with low-fat ingredients, see "Silent Substitutions," page 172.

Cheese — Be Aware and BEWARE

If you have been a cheese nibbler, your nibbling days are numbered. Snacking on cheese is a fattening habit because most of what makes cheese cheese is fat. One ounce of cheddar has 85 calories of fat; one ounce of Gruyère has 83; one ounce of Camembert cheese has 73; one ounce of Gouda has 70. Those chunks of Roquefort cheese in your salad dressing add 78 calories of fat to your greens (and to your thighs). If cheese is not one of your great loves, for the sake of your body, leave cheese off your shopping list. And, if it is, buy a small amount and fit it into your Fat Budget with great care. You will appreciate every bite.

"Lo" and "Light": Read the Label. Low-calorie and nonfat cheeses abound. Most "light" cheeses have 35–50 fat calories an ounce. While this may be half the fat of the cheese they are imitating, it still can add up to a lot of fat. Two slices may equal 70 or 100 fat calories. Read labels and choose carefully. If you really love cheese, you might want to skip the cheese imitations and budget in a small piece of the real stuff very occasionally.

We say "occasionally" because you want to move away from the high-fat taste of cheese. If you add a slice of low- or nonfat cheese to every sandwich and always eat your broccoli covered in melted cheese, you will never lose your high-fat taste.

Mozzarella Loses Its Reputation. Mozzarella and Parmesan cheese have gained the reputation of being low in fat. However, one ounce of whole milk mozzarella has 55 fat calories and one ounce of part skim milk mozzarella has 41. One ounce of Parmesan cheese has 77 fat calories. When you bite into that slice of pizza oozing with mozzarella and Parmesan cheese, think about all those fat calories being deducted from your budget and added to your hulk. For delicious, low-fat pizzas try *Focaccia* (page 286) — a cross between pizza and bread — or other pizza and calzone recipes found in *Eater's Choice Low-Fat Cookbook*.

Cottage Cheese: The Percent Tells All. Americans think of cottage cheese as the quintessential diet food. And low-fat cottage cheese really is. However, not all cottage cheeses will help you lose weight. That half-cup scoop of whole milk cottage cheese on the "diet plate" has 43 fat calories. Two percent cottage cheese is a better choice at 20 fat calories per half-cup. One percent is even better at a mere 10 fat calories per half-cup. Mix cut-up fruit with your 1% fat cottage cheese or scoop cottage cheese on half a cantaloupe for two super, low-fat lunches.

Dieters: Don't Give up Dairy Products

You are embarking on a new way of eating that will keep you slim and healthy for the rest of your life. Your goal is to lose weight but not by giving up nutritious foods. For you to succeed, your diet must be well balanced. And that means eating dairy products — nonfat and low-fat dairy products.

Dairy products supply your body with calcium (skim milk has more calcium than whole milk) to keep your bones and teeth strong. Females, especially teenagers and younger, need calcium to prevent osteoporosis in old age. Children need calcium for strong bones and teeth and to grow. Fill your diet with nonfat and low-fat dairy products — nonfat yogurts, low-fat cottage cheese, skim milk, buttermilk*. Cook with nonfat yogurt, buttermilk, and skim milk. Save high-fat dairy products for splurges.

*Check the nutrition label on your buttermilk carton. Fat calories of buttermilk range from 0 to 36 per cup. Try to use buttermilk with no fat.

Eggs: A Fat Surprise (page 319, Food Tables)

Before you fix yourself a four-egg omelet for a "light" dinner, be aware that each egg yolk contains 50 calories of fat. A four-egg omelet contains 200 fat calories. Make your omelet with one egg and three egg whites (0 fat calories) for a similar tasting but less fat-filled dish.

FATS

(pages 358–364, Food Tables)

The Vegetable Oil Contest

Olive Oil = Number One Choice. Why choose olive oil? Olive oil is considered the healthiest of all the oils because it lowers LDL-cholesterol (bad cholesterol) without lowering HDL-cholesterol (good cholesterol). It also protects arteries from atherosclerosis by preventing the oxidation of LDL-cholesterol (see *Eater's Choice* for a fuller discussion of diet, cholesterol, and heart disease). Epidemiological studies show that populations (such as Greeks and Italians) who use olive oil as the predominant source of fat in their diet have very low rates of heart disease. But before you begin to guzzle the nearest bottle of olive oil, be warned. Olive oil has 119 fat calories per tablespoon.

Second choice is canola oil. Canola oil is derived from the rapeseed plant after detoxification. While it has the lowest amount of saturated fat of all the oils and a lot of monounsaturated fat, canola oil has not been people-tested for a long enough time to assure us that it has no long-term adverse health effects. Why take a chance when you can easily and safely use olive oil?

Peanut oil is a poor choice. Scientific research has shown that peanut oil causes clogging of coronary arteries in some animals. Why not use olive oil instead?

Polys Have a Problem. Because consumption of large quantities of polyunsaturated fats have been implicated in the development of certain cancers in animals, we recommend limiting your use of polyunsaturated oil and using olive oil instead (even in baking breads and pie crusts). In addition, when vegetable oils are hydrogenated or partially hydrogenated, they become trans fats and raise LDL-cholesterol (bad cholesterol) levels.

Trans Fats: Bad Guys

For years margarine has been accepted as a good substitute for butter because it contains much less saturated fat and thus is assumed to be

Fat Facts

Total fat is the sum of three types of fat: saturated, monounsaturated, and polyunsaturated fat. While all three types of fat are fattening, each has a different effect on your heart-health.

Saturated fat
Monounsaturated fat
+ Polyunsaturated fat
Total fat

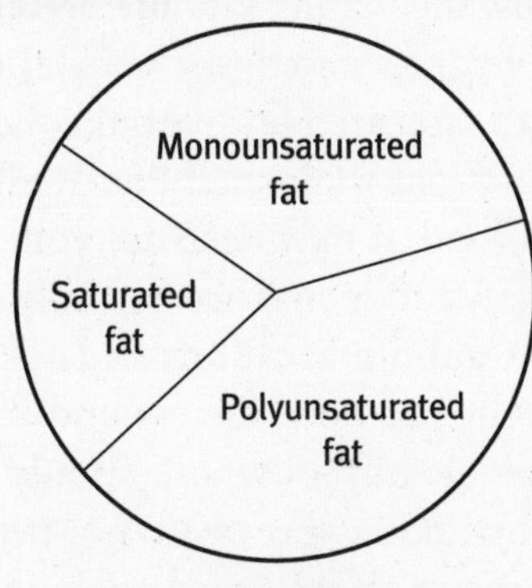

Saturated fat raises your LDL-cholesterol (bad cholesterol) and your risk of heart disease. Saturated fat is found predominantly in animal fats (beef fat, veal fat, lamb fat, lard or pork fat, chicken fat, turkey fat, and butterfat) and the following vegetable fats: coconut oil, palm kernel oil, palm oil, cocoa butter, and hydrogenated and partially hydrogenated vegetable oil. If you think these vegetable fats are rare, check out the labels on convenience and snack foods. Saturated fat is easily recognizable because it is solid at room temperature.

Monounsaturated fat lowers your LDL-cholesterol (bad cholesterol) and your risk of heart disease. An excellent source of monounsaturated fat is olive oil. Olive oil is considered the healthiest of all the oils (see page 92).

Polyunsaturated fat lowers LDL-cholesterol but may also lower your HDL-cholesterol. The major sources of polyunsaturated fat are vegetable oils such as soy, corn, cottonseed, and safflower.

heart-healthy. It turns out that margarine contains trans fats, which act like saturated fat in the body. So, while a somewhat better choice than butter, margarine is not heart-healthy.

Trans fats are rampant in our food system. Anytime you see "partially hydrogenated" on an ingredient list, you know the food contains trans fats. Trans fats are created when polyunsaturated fats and oils are hydrogenated (a chemical process that adds hydrogen to create a more saturated fat) to make them more solid at room temperature and to increase their shelf life. While hydrogenation increases the shelf life of a food, it may decrease yours, so you would be best advised to limit strictly foods that are partially or fully hydrogenated.

Caution: Most fats and oils contain 100–120 fat calories per tablespoon. For both dieters and the health conscious, fats and oils, even heart-healthy olive oil, should be limited in the diet. And remember that cooking sprays are not necessarily a saving in fat calories. To get the scoop on oil-based cooking sprays, see page 88.

Mayonnaise

(page 362, Food Tables)

One hundred percent of the calories of mayonnaise come from fat. One tablespoon of regular mayonnaise has 100 fat calories. At 30–45 fat calories per tablespoon, reduced-calorie mayonnaises are better, but may still represent a good percentage of your Fat Budget. Nonfat mayonnaise is a boon to mayonnaise lovers, but be careful. When you dump a cup of fat-free mayo into your potato salad, you are nurturing your high-fat taste. Use fat-free mayonnaise carefully.

Mayonnaise Salad Sandwiches. Although tuna, shrimp, and chicken are basically low-fat meats, when they are mixed with mayonnaise and made into salads, they lose their low-fat label. A tuna sandwich from a restaurant was analyzed as containing 387 fat calories. If you can't avoid ordering a meat salad sandwich, insist that no additional mayonnaise be spread on the bread.

If you must have mayonnaise on a sandwich, ask the sandwich maker to use no more than a teaspoonful — just enough to moisten the bread. For added taste, sprinkle on hot peppers. If you don't think you like mustard, try it anyway. You may surprise yourself.

Salad Dressings

(pages 360–364, Food Tables)

Although salads have acquired a reputation as the perfect diet food, salads are one of the worst choices for people who want to eat healthfully, and that is because of salad dressing. Most salad dressings are loaded with fat calories. One tablespoon of Russian or blue cheese dressing has about 70 fat calories as does one tablespoon of Italian. But who eats one tablespoon? Make that 8 tablespoons or 560 fat calories.

To reduce the damage, ask to have your dressing served on the side (or if serving yourself, find a small bowl or cup and pour in some dressing) so you can regulate the amount you use. Keep track of how much dressing you put on your salad. Measure by tablespoonfuls (or soup spoons in a restaurant) so that you don't use up your day's Fat Budget before you get to the main course. Better yet, use the Fork Method (see page 46) and get the taste of the dressing without the fat calories.

Low-fat and Fat-free Dressings: Dieters' Refuge. Good news! Many lo-cal and fat-free dressings are truly low- or nonfat. Read the label. Or make your own low-fat dressings. Try substituting nonfat yogurt in salad dressing recipes that call for sour cream. But caution: don't go overboard on your use of nonfat dressing. You want to be moving away from a high-fat taste.

DESSERTS

No dieter has to be told that desserts can cause a Fat Budget blowout. But what's the fun of living, after all, if you can never have a dessert? The beauty of *Choose to Lose* is that it allows you to fit in an occasional high-fat dessert. Just eat under your budget for several days to save up enough fat calories for a big splurge. (See "Splurging," page 181.) But use caution. Desserts are not going to make you thin.

Cakes and Pies — Sweet Temptation

(pages 511–556, Food Tables)

Think very carefully before that piece of cheesecake (270 fat calories), German chocolate cake (277 fat calories), or glazed doughnut (130 fat calories) passes between your lips. Even a little Hostess lemon snack pie has 220 fat calories. Check the Food Tables for the fat calories of your favorites. They can be very high. Try eating a sliver of cake. Eat it very slowly. Savor every bite.

You might also try baking scrumptious low-fat desserts from *Choose to Lose* such as *The York Blueberry Cake* (page 297) at 3 fat calories per slice or *Key Lime Pie* (page 300) at 36 fat calories per slice, or from *Eater's Choice Low-Fat Cookbook* such delights as *Lemon Cheese Cake* (10 fat calories per slice) or *Banana Cake* (6 fat calories a slice).

But before you bake a different low-fat *Choose to Lose* or *Eater's Choice* dessert for each day of the week, remember that overeating all these empty, sugar calories will stall your weight loss. Treat even low, low-fat or nonfat desserts as a splurge.

Cookies: Who's Counting? **(pages 527–537, Food Tables)**

Now that you are so fat-savvy, would you be surprised to learn that one small Fudge Covered Oreo cookie has 50 fat calories, one Pecan Sandie has 45 fat calories, and one Chips Ahoy! Chunky chocolate chip cookie has 40? Would you be shocked to find that one Archway Chocolate Chip 'N Toffee has 70 fat calories? Even one tiny animal cracker has 5 fat calories.

Generally small, cookies don't look fat-packed. But commercially baked cookies range in fat calories from a low of 0 fat calories for Nabisco Fat-Free Fig Newtons to a high of 70 for one Pepperidge Farm Chocolate Chunk Classic cookie.

Read food labels to find the fat calories of your favorites. The problem with cookies is that they are addictive and additive. One (18–70 fat calories) tastes so good, why not have two? or three? or ten (180–700 fat calories)? If you know that you are a cookie monster, leave temptation in the store and keep your home and office cookie jars filled with fruit.

For a discussion of fat-free food abuse, see page 122.

Frozen Desserts **(pages 540–549, Food Tables)**

Rich Ice Cream: Poor Choice. The moment you bite into a Dove Bar or let a spoonful of Häagen-Dazs settle in your mouth, you know. Rich ice cream is full of fat. A cup of Ben & Jerry's Wavy Gravy has 440 calories of fat. A cup of Häagen-Dazs Butter Pecan has 420 fat calories. A Milk Chocolate & Vanilla Dove Bar has 200 fat calories. All ice cream is not that high fat, but it is definitely not diet food. A cup of ordinary vanilla ice cream has about 160 fat calories. Read the label. Of course, you can have half a cup or a quarter of a cup or a tablespoon or two. But don't bring a gallon of ice cream into the house "for the kids" if you know that you are the kid the ice cream is for.

Light Ice Cream: Better Choice. For those who want an alternative to the megafat varieties and aren't quite ready for nonfat, there are some "light" ice creams that will not totally decimate your Fat Budget when eaten in small amounts and with thought. Be careful to read the labels to find the flavor with the lowest number of fat calories. Remember, the serving size is given as one half-cup. Double the fat calories to get a true picture of the cost. Some of the lowest are 40 fat calories for a cup of any Healthy Choice Ice Cream and 50 fat calories for a cup of Edy's Light chocolate fudge mousse or vanilla ice cream. Be aware that although lower than most ice creams, these "light" or "low-fat" varieties still have enough fat calories to demand extra vigilance. A cup of Edy's Light Chiquita 'n Chocolate contains 90 fat calories; a cup of Safeway Select Light Chocolate Chip Cookie Dough ice cream contains 100.

Fat Free Ice Cream: A Mixed Blessing. The word "free" adds to the appeal of nonfat food. "Free to eat as much as you want" is how we read it. But for people interested in losing weight, "free" means only "free of fat" and not "free from restraint." Forget those images of swimming pools full of nonfat ice cream. Treat fat-free ice cream as a splurge.

Frozen Yogurt: Sometimes the Best Bet. If you are an ice cream purist, no other frozen confection may satisfy you. However, a delicious, low-fat alternative does exist: frozen yogurt. Many frozen yogurts are quite low in fat. A typical small serving (5 fluid ounces) of Colombo low-fat frozen yogurt has 23 fat calories; for the same amount ICBIY has 32 fat calories and TCBY has 34.

But don't equate "frozen yogurt" with "fat-free" or "low-fat." Not all frozen yogurts are low-fat. Before you order the super-large size in three different flavors, be sure you ask the server the number of fat calories per serving and the serving size. If he doesn't know, find another yogurt parlor.

You also need to check the labels of the frozen yogurt you find in the freezer cases of your grocery store. Many are truly low- or nonfat. But some of the gourmet flavors rival ice cream for fat calories. A cup of Ben and Jerry's Vanilla Heath Bar Crunch has 100 fat calories; a cup of Chocolate Chip Cookie Dough has 80.

The Best Fat Buy: Nonfat Frozen Yogurt. The greatest treat is nonfat frozen yogurt which has 0 calories of fat and is truly delectable. Again, a caution: although the basic frozen yogurt may be almost fat-free, the almonds, Oreo cookie pieces, peanuts, M & M's, and coconut you load on top are not. Eat your frozen yogurt plain or with fruit toppings.

Caution two: don't eat frozen yogurt for breakfast, lunch, and dinner. Treat it like any other dessert and eat it in moderation. Those empty calories can add up quickly and stall your weight loss. Consider it a sweet, not a dairy. Some frozen yogurts contain little or no dairy products.

Candy Packs a Fat Wallop (pages 522–527, Food Tables)

A candy bar may be small in size but it's gigantic in its fat-calorie count. A 1.7-ounce Mr. Goodbar has 150 fat calories. Two little fun-size Mounds bars contain 90 fat calories. One little caramel has 25 calories of fat. When the taste is gone, much too quickly, all you have left is fat deposits on your frame and a few extra cavities in your mouth. However, if never eating candy would cause you deep remorse, budget candy into your diet . . . CAREFULLY.

Be sure to read the number of servings on the nutrition label. A candy bar you think you could finish off in a few bites may be four servings. Instead of eating the 50 fat calories you quickly noted on the nutrition label, you are eating 4 × 50 fat calories per serving = 200 fat calories.

Although candy has no redeeming nutritional value, some candy has little or no fat. Gumdrops, most hard candies, jelly beans, and marshmallows have no fat, so you can enjoy *a few* with no guilt. They are also devoid of any vitamins, minerals, or fiber, so you don't want to become a jelly bean junkie or marshmallow maniac.

SNACKS

(pages 487–506, Food Tables)

The problem with snacks is not just their high fat content. The problem with snacks is that they are addictive. You can't eat just one — one peanut or one cracker or one potato chip. Or even two. Or even three. Try fifty.

Sitting in front of the television set or among a group of friends, the snacker reaches into a bowl of nuts, crackers, chips, over and over and over: hand to mouth, hand to mouth. She barely tastes what she is eating much less thinks about it. Choosers to Lose must be thinkers. Here are some truths about snacks.

Popcorn

(pages 499–501, Food Tables)

Homemade Air-Popped Popcorn: The Best Buy. We start with popcorn because it has the potential to be one of the greatest snacks of all time. Popcorn that you pop in an air popper has almost no fat. It is chock full of fiber — half insoluble (important for effective bowel function and reducing risk of colon cancer) and half soluble (helps a little bit to lower blood cholesterol and risk of heart disease). It is a perfect snack. It fills you up and satisfies that hand-to-mouth craving at no punishment to your Fat Budget. Air-popped popcorn may taste a bit like bumpy cardboard at first, but in a short time you'll develop a yen for it. And because it contains so little fat and so much fiber, you can eat it until you pop.

Homemade Popcorn Popped in Oil: Not Such a Great Buy. The dietary benefits of popcorn diminish when other popping methods are used because it takes a lot of fat to pop popcorn. Each cup of popcorn prepared in a saucepan with sunflower or corn oil contains about 25 fat calories. That's 1 cup. Who eats 1 cup of popcorn? (To help put 1 cup into perspective, note that the smallest container of popcorn you can buy at a movie theater holds 5 cups and the largest bucket holds 20 cups.) And then, if you flavor it with melted butter or margarine, add 100 fat calories per tablespoon. And who uses just 1 tablespoon of butter or margarine?

Commercial Popcorn: Read the Label. It's a cruel world out there.

The food industry has us pegged. We want to think we are eating healthfully, so we search packages for reassuring buzz words, like "all natural" and "light." We associate "air-popped" with healthy, so we want our popcorn air-popped. But does "air-popped" necessarily mean healthy and does "lite" necessarily mean low-fat? Read the label.

Smartfood White Cheddar Cheese "fresh-tasting, light-textured, totally natural [read as healthy] air-popped" Popcorn contains 90 fat calories for 1¾ cups. The first ingredient is air-popped popcorn, the second is corn oil. Smartfood took a perfect weight-loss food and coated it with fat.

Neither Boston's Lite Popcorn (50 fat calories for 4 cups) or its *40% less fat* White Cheddar Popcorn (50 fat calories for a mere 2¾ cup) is a bargain. On the other hand, Bachman Air-Popped Lite Popcorn has only 15 fat calories for 5 cups. To ensure that you are *really* getting a low-fat buy, read the label. Better yet, air-pop your own popcorn and you will know for sure that the popcorn you are enjoying has very little fat and no risky oils and additives.

Microwave Popcorn: No Buy. Microwave popcorns do not save you money, time, or fat calories. Microwave popcorn costs about $3.65 a pound (versus 49¢ a pound for plain popcorn kernels), takes 4½ minutes to make (versus 3 for air-popped), and ranges from 3 to 33 fat calories per cup (versus 0 fat calories for plain air-popped popcorn). Look at the box below for a comparison of 10 cups of popcorn (the amount of air-popped popcorn we each eat every day) prepared in a variety of ways and packaged by different food producers.

	10 CUPS OF POPCORN	
POPCORN	**COST**	**FAT CALORIES**
Homemade air-popped	$ .07	0
Jolly Time Light Microwave	$ .58	100
Boston Lite	$ .83	125
Boston White Cheddar (40% less fat)	$1.21	182
Orville Redenbacher's Double Feature	$ .81	325
Weight Watchers White Cheddar	$8.90	350
Smartfood White Cheddar	$1.89	514

Movie Popcorn

AMOUNT	SIZE	POPPED IN COCONUT OIL CALORIES TOTAL	POPPED IN COCONUT OIL CALORIES FAT	POPPED IN COCONUT OIL WITH "BUTTER" CALORIES TOTAL	POPPED IN COCONUT OIL WITH "BUTTER" CALORIES FAT	POPPED IN CANOLA SHORTENING* CALORIES TOTAL	POPPED IN CANOLA SHORTENING* CALORIES FAT
5 cups	Small (A); Kids (C)	300	180	472	333		
7 cups	Small (C,U)	398	243	632	450		
	Small (A)					361	198
11 cups	Medium (A)	647	387	910	639	627	342
16 cups	Medium (C,U)	901	540	1221	873		
	Large (A)					850	468
20 cups	Large (all)	1161	693	1642	1134		

A = AMC*; C = Cineplex Odeon; U = United Artists

Sources: *Nutrition Action Healthletter;* Center for Science in the Public Interest; SGS Control Services, Inc.
*Some AMC theaters use canola oil.

Movie Popcorn: BYO. Beware of movie popcorn. It is almost always prepared in coconut oil, the most saturated (heart-risky) oil there is. Don't even consider buying buttered popcorn unless you want a total Fat Budget wipe-out for several days. Bring your own air-popped popcorn. If the usher questions you, tell him you're following doctor's orders.

Nuts to Nuts

(pages 439–442, Food Tables)

Nuts make a delicious snack, add zest to entrées and desserts, possess delectable taste and crunch, and (sigh) are brimming with fat. For example, the little (very little) bag of peanuts the stewardess brings with your drink contains 65 fat calories. Nuts range from cashews at 118 fat calories per ounce to macadamia nuts at 196 fat calories per ounce. Considering that 1 ounce of nuts isn't many, it turns out to be a lot of fat calories. And they are habit-forming. You might want to save nuts for cooking and then, when a recipe calls for a half-cup of peanuts (323 fat calories), use only a quarter-cup (161 fat calories) or 3 tablespoons

(121 fat calories). If having nuts around for any reason demands more self-control than you are able to muster, leave them on the shelf in the store.

Nuts to Coconut. When the soda jerk asks you if you want your dish of nonfat frozen yogurt covered with shredded coconut, you might offer him a lecture on nutrition. There are two reasons to avoid coconut: First, coconut is a high-fat food and almost 100 percent of the fat is saturated. (Saturated fat raises blood cholesterol, which raises your risk for heart disease.) Coconut is what scientists feed rats (which are quite resistant to heart disease) to give them heart disease. Second, and of most importance to the new thin you, coconut is incredibly high in fat calories. One cup of shredded coconut has 241 fat calories. One-half cup of coconut cream has 375 fat calories. One-half cup of coconut milk has 258 fat calories. A little dab will undo you.

Nuts to Seeds. Seeds represent another sneaky fat carrier. They are so small and seem so healthy, but just a few handfuls can deplete your whole Fat Budget. Whole pumpkin seeds contain 50 fat calories per ounce. One tablespoon of sesame seeds contains 39 fat calories. Shelled sunflower seeds contain 127 fat calories per ounce; 39 fat calories per tablespoon. The little package (1⅛ oz) of shelled sunflower seeds you pick up at the grocery store check-out line has 143 fat calories. And they are covered with salt so you want to eat more and more. If you love to nibble on sunflower seeds, apportion yourself 1 tablespoon and consider that a hefty treat.

For a real education, check the **NUTS AND SEEDS** section of the Food Tables.

Trail Mix. Trail Mix (mixed nuts, seeds, and dried fruit) conjures up the aura of outdoor vitality and good health. Out on the trail, hiking up a mountain, fording a stream (or, perhaps, sitting in your den watching TV), you gobble up handfuls of this tasty mix. STOP. Consider the contents — almonds (132 fat calories per ounce), cashews (118 fat calories per ounce), peanuts (126 fat calories per ounce), sunflower seeds (127 fat calories per ounce), flaked coconut (82 fat calories per ounce), dried fruit (0–13 fat calories per ounce). This "healthy" snack food could contain anywhere from 85 to 120 fat calories per ounce.

Just Say No to Chips

(pages 487–490, Food Tables)

If you are like most normal human beings, you will find it impossible to keep from eating potato chips if they are in your house. Since 1 ounce (about 14 chips) has 80–100 fat calories, you might want to avoid bringing temptation home from the store. If you have an uncontrollable urge to eat potato chips, buy the smallest deli bag (80 fat calories). You may find it worth a big chunk out of your Fat Budget every once in a while.

Any chips — Fritos, Chee-tos, nachos — should be eaten with considerable thought. Chips are basically fat stiffened with a bit of vegetable. In fact, the first ingredient of French's French Fried Onions is partially hydrogenated soy oil. The second is onions!* And the fat calories show it. A tiny (2.8-oz) can contains 330 fat calories. A 3-ounce deli bag of Fritos corn chips contains 270 fat calories. Snack chips are truly addictive and can be consumed by the hundreds and thousands. The last one always invites another. Our strong recommendation is to leave temptation out of the house by not bringing chips in. No one needs them.

New Chips in Town. As if potato chips, cheese twists, and corn chips weren't bad enough, bagel chips and pita chips were invented to add even more temptation to an already gigantic selection. At first glance, you may even have thought that bagel chips and pita chips were healthy and low in fat.** Ha! Three bagel chips — a whiff and they're gone — cost 35 fat calories. More likely you'll eat the whole bag for 280 fat calories.

If you are swayed to buy apple chips or plantain chips for a healthy snack, read the package label, shake your head in disgust, and buy the real fruit instead. Twelve Seneca Apple Chips contain 65 fat calories, and 1 ounce of Plantain Chips contain 100. One-third cup of banana chips contains 60 fat calories.

For a discussion of fat-free chips see page 124.

---

*Ingredients on a food label are listed in descending order by weight.

**A truly low-fat bagel chip recipe appears on page 213 of *Eater's Choice Low-Fat Cookbook*.

Crackers (pages 490–495, Food Tables)

Crackers are a high-fat food and, like potato chips and peanuts, are consumed in great numbers with nary a thought. You could guess that six small cheese and peanut butter crackers would have a lot of fat — 100 fat calories — because cheese and peanut butter are high fat. But even innocent sounding crackers like Vegetable Thins have 80 fat calories for 14 little crackers. It is very easy to eat 14 crackers or even 28 crackers (160 fat calories). Choosers to Lose must pay attention. The best policy is to avoid the cracker section of your grocery store and the cracker tray at a party. But if you must have crackers, keep track of every one you eat. Perhaps, as a treat, have two crackers a day. You know yourself. Will the box of crackers be so tempting that you will eventually weaken and gobble them all up? If so, leave the crackers in the store.

Reduced Fat and Fat-Free. You can find low-fat and reduced-fat crackers on your grocery shelves, but any kind of cracker is a poor choice. First, reduced fat does not mean nonfat, so the 25 fat calories you pay for an ounce of reduced-fat crackers can quickly become 50 fat calories for 2 ounces. Even the few truly nonfat brands like Devonsheer's Melba toast and Melba Rounds and Wasa's Crispbreads should be eaten with sense.

Eat them plain or with jelly. Nonfat crackers covered with cheese, cream cheese, or some other high-fat spread join the high-fat hit parade. The warnings that apply to eating too many nonfat cookies also apply here. See page 122 for a discussion of fat-free food abuse.

THE GOOD NEWS

Dry your tears. You can still fit any of these fat-laden foods into your Fat Budget. But now you know how to choose carefully and wisely. You know that you must always read the nutrition information box and determine fat calories per serving and the number of servings you are eating. You won't be duped by misleading hype. You'll be able to make better choices.

You might ask yourself some questions. Why eat a can of solid white tuna packed in oil at 295 fat calories when a can of solid white tuna packed in water has only 30 fat calories? Do I really want to spend 130 fat calories on one cup of Ramen Noodles? Will a bite of apple pie (about 16 fat calories) instead of a whole slice (160 fat calories) satisfy me? Refer to this chapter and to the Food Tables in the back of the book to make educated choices that will produce a streamlined and glorious you.

Coming next . . . finding out the great foods YOU can eat.

Remember:

1. Americans like to eat high-fat foods, and the foods available in supermarkets, restaurants, and everywhere cater to this high-fat taste. In order to become and stay thin and healthy in this high-fat world, you need to know not only where fat lurks but in what amounts.
2. Always judge the fat content of foods with your Fat Budget in mind.
3. Use the Food Tables to compare the fat contents of foods and to find lower-fat alternatives to high-fat favorites. Remember to adjust total and fat calories for the actual amount of a food you eat.
   a. Red meats, such as beef, veal, lamb, and most pork products, are high in fat. The higher the grade of meat, the more fat it contains.
   b. Turkey and chicken with the skin removed (always remove the skin before cooking chicken), shellfish, and many fish are low in fat. But these low-fat foods are easily made high-fat foods by the way they are prepared.
   c. Most sausage and luncheon meats are extremely high in fat (in addition to containing cancer-risky nitrites and blood pressure–raising sodium).
   d. Many frozen dinners and convenience foods are riddled with hidden fat. Read the food labels to see how (or if) you can fit these foods into your Fat Budget.

e. Most fast foods are loaded with fat (and sodium). Do your diet and health goals a favor and approach fast-food restaurants with extreme caution or not at all.

f. Dairy products are excellent sources of calcium, but many are high in fat. A glass of whole milk contains 75 fat calories. An ounce of cheddar cheese contains 85 fat calories. Choose low-fat dairy products such as low-fat or nonfat yogurt, low-fat cottage cheese, and skim milk. Fit cheeses into your budget sparingly, and you will really enjoy them.

4. Remember that your Fat Budget is a ceiling. You should not eat above it and preferably should eat well within it. But, for your psychological health, be sure to include an occasional high-fat splurge.
5. Your minimum total calories are a floor and you must eat well above this amount to ensure that you provide enough calories to fuel your physical activity, maximize your basal metabolic rate, and consume adequate amounts of vitamins, minerals, and fiber.

5

Low-Fat and Delicious

"For the first time I think of what I *can* eat instead of what I *can't* eat."

— Dawn Monroe, Bolton, Ontario

EVERYONE IS ALWAYS TELLING YOU what you can't eat. You can't eat chocolate. You can't eat French fries. Never touch a baked potato. Forget jelly unless it is 100 percent fruit. The list of no-no's is encyclopedic. In fact, knowing that they will have to give up everything they love often makes people hesitant to change their high-fat ways.

But when you become a Chooser to Lose, you will learn you don't have to give up enjoying food — quite the contrary. You will be able to eat more delicious food than you ever ate before.

Making Educated Choices

Choose to Lose will help you in two ways. First, since you now know how much food costs in fat calories, you can decide if a certain food is worth the splurge. Of course, it won't be at every meal or every day, but you won't feel guilty when you fit it in. Second, in this chapter, you will find out which foods that you once avoided should be relished and which "no-no's" are really "sometimes." You will learn that there are many delicious foods that you should eat a lot of. You need never be hungry.

Eating Is a "Yes-yes"

It is crucial that you eat enough calories — nutrient-dense/fiber-rich calories. You figured out your minimum daily total caloric intake in

Chapter 2, "The *Choose to Lose* Plan." Remember, this is a FLOOR. It is the number of calories you need to satisfy your basal metabolic rate. Make sure that you are eating *more* than this amount. You need additional calories to provide energy for your physical activity. When you reduce fat in your diet, there is a real danger that your total caloric intake will fall too low. You must replace some of the fat you eliminate with foods rich in complex carbohydrates — whole grains, vegetables, fruits, and low-fat meats. You need the vitamins, minerals, and fiber these foods provide.

Overcoming Undereating

One of the most difficult aspects of *Choose to Lose* is getting people to eat a lot of food. Many people are terrified of food. Food is the enemy. To lose weight, they reason, you need to eliminate the enemy. Besides, they do not think a "diet" is worth its salt unless they are suffering. Only if they are eating lettuce with lemon juice for breakfast, lunch, and dinner will they lose weight. Starvation diets have an advantage of being over quickly because you can't follow them for long. Within a few days or weeks you can return to old high-fat eating habits.

In addition, many people are not used to eating a lot of food. They eat a tremendous amount of fat, but very little bulk. A little sandwich with 2 slices of bologna and 1 slice of cheese and a tablespoon of mayonnaise could contain 350 fat calories. Not much food but a heap of fat. Contrast this with a white-meat turkey sandwich with mustard, tomato, and hot peppers on whole-wheat toast, an apple, an orange, carrot sticks, 6 cups of air-popped popcorn — all for less than 40 fat calories. The person eating the bologna and cheese sandwich can truly complain that she eats little food but she shouldn't be surprised that she is fat. The person eating the turkey sandwich is full, satisfied, and losing weight.

Conquering Your Fear of Eating

First, you have to get over your fear of eating. It's not the amount of food you eat but the amount of fat that is fattening. Second, you have to replace your old ideas about which foods are fattening.

CARBOHYDRATES DON'T MAKE YOU FAT

Carbohydrates used to be (and remain to a great extent) the Number One Forbidden Nutrient in every diet. You were advised to eat the 6-ounce hamburger and toss the bun, eat the 5 ounces of veal breast and chuck the baked potato. Let's see what that means in fat calories. Keep the hamburger (348 calories of fat) and toss the bread (20 calories of fat). Keep the veal (270 fat calories) and chuck the potato (0 fat calories). Huh? They must be kidding.

FOOD GUIDE PYRAMID

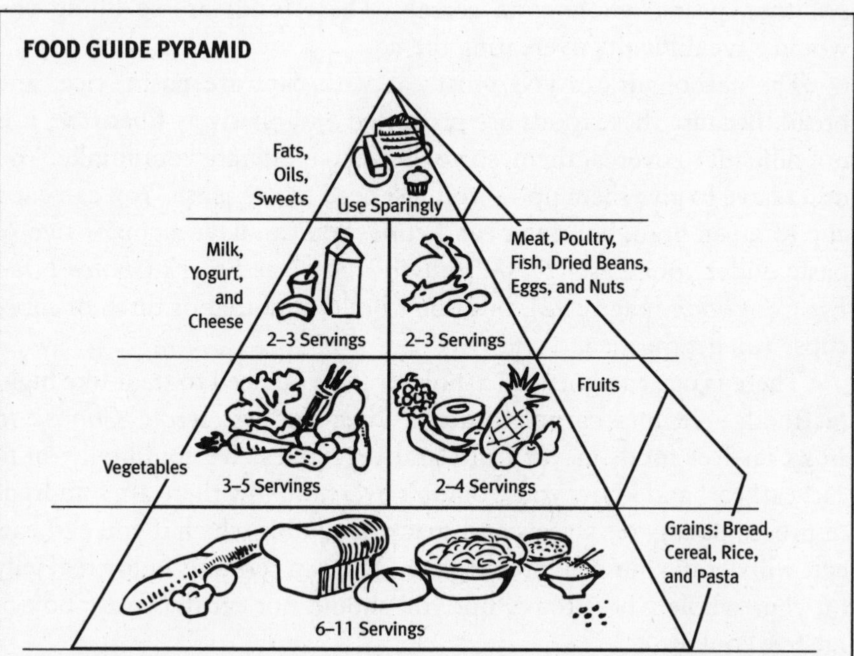

This is the Food Guide Pyramid developed by the Department of Agriculture to give Americans an easy way to see which foods they should be eating for long-term health. You will notice that the two lower levels consist of carbohydrates. These foods — grains, vegetables, and fruits — should be eaten in the greatest amounts because they are most beneficial to your health.

Here's the great news. You may eat bread. You may eat potatoes. You may eat rice. You may eat spaghetti. You may put jelly on your toast, real jelly with sugar, not all-fruit jelly. You may eat a *whole* banana and an apple and pear and orange and grapes. The USDA Food Guide Pyramid recommends that you eat *at least* 2 to 4 servings of fruits, 3 to 5 servings of vegetables, and 6 to 11 servings of grains each day to maintain your health.

As you can see from the Food Guide Pyramid, carbohydrates are our nutritional base. The carbohydrates you can eat with abandon are the ones that have lots of nutrients and fiber — like carrots, broccoli, potatoes, pears, whole-grain cereals. **These foods are so filling you would have difficulty overeating them.**

The carbohydrates you must eat with care are pasta, rice, and bread. Because these foods are processed and relatively fiber-free, it is not difficult to overeat them, so you have to moderate your intake. You don't have to give them up — you just need to use sense. You can eat a cup of raisin bran, but don't eat 3 cups. You can have a cup or two of pasta under your *Zesty Fresh Tomato Sauce* (see *Eater's Choice Low-Fat Cookbook,* page 192), but don't build a mound of three or more cups. You get the idea.

There is one category of carbohydrates you need to treat like high-fat foods — empty carbohydrates. When we first wrote *Choose to Lose,* fat-free foods meant fruits and vegetables. We could tell you to eat! eat! eat! and know you wouldn't overeat. Now there are hundreds of processed fat-free sweets and snacks available which if you eat! eat! eat! will derail your weight loss goals. You can certainly enjoy real jelly on your whole-wheat toast, but you should not eat the entire box of fat-free cookies.

Adding Carbohydrates to the Fire

Contrary to what many current diet fad books contend, carbohydrates are *not* the bad guys. (See Chapter 15, "Frequently Asked Questions.") Scientific evidence shows that carbohydrates eaten in moderation (even empty carbohydrates) do not turn to fat.

Let's review the reasons why this is true. Except for the small amount stored as glycogen, carbohydrates are entirely burned to provide energy to fuel your body functions and your physical activity. Any carbohydrate eaten in excess of your needs is burned and the energy is wasted as heat. **It is not converted into fat under normal conditions.** By eating lots of carbohydrates (such as fruits, vegetables, and whole grains) you keep your basal metabolic rate chugging away at a higher rate.

Putting Fat into Cold Storage

In contrast, the fat that you eat is immediately added to your fat stores. It doesn't get burned up. It doesn't add fuel to the fire. It is as if you took a syringe and injected the ice cream sundae that you just ate directly into your blubbery thighs.

COMPLEX CARBS: WEIGHT-LOSS STARS

Here are foods you no longer have to spurn.

Bread: A Yes-Yes

Hey! You can eat BREAD! Consider bread diet food. A slice of commercially baked bread has 10 fat calories. (Check labels — some have more.) You can eat it for a snack. Shun open-face sandwiches. Give your sandwich a top and a bottom. You need those calories of carbohydrate. Besides, you will feel fuller and more satisfied eating both slices of bread.

Choose whole-grain breads — they are healthier than white because they are rich in fiber, protein, thiamin, riboflavin, niacin, folic acid, vitamin E, iron, phosphorus, magnesium, zinc, and other trace minerals. Check labels. The first ingredient should be whole-wheat flour. Some breads labeled "wheat bread" may contain no whole wheat. Some bread manufacturers substitute caramel coloring for whole-wheat flour.

And, of course, even with whole-grain breads, don't overdo. A whole loaf of bread (1020 total calories and 170 fat calories) is a lot of calories for your body to burn and a lot of fat to go to your fat stores.

BREAD + Fat: A No-No

When bread is toasted with butter and garlic or cheese, it quickly loses its healthy carbohydrate rating. A slice of New York Brand Texas garlic toast contains a whopping 90 fat calories. A 2½-inch slice of Pepperidge Farm Mozzarella Garlic Cheese Bread also contains 90 calories of fat. A croissant may contain 110 fat calories, and muffins often contain more than 150 fat calories. Consider these "breads" fats and eat them as splurges.

Eat Potatoes to Avoid Looking Like One

Can you believe it? Eating potatoes will help you lose weight. Bake them, boil them, broil them. Fill them with nonfat yogurt, salsa, or eat them plain. Bake *Potato Skins* (page 284). At 7 fat calories a serving (or 0 if you use no oil), you can eat them with zest. Peel, slice, and boil sweet potatoes until tender. Then whip them with orange juice for a wonderful supper treat. Bake an extra potato when you bake potatoes for dinner and eat it cold as a snack.

Your mouth drops open in disbelief. How can potatoes be diet food? Potatoes are almost completely free of fat. And they are a good source of vitamin C, thiamin, niacin, and iron. Sweet potatoes are loaded with vitamin A in addition. Screw up your courage and eat a potato or two every day. See if you don't look less and less like a potato.

Low-Fat Potato + Fat = High-Fat Potato

Of course, if you stuff your potato with a fat (butter, margarine, cheese, sour cream) or slice and fry it into French fries or slice it even thinner and fry it into potato chips, you can no longer consider potatoes diet food. French fries and potato chips are among the worst offenders on the Most Unwanted Fat list.

Pastas + Rice = Fat-Loss Success

Two more complex carbohydrates to add to your diet list — pasta and rice. Don't be shy. Pile a heap of steamed rice on your plate and top it with a low-fat chicken or vegetable dish. Fill a large bowl with spaghetti and cover it with a low-fat tomato sauce. Eat more of the pasta or rice and less of the topping.

Are you having difficulty imagining a cup or two of rice or spaghetti on your plate? And actually eating them? Those who have always considered a quarter cup of rice or spaghetti extravagant must make an effort to regard them in a positive light. Of course, use sense. One to two cups are fine; four are more than you need.

Pasta + Cream Sauce = Weight Gain

Of course, pasta and rice are not always reducing foods. Mix in butter or margarine or cover them with high-fat sauces or cheese and you destroy their beneficial weight loss properties. Prepared pastas such as Contadina Chicken and Prosciutto Tortelloni (110 fat calories for 1 cup) don't even need added sauces to give your Fat Budget a nervous breakdown.

You must also be sure that the basic pasta is really nonfat. Pastas such as Oodles of Noodles and Top Ramen Noodles sound like great nonfat choices. They are not. A quick bath in sizzling hydrogenated vegetable oils has made them into high-fat no-no's.

Cereals: Breakfast of Champions

What a treat it is to eat a large bowl of cereal without feeling like a sinner. Eat cereal for breakfast and eat it for a snack. Of course, the milk should be skim and the cereal should have no more than about 10–20 fat calories per serving. The more fiber the healthier. Check the fiber listing on the box. (Read on to find the importance of fiber in your diet.) Cereals filled with cookies or candies or marshmallows should be considered dessert and eaten in limited amounts.

VEGGIES

No one needs to tell you that vegetables are an important part of a diet. You have probably followed diets in which the fear of turning into a rabbit was justified. But don't disparage vegetables. Not only are they low in fat, high in vitamins, minerals, and fiber — they are also just plain delicious. Eat them raw. Or cut them into matchsticks, steam them until just tender, and press garlic or grind pepper over them or sprinkle them with herbs. Create a vegetable dip by mixing Dijon mustard into nonfat yogurt. Try vegetables you have never tried before. Have you ever baked spaghetti squash? When you scrape out the cooked interior, it separates into strands, exactly like spaghetti but more nutritious and delicious. Top it with a low-fat tomato sauce. Bring home a cartful of vegetables each time you shop. Then you'll have cauliflower for *Cauliflower Soup* (page 265), green beans for *Indonesian Chicken with Green Beans* (page 272), and sweet potatoes, spinach, zucchini, carrots — a panoply of vegetables — to enhance your dinners.

FRUIT

Food of the Gods

Fruit makes a great snack, a great lunch, and a great dessert. In fact, we should emulate the Europeans and top off our meal with fruit instead of rich, high-fat desserts. In season, cantaloupe, honeydew melon, watermelon, strawberries, raspberries, peaches, nectarines, pears, and plums taste divine. Packed with vitamins, minerals, and fiber, fruit is nutritious and will put no strain on your Fat Budget. You can eat fruit until you feel like a human fruit bowl (except for avocado, which has 270 fat calories). Eat a *whole* banana. Eat *two bananas*! Bring a bag of fruit to the office and dig in whenever hunger strikes.

Fattening Fruit

Of course, even fruit can be ruined. Vanilla ice cream covered with strawberries, raspberries in cream, or blackberries in sour cream may taste delicious, but they won't keep buttons from popping off your slacks. Substitute nonfat yogurt or nonfat frozen yogurt for a similar but non-fattening effect.

FIBER

By eating all these wonderful fruits and vegetables and whole grains, not only are you getting a heaping amount of vitamins and minerals, you are also fulfilling your requirement for fiber. Fiber is essential for good health. A high-fiber diet reduces the risk of colon cancer and may reduce the risk of breast cancer. It helps prevent diverticulosis and hemorrhoids as well as constipation. It can even lower blood glucose levels.

The fiber chart (page 116) shows you the amount of fiber in a variety of foods. The Food and Drug Administration recommends eating 25 grams of fiber a day. Most Americans eat less than 10 (which may be the reason why, along with a high-fat diet, we have such high rates of colon cancer). But don't overdo a good thing — eating more than 50 to 60 grams of fiber a day may decrease the amount of vitamins and minerals your body absorbs.

When you are making choices, keep fiber in mind. Eat the fruit rather than just drinking the juice. Eat the skin on your potato. Choose a cereal with a high-fiber content.* Eating the *Choose to Lose* way, you will automatically be eating enough fiber. See Chapter 13, "Ensuring a Balanced Diet," for more details on the fruits, vegetables, and whole grains you need to eat to ensure an adequate intake of fiber.

EAT THE REAL THING

Get your antioxidants right here. You don't need megadoses of vitamin A, vitamin E, or vitamin C supplements. You need to get your vitamins from real food — fruits, vegetables, and whole grains, low-fat meats, and non- and low-fat dairy. It is the antioxidants, phytochemicals, fiber, vitamins, and minerals in real foods that work together to help reduce your risks for cancers and heart disease.

---

*Nutrition labels of commercial foods list dietary fiber. See page 145.

Dietary Fiber in Foods

FOOD	AMOUNT	DIETARY FIBER (G)
Vegetables, cooked		
Broccoli	½ cup	2.0
Brussels sprouts	½ cup	3.4
Potato, baked w/skin	1 medium	3.6
Spinach	½ cup	2.0
Sweet potato	½ medium	1.7
Zucchini	½ cup	1.3
Vegetables, raw		
Carrots	1 medium	2.3
Celery	1 stalk	0.6
Cucumber, sliced	½ cup	0.5
Lettuce, romaine	1 cup	1.0
Mushrooms, sliced	½ cup	0.5
Spinach	1 cup	1.5
Tomato	1 medium	1.6
Fruits		
Apple (with skin)	1 medium	3.0
Banana	1 medium	1.8
Blueberries	½ cup	1.7
Cantaloupe	¼	1.1
Figs, dried	2	3.5
Orange	1 medium	3.1
Peach (with skin)	1	1.4
Pear (with skin)	1 medium	4.3
Prunes, dried	3	1.8
Raisins, seedless	¼ cup	1.9
Strawberries	1 cup	3.9
Legumes, cooked		
Baked beans, canned	½ cup	9.8
Kidney beans	½ cup	7.3
Lentils, cooked	½ cup	3.7

Dietary Fiber in Foods (continued)

FOOD	AMOUNT	DIETARY FIBER (G)
Breakfast cereals		
All-Bran	⅓ cup	8.5
Bran Chex	⅔ cup	4.6
Bran flakes	¾ cup	5.3
Cornflakes	1¼ cup	0.6
Granola	¼ cup	3.2
Oat bran, raw	⅓ cup	4.9
Oatmeal, cooked	¾ cup	1.6
Raisin bran	¾ cup	4.8
Breads, grains, & pasta		
Bagel	1	1.2
French bread	1 slice	0.8
Pumpernickel bread	1 slice	1.9
Rice, brown, cooked	½ cup	1.7
Spaghetti, cooked	½ cup	1.1
Whole-wheat bread	1 slice	1.9
Snack foods		
Popcorn, air-popped	1 cup	0.9

Source: J.A.T. Pennington, *Bowes and Church's Food Values of Portions Commonly Used,* 15th ed. Philadelphia: J. B. Lippincott, 1989.

CHOICE MEATS

Chicken: Infinite Variety

Chicken can be prepared in umpteen million delicious ways without harming your diet. Choose the breast, as it's the part lowest in fat. Always remove the skin *before* you cook it. Steam, bake, broil, or stew to minimize added fat. Use spices for taste. Substitute nonfat yogurt, skim milk, or buttermilk for cream, sour cream, or whole milk to create tasty, low-fat dishes.

Spices Get a Bum Rap

If you were to believe all television commercials, you would blame spices for giving heartburn to all those actors in the Pepcid AC, Tagamet, Alka Seltzer, Maalox, and Mylanta ads. It just ain't so. People get heartburn from eating fat, not from eating spices. It's the greasy cheeseburgers and fries, not the *Choose to Lose Cajun Chicken.* We have heard from countless Choosers to Lose that they no longer suffer heartburn or reflux since they moved to a low-fat, spice-filled diet. They feel great!

Turkey Without Skin: Low-Fat Treat

White turkey meat has only 2 fat calories per ounce. You can't do much better than that. For a low-fat treat, fill a whole-wheat pita or French bread baguette to overflowing with turkey breast meat and dress it with mustard, tomato, lettuce, and hot peppers. Try *Turkey Mexique* (page 277) for a tasty treat. Throw cooked turkey into low-fat sauces. Roast a big tom and you'll have turkey leftovers to freeze and use for weeks.

Seafood: Lean Cuisine

Many fish are irresistibly lean and delicious — 2 fat calories per ounce for raw cod, dolphin fish, haddock, lobster, pike, pollock, and scallops; 3 for grouper, snapper, and sunfish; and 4 for flounder, monkfish, perch, rockfish, shrimp, and sole. Just bake it, grill it, poach it, or broil it. Skewer it with vegetables. Cover it with low-fat sauces. Eat away.

Warning: Chicken, turkey, and fish can be exceedingly fattening if prepared incorrectly. Read the Chicken, Turkey, and Fish sections in Chapter 6, "Where's the Fat?" to learn how easily these low-fat meats can be pumped up with fat.

Treat Yourself: Turn Chicken, Turkey, and Fish into *Choose to Lose* and *Eater's Choice* Entrées

Nancy Goor loves to eat. She's not wild about cooking. She does it because she loves to eat so much. Most of the recipes she has created in *Choose to Lose* and *Eater's Choice Low-Fat Cookbook* are quick and easy to prepare because she doesn't want to spend her life cooking. She has other things to do and she knows you do, too. If you want to lose weight, you (or someone in your household) should cook. You need to be satisfied. You need to eat full dinners. It is not enough to eat only steamed broccoli and three carrot sticks for dinner. How long will you stay on a low-fat diet if you eat so little? But if you eat them with *Ginger-Carrot Soup* (page 265), *Black Bean and Corn Salad* (page 290), *Bar-B-Que Chicken* (page 270), and *Curried Whipped Potatoes* (page 283), you will be full and content. Eating lots of delicious, low-fat food is the secret to staying thin. When you start changing your eating habits and eat wonderful low-fat foods, you will find life worth living. Not only will your fat melt away, your whole life will be richer. You'll see.

SNACKS

Popcorn

Air-popped popcorn — that is, *true* air-popped popcorn that you pop yourself (as opposed to commercial packaged brands, which may add 100 calories of fat per ounce) — has almost no fat. If you ate the same amount of air-popped popcorn with no added fat as the large container of coconut oil–popped movie theater popcorn — 22 cups — you would not gain weight. (However, your jaws might never work again.)

Pretzels

Pretzels make a great diet snack if eaten in moderation. (We eat one or two for dessert.) Many, such as Snyder's of Hanover Sour Dough Pretzels, are fat-free. Many are very low in fat. Twelve Utz Pretzel Stix have only 5 fat calories. However, before you invest in a warehouse of pretzel tins, note that most brands are heavily encrusted with salt (sodium raises blood pressure in some people and causes fluid retention) and, as your aim is to eat primarily nutritious (not processed) foods, you don't want to pig out on even the lowest sodium nonfat pretzels.

Don't take for granted that all pretzels are low- or nonfat. Snyder of Hanover's Honey Mustard and Onion pretzels contain 60 fat calories for ⅓ cup. Combos Cheddar Cheese Pretzels contain 45 fat calories for ⅓ cup. Read the labels.

Yogurt: Grade-A Snack

Nonfat flavored yogurts are truly a treat. Colombo is our favorite. Choose coffee, lemon, vanilla, and a variety of fruit flavors. Even if you don't think you like yogurt, check these out. They make good snacks and desserts. Try the Roland Lippoldt special for a nonfat dessert delicacy: mix apple sauce and nonfat plain yogurt and top with a sprinkle of pumpkin pie spice; add a cut-up apple for extra crunch.

SUGAR GUILT

Sugar guilt is intense in dieters. Intense but misguided because sugar by itself will not make you gain weight. It is the fat that accompanies the sugar that makes you fat. Think of all those wonderfully rich desserts you have been avoiding. They contain sugar *and* butter, chocolate, whipped cream, cream, or sour cream. The big "and" makes the dessert fattening.

So, although you can only occasionally fit high-fat desserts into your Fat Budget, you can give yourself a daily treat by using *real* jelly or jam on your English muffin without a twinge of conscience. (In fact, you should use jelly and forget about the margarine or butter.) You can suck an occasional peppermint, jelly bean, or hard candy without guilt.

Sweet Treat

No, you should not binge on sugar, either. Sugar is just a lot of empty calories containing no vitamins, minerals, or fiber. If you are diabetic or have high triglycerides you should avoid sugar. Otherwise, treating yourself to an occasional hard candy may give you a lift. (Be sure to check labels for ingredients; some hard candy, such as butterscotch, is made with heart-risky, fattening oils.) If you ordinarily skimp on maple syrup on French toast, pouring a little more of the maple syrup and eating a lot less of the French toast will do both your psyche and your chassis a lot of good. A cup of vanilla nonfat yogurt may have about 120 calories of sugar, but since it has 0 calories of fat — enjoy the treat.

Alcohol — NOT a Fat-Free Panacea

Although alcohol has no fat, its consumption should be limited for other reasons. Not only are alcohol calories empty since they provide no nutritional benefits — drinking too much alcohol can wreak havoc on your health (and if you drive drunk, on the health of others). Alcohol can weaken your inhibitions and you may end up eating more fat than a sober you would have allowed. In addition, drinking too much alcohol will slow down or stall your weight loss. The body can't store alcohol so it burns it off immediately instead of burning fat from your fat stores. Limit alcoholic beverages to a maximum of 1–2 drinks a day.

DON'T FORGET: THIS IS A HEALTHY WAY TO EAT . . .

. . . Not a License to Overdo

Now that you are feeling free as a bird, we repeat our warning. You need not focus on limiting carbohydrates, but you must use sense. For example, many people who start following *Choose to Lose* become bagel-monsters. Bagels taste good and they are so convenient. However, while one large bagel at 300 total calories and 0 to 15 fat calories is a fine snack, two at 600 total calories or three at 900 total calories are an excessive amount of basically empty calories.

Don't eat a food without paying attention just because you *think* it is a carbohydrate and low-fat. The oat bran muffins you greedily devour without guilt or thought may contain anywhere from 75 to 250 fat calories each. Think first. Don't make assumptions. Always check the Food Tables or food labels.

Watch Simple Sugars

Choose to Lose allows you to eat jelly on your bagel or pop a lifesaver into your mouth. But don't abuse the system by consuming a can of hard candy and two six-packs of soda. Eating too much simple sugar will do nothing for your health, could push out nutritious foods you should be eating instead, and will stop weight loss by sparing the fat in the fat stores. Don't deprive yourself, just use sense.

FAT-FREE FOOD ABUSE

Free May Cost You

When many new Choosers to Lose find out that carbohydrates aren't the devil, they race to their supermarket. They whiz by the produce department and slide into the baked goods section, where they pile their baskets full of fat-free and low-fat goodies. They think they are following *Choose to Lose* by keeping their calorie count high with loaves of nonfat cakes and boxes full of fat-free cookies and zero-fat crackers. WRONG! Choosers to Lose need to eat lots of food but not just *any* food. They need to eat nutritious foods like fruits, vegetables, whole grains, nonfat dairy, low-fat poultry, and fish. A nonfat cookie or two after dinner, a fat-free pretzel or two with lunch, is a treat. But when your menu consists of mainly ersatz nonfat food instead of nutritious food, not only do you deprive yourself of necessary vitamins, minerals, and fiber for good health because you are forcing out nutritious foods, you will find it difficult to lose weight.

Fat Free Means Free of Fat

The term "fat free" is a great marketing coup. Not only does FREE describe a food that contains no fat, it gives you the feeling you are free as a bird to eat it. You are free to eat and eat and eat and kick up your heels as you do it. Alas, the foods that are marketed as fat-free are also free of fiber and nutrients. Eating them freely holds your weight loss/health goals hostage.

Weight Loss Standstill

Because these fiber-free, high-empty-calorie foods are so easy to eat, hundreds of these empty carbohydrate calories slide right down the ol' gullet. The calories quickly add up to huge amounts. Eat 4 slices (easy to do) of an Entenmann's Marble Loaf and you will have consumed very little fat but a hefty 520 total calories. Add 3 slices of an Entenmann's Louisiana Crunch Cake — hardly a dent in your appetite — and you will have consumed another 630 total calories. Add 2 cups of Safeway Select Fat-Free ice cream — no fat, but 320 total calories.

Your body can't store all those carbs, so it must burn them. But because you have so many calories to burn, your body burns them and never gets around to burning what you want to lose — the fat in your fat stores. You won't gain weight (unless you eat more than 2200 calo-

You could not possibly eat five plain baked potatoes (500 total calories) at one time, but you could inhale twelve Entenmann's fat-free oatmeal-raisin cookies (500 total calories) . . . and twelve more (500 total calories) and twelve more (500 total calories). Nutrient-dense/fiber-rich foods are self-limiting, but fiber-free foods make no impression on your appetite. You need to treat empty calories as high-fat splurges — fit them in rarely.

ries of pure carbohydrates in addition to your normal diet for five or six days — not easy to do), but your weight loss will be completely stalled.

New Kid on the Block

Your weight loss will be stalled and, in addition, you may have a severe case of stomach cramps if you eat chips or crackers made with the newest fat-free product on the market — olestra (Olean). Olestra is a man-made fat molecule composed of sugar and vegetable oil. What makes it so desirable to fat lovers is that it is too large to be absorbed by the intestine. So instead of being absorbed and then distributed to the fat stores, it just passes through. Thus the consumer gets to eat fat without the fat staying around to become blubber.

If it seems too good to be true, it is. Olestra comes with more than a few problems. First, there are the gastrointestinal problems just mentioned, which will negatively affect some consumers. In addition, because of its structure, when olestra passes through the body it takes the fat-soluble vitamins A, E, D, and K and carotenoids with it. The manufacturer of Olean®, Procter and Gamble, fortifies its products with the vitamins, but not with the carotenoids. The loss of carotenoids may have long-term health consequences because carotenoids appear to reduce the risk of cancer, heart disease, and macular degeneration (a common cause of blindness in the elderly). For those people who take a blood-thinning medication such as coumadin, olestra may be dangerous.

Equally as important, the long-term effects of eating olestra have not been measured. According to Dr. Henry Blackburn, a professor of public health at the University of Minnesota and one of the five advisors on the FDA Advisory Committee of 20 who voted against recommending olestra's approval, the lack of long-term clinical trials with adequate numbers of humans and the insufficient follow-up on cancer findings in one mouse study made him "unable to arrive at a reasonable certainty of no harm."

Even if olestra did not have potential health problems, consumers need to carefully monitor their intake of olestra-filled fat-free snack foods. Just like overdosing on high-sugar/fat-free foods, eating too much of these foods will push out more nutritious foods and the hundreds of empty calories consumed will be burned in preference to stored fat. In addition, these fat-free fatty tasting foods will keep up consumers' high-fat taste so they will never lose their taste for high-fat foods.

Fat-Free Mislabeling

Are you sure that what you see labeled as fat-free is truly free of fat? In its May 1994 issue, *New York* magazine had a variety of low- and nonfat foods sold in local food shops analyzed for their fat content. A tofu cheesecake billed as fat-free actually contained 221 fat calories; a fat-free, sugar-free diet corn muffin, 204. Tasti D-lite Frozen dessert (peanut butter*) claimed less than 9 fat calories per small serving, when in truth an 8.9 fluid ounce serving contained 58 fat calories. Ask to see the nutrition label. If that doesn't exist, ask to see the ingredient list. If there is none, leave the premises empty-handed.

Nonfat High-Fat Taste

The reason the American food supply is so high in fat is because Americans have developed a fat tooth. Even the nonfat and low-fat foods are doctored to appeal to our high-fat taste. Look at the low-fat frozen dinners — Fettuccine Alfredo, Lasagne Florentine, Veal Patty Parmigiana, pizza. They almost all include cheese or cream. An abundance of nonfat cheeses, nonfat cream cheeses, nonfat sour creams, and nonfat mayonnaises allows us to continue our high-fat taste without the fat calories. But will these substitutes help you achieve your long-term fat-loss goals? The most successful Choosers to Lose are those who use the Fat Budget to wean themselves from fatty foods and replace them with high-fiber, nutritious foods. While there is nothing harmful about spreading a tablespoon of fat-free cream cheese on your bagel or pouring a few tablespoons of nonfat dressing on your salad, those who overdo their use of creamy and cheesy-tasting, low-fat and nonfat er-

---

*Vendors sometimes label yogurt nonfat even if it contains peanut oil, cookies, or nuts.

satz foods are less likely to be successful because they never give up the taste and craving for fatty foods. If you are a nonfat fat addict, you must ask yourself if, once you are no longer keeping track of what you eat, you will slip back to your old high-fat habits because you never really let them go.

Scientific experiments have shown that it takes about 12 weeks to change to a low-fat, tasty taste *if* you don't keep subjecting yourself to your old high-fat tastes. Give yourself a chance to let the change take place.

RECAP

Choose to Lose is about eating lots of wonderful, low-fat/high-fiber food. To avoid high-fat or fat-free overload and to be successful in your quest to be lean and healthy, we repeat: eat at least the minimum number of servings of foods from the Food Guide Pyramid (see page 218). Eat real, full meals. If you are eating nutrient-dense/fiber-rich fruit, vegetables, and whole grains, you will be full, satisfied, and healthy.

Remember:

1. You must eat foods rich in nutrient-dense/fiber-rich carbohydrates to lose weight.
2. Carbohydrates when eaten in moderation are burned and not stored as fat.
3. Some foods you will want to eat in abundance are potatoes, other vegetables, fruit, nonfat dairy products. Foods you no longer need to fear, but which you need to eat in sensible amounts, are bread, pasta, rice, low-fat meats and fish.
4. Chicken and turkey without skin and most seafood are naturally low in fat and can be prepared in a large variety of delicious low-fat dishes. But they become very fattening if prepared incorrectly.
5. Eat fat-free ersatz food as if they were high-fat splurges — in small

amounts, rarely. They have the following drawbacks when eaten in large amounts:

a. they have no vitamins, minerals, or fiber;
b. they crowd out more nutritious foods;
c. they stall weight loss because the body burns them in preference to burning fat;
d. high-fat-tasting/fat-free foods perpetuate a high-fat taste and make switching to a low-fat cuisine difficult.

6

Putting *Choose to Lose* to Work for You

> "Just tonight a friend was praising me for my commitment to *Choose to Lose* to his sister, telling her that I was on a new 'diet' and he was very proud of me. I gently corrected him. 'I'm not on a diet,' I explained. 'I've experienced a lifestyle change.'"
>
> — Pat Frohe, Tallahassee, Florida

> "This is not like struggling with a diet but is becoming more and more a permanent part of our lifestyle."
>
> — Mildred Clark, Mentor, Ohio

ARMED WITH YOUR OWN FAT BUDGET, insight into the sources of fat in foods, and a desire to be light as a feather, you are ready to . . . begin making changes!

SLASHING THE FAT

Take Your Time

But whoa! Don't expect to revamp your entire eating repertoire today and lose 25 pounds tomorrow. (Ah, were it so easy.) You have been eating the way you have been eating for a long time. Changes made too quickly are rarely permanent, and you want to be permanently slim. Take a few weeks to work into your new eating plan. You'll even lose during this time because you will be making changes — enduring changes.

Or Go Cold Turkey

For those of you who want to cut your fat intake to budget level immediately and know that you can stick to these big changes, disregard the last paragraph. Each to his or her own style.

Collect Data

To analyze and change your eating habits, you need to know what they are. If you haven't done it already, order yourself a *Choose to Lose* Passbook (see the ad in the back of the book) and read the Appendix, "The Nitty-Gritty of How to Keep a Food Record." Keep a three-day (two weekdays and one weekend day) baseline food record. It will be extremely helpful to have this information in hand as you read on.

Paring Down by Stages

One approach to making gradual changes is to focus your attention on the worst culprits. Can you eat these foods less often, in smaller amounts, or eliminate them entirely? Go for the worst first. Make changes in stages. Don't go from whole milk to skim; go from whole milk to 2% milk (the taste is very similar). Then go from 2% milk to 1% milk (the taste is close); then, from 1% milk to nonfat skim milk. You made it.

Choices and Changes

First, you want to target your high-fat food choices. Make four lists:

1. High-fat foods you can eat in smaller amounts
2. High-fat foods you can eat less often
3. High-fat foods you can eliminate entirely
4. High-fat foods you can replace with low-fat substitutes

Study this list and use it to analyze your food records. Perhaps you find that margarine, potato chips, cheeseburgers, and waffles made with whole milk are your high-fat downfall. (If you can pinpoint just a few offenders, you are lucky. For most people no one item is making them fat — fat is riddled throughout their diet.)

Choose Smaller Amounts. You decide that margarine is a high-fat food that you are willing to eat in smaller amounts. Check the **FATS AND OILS** section of the Food Tables. The 1½ tablespoons of margarine (150 fat calories) you use to butter your bagel inflicts major damage on your Fat Budget. Although not insignificant fat-calorie-wise, 1 teaspoon (30 fat calories) would still give you the taste and texture of margarine, but would only create a minor wound. Best, of course, would be to use jelly and no margarine.

Choose Less Often. Your food records show that you ate three cheeseburgers in a three-day period. According to the **MEATS, DAIRY AND EGGS**, and **GRAINS AND PASTA** sections of the Food Tables (and your food diary) a cheeseburger has 288 fat calories (4 ounces of lean hamburger = 188 fat calories, 1 ounce of American cheese = 80, a hamburger roll = 20). Realizing that you are not *that* crazy about cheeseburgers — certainly not at that price — you choose to cook them rarely and instead fix low-fat dishes that are just as simple and quick to prepare.

Choose Not at All. How do you want to treat your third high-fat booby trap, potato chips? You look at the Food Tables under **SNACK FOODS.** Eliminating potato chips at 90 fat calories for a mere 15 chips (and who eats only 15 potato chips?) is a change you can make with no pain.

Choose Low-Fat Alternatives. You love making waffles Sunday morning. Light streaming through the kitchen window, the aroma of coffee filling the air, maple syrup flowing over a pile of light brown crispy waffles — a scene to warm the heart. Why start such a beautiful day with a dent in your Fat Budget? Why not replace the ¾ cup of sour cream (276 fat calories) and ¾ cup of whole milk (55 fat calories) with 1½ cups of nonfat buttermilk (0 fat calories)? Why not replace the half cup of melted butter (800 fat calories) with 1 tablespoon of olive oil (119 fat calories) or no fat at all (0 fat calories)? The waffles will still be incredible.

Use the Food Tables to find alternatives. For example, if you normally fry up a rasher of bacon for breakfast, check the **MEATS (Pork)** section in the Food Tables for a lower-fat substitute. You'll find that for about one-third the fat calories, you'll get twice as much Canadian bacon.

A DAY TO BE RECKONED WITH

Before you analyze your own food records, let's look at Ellen's and see what modifications she can make so that her diet is nutritious, delicious, and filling.

A Look at Ellen

It's amazing how total calories and fat calories accumulate while you're not paying attention. Without even eating much food, Ellen has accumulated 1315 fat calories (Chapter 3, "Self-Discovery"). Ellen needs to look over her meal plan and pinpoint the worst culprits. Knowing the bad news, she can then make the changes that will bring her fat intake in line with her Fat Budget.

Let's look at Ellen's baseline food record (pretend it's one day of your food record). Turn to page 134 of this chapter. Start with breakfast. Which is the biggest fat offender? The large blueberry muffin contributes 140 calories of fat.

A Better Breakfast. She could have had:

- a bowl of Wheat Chex (10 fat calories)
  with ¾ cup of skim milk (0 fat calories)
  and half a banana, sliced (2½ fat calories)
- a slice of whole-wheat toast (10 fat calories)
  with fancy preserves (0 fat calories)
- a half cup of 1% cottage cheese (10 fat calories)
  mixed together with slices of the other half banana (2½ fat calories) for a total of 35 fat calories instead

A Better Morning Snack. Chuck the glazed doughnut at 130 fat calories. What about one of the following:

- a bagel toasted in the office microwave (10 fat calories)
  with fancy preserves (0 fat calories)

- a carton of flavored nonfat yogurt (0 fat calories)
- a peach, apple, orange, or pear

Her snack would then cost 10 fat calories — a saving of 120 fat calories. However, having eaten such a big breakfast, Ellen doesn't even want a snack.

A Better Lunch. The 280 fat calories of Caesar dressing can easily be reduced to 0 fat calories with a nonfat dressing or about 35 using the fork method (see page 46). The 130 fat calorie biscuit with butter (36 fat calories) could be replaced with an English muffin with jelly at 10 fat calories. But what kind of lunch is a dinky salad and biscuit, anyway? Instead of submitting herself to fast-food temptation, she could bring from home:

- a sandwich of sliced deli turkey breast (8 fat calories per ounce) with mustard, sliced tomato, sprouts, and hot peppers on mixed-grain bread (a total of 36 fat calories)
- or a few pieces of last night's *Chinese Chicken* (13 fat calories), eaten cold
- carrots and an orange
- a dish of nonfat frozen yogurt (0 fat calories)

Ellen's diet cola can be replaced with a healthy drink — skim milk or orange juice. Instead of spending 485 fat calories for a few boring bitefuls, she will be spending at most 36 fat calories for a full, satisfying lunch.

A Better Afternoon Snack. Ellen suffered every time she thought about her snack fiasco. Never again will she consume 130 fat calories without knowing exactly how much fat she is eating. She will choose the optimal way to spend her fat calories. Instead of eating Ramen Noodles, which gave her little satisfaction but a lot of fat, Ellen might have had a carton of strawberry or peach nonfat yogurt (0 calories) or a bowl of blueberries (0 calories).

A Better Dinner. For dinner Ellen spent 80 fat calories for the Breaded Country Chicken. Why waste 80 fat calories on such a skimpy meal — about 2 ounces of meat*? She doesn't have to punish herself this way.

Try this scenario on for size. Before she goes to work or the night before, Ellen takes five minutes to make *Cucumber Soup* (page 266) at 0 fat calories or 10 fat calories with the walnut garnish. When she returns home at the end of the day, the soup is chilled and ready to eat. Then she starts the rice (0 fat calories) so it will be cooking while she whips up *Apricot Chicken Divine* (page 269) at 23 fat calories. While the chicken is cooking she puts a sweet potato (0 fat calories) into the microwave. She cuts off the flowerets from a stalk of broccoli (0) and sticks them in a saucepan. She'll steam them so they can cook while she eats her soup.

"Yummy," she says as she takes the chicken from the oven. What a feast! Soup, an entrée over rice, baked sweet potato, steamed broccoli — and then a bowl of sliced strawberries for dessert — all for a mere 33 fat calories. After completing dinner, she is too stuffed even to think about food.

A Better Evening Snack. At 320 fat calories, the cup of peanuts was a blunder of earth-shaking magnitude. However, Ellen has no reason to ever repeat this mistake. The solution is simple. If nuts are your weakness, don't keep them in the house. Even if they are not your weakness, why keep them in the house? Ellen could have enjoyed 6 cups of air-popped popcorn with 0 fat calories, loads of soluble and insoluble fiber, and lots of crunch.

With all these changes she has brought her fat intake down to 103 fat calories — way below her Fat Budget of 315.

Adding Bulk to Reduce Bulk

Look at the difference in the amount of food Ellen ate. On her normal diet, she ate about 2.5 pounds of food. On the modified diet, she would

*Ellen determined the amount of meat by looking at the amount of cholesterol (45 mg) on the nutrition label. She knows there are about 18–25 mg of cholesterol per ounce of any meat, so she divided the amount of cholesterol on the package by 22 to get an approximate number of ounces. 45 mg ÷ 22 = 2 oz.

Ellen's Food Record

BASELINE			REPLACEMENT MEAL PLAN		
	CALORIES			CALORIES	
FOOD ITEM	TOTAL	FAT	FOOD ITEM	TOTAL	FAT
Breakfast			***Breakfast***		
1 large blueberry muffin	400	140	1 cup Cheerios	110	15
1 cup coffee	0	0	¾ cup skim milk	65	0
1 tbsp half-and-half	20	15	1 banana	105	5
			1 slice whole-wheat toast	70	10
Snack			1½ tsp jelly	27	0
1 large glazed doughnut	240	130	½ cup 1% cottage cheese	80	10
1 cup coffee	0	0	1 cup coffee	0	0
1 tbsp half-and-half	20	15	2 tbsp 1% milk	10	0
Lunch			***Lunch***		
2 cups mixed salad	30	0	1 turkey sandwich		
4 tbsp Caesar dressing	300	280	2 oz sliced turkey breast	60	16
1 tbsp sunflower seeds	50	39	3 slices tomato	4	0
1 diet cola	0	0	2 slices multi-grain bread	140	20
1 biscuit	280	130			
1 pat butter	36	36	1 carrot	31	0
			1 orange	60	0
			8 oz skim milk	80	0
			5 oz nonfat frozen yogurt	100	0
Snack			***Snack***		
1 cup Ramen Noodles	300	130	8 oz nonfat peach yogurt	190	0
Dinner			***Dinner***		
Breaded Country Chicken (Healthy Choice)	280	80	1½ cups *Cucumber Soup*	117	10
			Apricot Chicken Divine	215	23
¼ head iceberg lettuce	20	0	¾ cup white rice	168	0
4 tbsp fat-free dressing	60	0	1 cup steamed broccoli with balsamic vinegar	24	0
11 oatmeal raisin cookies (Entenmann's fat-free)	550	0			
			1 baked sweet potato	95	0
			1 cup strawberries	45	0
Snack			***Snack***		
½ cup peanuts	420	320	6 cups air-popped popcorn	180	16
TOTAL:	3006	**1315**	**TOTAL:**	1976	**125**
Food Weight: 2.5 lb			**Food Weight:** 4.1 lb		
Fruits = 0; vegs = ½; grains = 0; dairy = 0			**Fruits = 3; vegs = 3; grains = 7; dairy = 4**		

eat 4.1 pounds. Imagine how much fuller and satisfied she will feel eating the healthier diet.

An aside: One of the biggest complaints people make about *Choose to Lose* is that they can't eat enough food. (What a problem!) Our advice is to take it slowly. Start adding vegetables to your dinner. Add fruit to lunch. Begin making real, full meals. Eventually you will have no trouble eating a healthy diet that is 300–400 total calories above your total calorie floor.

Calories Come Into Line. Even though Ellen ate much more food, by reducing the fat in her diet, she brought her total calorie intake into line. Her total calorie intake went from 3006 calories to 1976 calories. With her modified diet she consumed almost exactly 400 more calories than her total calorie floor of 1579.

No pain. Making these changes didn't hurt Ellen a bit. You'll notice that the modified menu is filled with much more food — more interesting tastes, more bulk, a more balanced and healthier combination of foods.

First Cut. In your first round of fat modification you may not be able to make as many changes as Ellen has. If so, start slowly. For example, if giving up a blueberry muffin would make you cry, you might eat it today and substitute an English muffin or whole-wheat toast with jam tomorrow. If you can't imagine eating a baked potato plain, cut the 4 tablespoons of sour cream you normally eat to two. Next time, you can reduce the amount even more or try nonfat yogurt or a nonfat sour cream substitute. When you feel ready, perhaps in a few days, try to modify more of your high-fat choices.

The *Word*

Always keep in mind that you are trying to lose weight, and the less fat you eat, *without feeling like a martyr,* the faster you will lose. You can easily fit many wonderful low-fat foods into your budget. You just have no room for a lot of greasy, ugly high-fat foods.

Treat this time of transition as an adventure. Explore new foods, new tastes, new textures — experiment.

Try a Little Harder. Next round, take another look at your food record to see if there are other foods you wouldn't mind eating less often or in smaller amounts or eliminating entirely. Do you need half-and-half in your coffee? Why not give 1% milk a chance? Take the skin off your baked chicken. It will make a 56-fat-calorie difference. Better yet, cover the chicken in spices, and create a more interesting dish, *Blackened Chicken.*

ADDING BACK CARBS

Ellen analyzed her food record to find the biggest fat donors so she could make better choices. She can't stop there (and neither can you). It is equally important that she review her day's intake to see if she is satisfying the minimum basic nutritional requirements according to the Food Guide Pyramid. You must eat a minimum of 2 to 3 servings* of fruit, 3 to 5 servings of vegetables, 6 to 11 servings of grains (preferably whole grains), and 2 to 3 servings of low-fat or nonfat dairy every day to maintain your health and to ensure that you are eating a nutritious diet.

Important: The beauty of this recommendation is that when you combine it with eating a low-fat diet, you get the perfect prescription for weight loss.

A Big Tip

The easiest way to reach and maintain your goal weight is to:

- eat below your Fat Budget
- eat a lot of nutrient-dense/fiber-rich total calories — more than the minimum total caloric floor you used to determine your Fat Budget
- **make sure these calories include foods to satisfy your minimum basic nutritional requirements**
- exercise aerobically every day for 20–30 minutes.

Soon people who don't even know you will comment on how great you look.

*See page 223 for definitions of what constitutes a serving.

Check Out How Ellen's Baseline Intake Meets the Food Guide Pyramid: Really Poorly

Recommendations

Fruit: 2–3 servings: 0 servings. One quick glance shows you that Ellen is in the hole concerning fruit. She is supposed to eat at least 2 servings and she has eaten none.

Vegetables: 3–5 servings: ½ serving. Lettuce has little to recommend it nutritionally. Perhaps with the skimpy pieces of cucumber, tomato, and carrot included, she can consider this half a serving of vegetables. She is short 2½ vegetable servings.

Grains: 6–11 servings: 0 servings. Grains are a tricky category. The idea is to eat nutritious whole grains, not a little flour holding together a lot of fat. Use this rule of thumb: If a food is high fat, do not count it as a grain. By this criterion, Ellen ate no grains.

Dairy: 2–3 servings: 0 servings. Ellen has eaten no dairy.

Ellen's Replacement Meal Plan Gets an A+

Ellen has easily met the nutritional requirements in her new meal plan. Her choices are filled with vitamins, minerals, and fiber. By fulfilling these recommendations, she will be so stuffed at the end of the day she won't have room to pig out on empty and/or fat calories.

Look over her new meal plan yourself and see how she satisfies each requirement.

Fruit: 2–3 servings: 3 servings. Ellen ate a banana at breakfast, an orange at lunch, and strawberries at dinner.

Vegetables: 3–5 servings: 3 servings. The carrot, cup of broccoli, and sweet potato add up to 3 servings.

Grains: 6–11 servings: 7 servings. Except for the white rice, all of Ellen's grains are whole grains. Her bowl of Cheerios, slice of whole-wheat toast, and ¾ cup of rice each count as one grain. Her two slices of multi-grain bread and 6 cups of air-popped popcorn each count as 2 servings. Popcorn is considered a cereal grain. 3 cups = 1 serving.

Dairy: 2–3 servings: 4 servings. Ellen's ¾ cup, plus 1 cup, plus 2 tablespoons add up to almost 2 cups of skim milk. Add ½ cup of cottage cheese and the 8-ounce container of nonfat yogurt, and she has more than fulfilled her dairy requirement. Her frozen yogurt (even nonfat) does not fulfill a dairy requirement because the amount of dairy in frozen desserts is often negligible. Again, the idea is to satisfy your minimum recommendations with nutrient-dense foods, not with processed, ersatz fluff.

For more on eating a balanced diet, see Chapter 13, "Ensuring a Balanced Diet."

Educated Choices

We gave you our modification of Ellen's meal plan, but had it been your meal plan, you could have modified it in an infinite number of different ways to suit your taste. It is your choice. But now, when you make choices, you will think before you choose. On the one hand, you'll think about fat. You might ask: How many fat calories are in this "All Natural" dressing? Is this turkey burger really low in fat? Do I really want a cup of ice cream that demolishes my entire Fat Budget for the day?

On the other hand, you'll think about healthy food. You'll think about digging into that cold, sweet piece of watermelon. You'll try to figure out which wonderful recipes you can make using the broccoli and red peppers you see in the produce department. You'll think vitamins, minerals, and fiber when you choose whole-wheat toast over white bread.

Don't worry. Thinking about what you eat or what you don't will not ruin your life. You'll still be able to enjoy and love food — maybe even more than before. You'll be able to make more informed choices. When you do choose that hot apple turnover with vanilla ice cream after a month of doing without, you'll love every bite. You'll even taste every bite. Or, perhaps, you'll take a bite and push it away because it isn't worth 350 fat calories.

Remember:

1. For some people, gradual changes are more likely to be lasting.
2. Four ways to reduce your fat intake:
   a. choose some high-fat foods in smaller amounts

b. choose some high-fat foods less often
c. eliminate some high-fat foods entirely
d. replace some high-fat foods with low-fat alternatives

3. Use the Food Tables to choose lower-fat alternatives.
4. By reducing the fat in your diet and adding low-fat nutrient-dense/fiber-rich foods, you will bring your total calorie intake into line.
5. You will be eating much more food when you reduce the fat and increase the low-fat nutrient-dense/fiber-rich foods, thus ensuring you will be full and satisfied.
6. Be sure you are meeting your daily minimum basic nutritional requirements:
   a. 2–4 servings of fruit
   b. 3–5 servings of vegetables
   c. 6–11 servings of grains, preferably whole grains
   d. 2–3 servings of low- or nonfat dairy
7. Eating these recommended foods ensures an adequate supply of vitamins, minerals, and fiber and will keep you full and happy.

7

Decoding Food Labels

"I didn't realize how important label reading was until I went grocery shopping the next time after reading your book. I found myself putting back things after taking them off the shelf just because I finally knew what the number of grams of fat really meant. . . . Your book has also opened up a dimension in my taste buds and I really seem to enjoy food more now than ever before, without the added butter or other fats."

— Diana Firey, Sand Springs, Oklahoma

YOU NOW KNOW that you want to reduce fat and you want to increase nutrient-dense/fiber-rich foods, but that is not enough. You have to know exactly how much fat is in foods so that you can make choices. The U.S. Government has made it much easier for us to do so by regulating the nutrition information that goes on a food label. This information is invaluable, for it gives you the power to make choices, because the fat calories per serving are clearly listed on every product. There are also a lot of other numbers and percentages that may have befuddled you in the past. Good-bye confusion; total understanding is on its way. In this chapter you will learn the fine art of reading food labels.

THE FINE ART OF READING FOOD LABELS

In the following section we will show you how to become a label-reading expert. Before you read on, please select a can or package from your kitchen or pantry so that you can examine the label.

Ingredient List

Find the list of ingredients on the container you chose. Although ingredient lists do not give you quantitative information, they may give you facts that influence your choice. Do you want a product with coconut oil as its main ingredient? Find the ingredient list on your package. What ingredient is listed first? The ingredients are always listed in descending order by weight.

Tomato Soup

INGREDIENTS: TOMATOES (WATER, TOMATO PASTE), HIGH-FRUCTOSE CORN SYRUP, WHEAT FLOUR, SALT, VEGETABLE OIL (CORN, COTTONSEED OR PARTIALLY HYDROGENATED SOYBEAN OIL), SPICE EXTRACT, VITAMIN C (ASCORBIC ACID) AND CITRIC ACID.

The ingredient list for this tomato soup tells you that tomatoes account for more weight than any other ingredient. On second glance you see that the first ingredient isn't *really* tomatoes. It's water and tomato paste. Now you know there is more water and tomato paste than any other ingredient. Unfortunately, since the actual weights of the ingredients are not provided, you have no way of knowing if these two ingredients comprise 50 percent of the soup or 40 percent or 10 percent.

Read on and you will see that high-fructose corn syrup is the ingredient second in weight, wheat flour is the third, and so on.

Notice that the listing for vegetable oil is a choice between corn, cottonseed, or partially hydrogenated soybean oil. Manufacturers typically list a variety of oils, choosing whichever is cheapest at the time. Of course, you have no idea which oil was used. But the choice of oil (fat) makes a big difference to your arteries. (See page 93 for a discussion of the health effects of saturated, monounsaturated, and polyunsaturated fats.)

What does the ingredient list on *your* package tell you?

Nutrition Facts Box

Thanks to the Food and Drug Administration's food label regulations, every food package should have a clear, easy-to-read label. The format is standardized so that every product must include the same nutrition information.

In many ways the nutrition labels are a great improvement over the past haphazard labeling. Calories from fat are boldly listed near the top. Saturated fat is also listed (albeit in grams, not calories). Fiber is an important new category. These are the pluses. The minuses concern the Daily Value and % Daily Value concepts. We will discuss them later.

DISSECTING A FOOD LABEL

We are going to analyze food labels from top to bottom so that you will understand what every part means. However, because you are a Chooser to Lose and have a Fat Budget, most of the numbers and percentages on the label need not concern you.

The only two numbers you really need to know are the calories from fat and the serving size. With this information you can determine the number of fat calories for the amount of food you want to eat. Then you can decide if you want to fit the food into your Fat Budget.

An Enlightening Example

Let's dissect the Marie Callender's White Meat Chicken & Broccoli Pot Pie label on page 143. You might have pulled the package from the freezer compartment because it is white-meat chicken (and thus, you figured, low-fat), and it contains broccoli, so it is healthy. You'll soon see.

Information Is for One Serving

As with all nutrition labels, all of the information — calories, calories from fat, total fat, cholesterol, and even the amounts of vitamin A and vitamin C — is for one serving of the food. The serving size is listed at the top of the Nutrition Facts box. First, find the information that is most important to you:

Calories from Fat 440. At the top right of the Nutrition Facts box you can see the listing of calories from fat. Calories from fat! A boon for us Choosers to Lose. Now you know exactly how much fat you are getting. This little pie contains 440 fat calories for one serving. (You will find another listing of fat, but it is in grams (a weight) and not calories: **Total Fat 49 g.** You can ignore this listing. To determine the grams of fat, the manufacturers determine the fat calories and divide by 9. Then they round to the nearest whole number. This number is less specific and accurate than the listing of calories from fat.)

Chicken & Broccoli Pot Pie

Nutrition Facts	
Serving Size 1 Cup (290g)	
Servings Per Container About 2	
Amount Per Serving	
Calories 710 Calories from Fat 440	
	% Daily Value*
Total Fat 49g	**75%**
Saturated Fat 13g	**67%**
Cholesterol 25mg	**8%**
Sodium 1060mg	**44%**
Total Carbohydrate 51g	**17%**
Dietary Fiber 4g	**16%**
Sugars 5g	
Protein 17g	

Serving Size: 1 Cup (290 g). Look under the heading "Nutrition Facts" for the serving size. In this case, the serving size is one cup. Thus, one cup of this chicken pot pie costs 440 fat calories. That's an earth-shattering amount of fat for a little convenience. How does this dish fit into your Fat Budget or vice-versa?

Servings Per Container: About 2. Many people read the Calories from Fat and then stop reading, but they don't stop eating. The food label tells you the Calories from Fat for *each* serving. In this case, there are about 2 servings per container. That means that if you eat the whole pie, an easy mission to accomplish, you are eating twice as much fat, sat-fat, and total calories as the information printed on the label for one serving. As you can see, ignoring the serving size can have calamitous consequences.

Actual Consumption

We were shocked when we read "Calories from fat (per serving): 440." But when we realized that the whole little 6-inch chicken pot pie contains 2 servings or 880 fat calories, we keeled over in a dead faint. Aaaggghhh!

Perhaps the 440 fat calories per serving was so much fat you would have dropped the chicken pot pie like a hot potato before you took time to look at the number of servings, but there are other products where a glance at the Calories from Fat alone may induce you to consume food you should pass up.

If you want your suits to drop three sizes and your cholesterol level to fall 50 points, always take note of serving size and the number of servings you are eating or you may seriously underestimate the amount of fat you are consuming.

Making Sense of the Rest of the Label

These listings may also be of interest to you.

Saturated Fat 13 g. Everyone who is interested in a healthy diet, but particularly people with elevated blood cholesterol, should limit their intake of saturated fat. Read page 93 about saturated fat (for *everything* you need to know about saturated fat, also read *Eater's Choice*). To determine the calories of saturated fat in a serving of the pot pie, multiply the grams by 9. There are 9 calories per gram of saturated fat (13 grams × 9 calories/gram = 117 sat-fat calories). To help put the 13 grams or 117 sat-fat calories into perspective, a sedentary female who wants to weigh 140 pounds has a Sat-Fat Budget of 176. If she eats the whole pot pie, the 2 servings total 234 sat-fat calories, almost 1⅓ times her Sat-Fat Budget for the whole day.

Cholesterol 25 mg. This number is useful, not because you should be concerned with cholesterol (dietary cholesterol has a very minor effect on raising blood cholesterol) but because the cholesterol content can help you determine the amount of meat in a product. For example, the pot pie lists 25 mg of cholesterol. Since 1 ounce of chicken contains about 22 mg of cholesterol*, this meal contains about 1 ounce of chicken. (We measured it. One serving contains even less than that. It contains ⅘ of an ounce.) Although it is healthier to eat less meat, you might be a bit annoyed that you paid so much for so little meat.

The trend in frozen foods today is to make the portions larger. However, if you check the cholesterol content, you'll see that a dinner

---

*All meats (beef, veal, and lamb), poultry, and seafood (except for shrimp at 43 mg per oz) contain 18–28 mg of cholesterol per ounce. To determine the amount of meat in a product, divide the amount of cholesterol on the package by 22.

that may be a third larger than the normal meal still has the same paltry amount of meat.

Dietary Fiber 4g. Because the FDA considered dietary fiber an essential element of a healthy diet, it was added to the new nutrition labels. What a great addition. With a diet of fast foods, frozen dinners, and convenience foods, Americans eat too little fiber (the current average American intake is 10 grams), and it shows in our massive health problems. (See page 115 for a discussion of the benefits of consuming dietary fiber.)

You're in luck. By following *Choose to Lose* and eating a diet high in vegetables, fruit, and whole grains, you will be eating a high-fiber diet. (To be sure you are meeting your minimum daily nutritional requirements and thus getting adequate fiber, see Chapter 13, "Ensuring a Balanced Diet.") The FDA recommends eating 25–30 grams of fiber a day. (But too much of a good thing is not necessarily good: more than 50 or 60 grams of fiber per day may decrease the amount of vitamins and minerals your body absorbs.)

Special K Cereal

Nutrition Facts
Serving Size 1 Cup (31g/1.1 oz.)
Servings per Container 6

Amount Per Serving	Cereal	Cereal with ½ Cup Vitamins A & D Skim Milk
Calories	110	150
Fat Calories	0	0
	% Daily Value **	
Total Fat 0g*	**0** %	**0** %
Saturated Fat 0g	**0** %	**0** %
Cholesterol 0mg	**0** %	**0** %
Sodium 250mg	**11**%	**13** %
Potassium 60mg	2 %	7 %
Total Carbohydrate 22g	**7** %	**9** %
Dietary Fiber 1g	**3** %	**3** %
Sugars 3g		
Other Carbohydrate 18g		

All-Bran Cereal

Nutrition Facts
Serving Size 1/2 cup (31g/1.1 oz.)
Servings per Container 13

Amount Per Serving	Cereal	Cereal with ½ Cup Vitamins A & D Skim Milk
Calories	80	120
Calories from Fat	10	10
	% Daily Value **	
Total Fat 1.0g*	**2** %	**2** %
Saturated Fat 0g	**0** %	**0** %
Cholesterol 0mg	**0** %	**0** %
Sodium 280mg	**12** %	**14** %
Potassium 350mg	**10** %	**16** %
Total Carbohydrate 23g	**8** %	**10** %
Dietary Fiber 10g	**40** %	**40** %
Insoluble Fiber 9g		
Sugars 5g		
Other Carbohydrate 8g		
Protein 4g		

A good way to use dietary fiber information is to comparison-shop at the cereal aisle in your grocery store. Look at the amount of dietary fiber in the Kellogg's Special K on the left on page 145. Compare it with the dietary fiber in the Kellogg's All-Bran to its right. At 1 gram per serving, the Special K is almost fiber-free. At 10 grams per serving, the All-Bran cereal meets almost half your daily fiber needs. Before you choke down a spoonful of a high-fiber cereal as if it were medicine, remember that you can get lots of fiber in other foods. But if a cereal tastes good and also contains 10 grams of dietary fiber per serving, why not choose it over another favorite that has only 1 gram?

Wheat Snack Crackers

Nutrition Facts
Serving Size 15 Crackers (30g)
Servings Per Container about 7

Amount Per Serving	
Calories 140	Calories from Fat 50
	% Daily Value*
Total Fat 6g	**9%**
Saturated Fat 1.5g	**8%**
Cholesterol 0mg	**0%**
Sodium 320mg	**13%**
Total Carbohydrate 16g	**5%**
Dietary Fiber 2g	**6%**
Sugars 2g	
Protein 3g	

Vitamin A	0% •	Vitamin C	0%
Calcium	4% •	Iron	4%
Thiamin	8% •	Riboflavin	4%
Niacin	6% •	Folate	4%

*Percent Daily Values are based on a 2,000 calorie diet. Your daily values may be higher or lower depending on your calorie needs:

	Calories:	2,000	2,500
Total Fat	Less than	65g	80g
Sat. Fat	Less than	20g	25g
Cholesterol	Less than	300mg	300mg
Sodium	Less than	2,400mg	2,400mg
Total Carbohydrate		300g	375g
Dietary Fiber		25g	30g

Calories per gram:
Fat 9 • Carbohydrate 4 • Protein 4

We would not recommend that you choose frozen dinners by their dietary fiber listings, for the numbers seem to be greatly exaggerated. For example, the chicken & broccoli pot pie discussed earlier lists 4 grams of dietary fiber per serving. The serving contains 1 tablespoon of broccoli and 1 tablespoon of carrot slices and some white flour. According to our fiber table (page 116) the fiber would add up to at most 2 grams.

Daily Value: Universal Nutrition Budget

Woe to the people who have no *Choose to Lose* Fat Budget. The Food and Drug Administration (FDA) was worried that most Americans had no way of judging food labels to make healthy food choices. After many months of hearings and debate, they came up with a solution: give all Americans a nutrition budget — they called it a Daily Value

(DV) — for fat, saturated fat, cholesterol, sodium, total carbohydrate, and dietary fiber. Look in the circled area of the Wheat Snack Crackers label. These Daily Values are based on a 2000-calorie diet for everyone, regardless of sex, weight, shape, or size. For those who think this basic caloric intake is too small, a 2500-calorie diet is listed in the last column on the right. You can see that the DV for cholesterol is less than 300 mg, the DV of sodium is less than 2400 mg, and so on.

Daily Value Total Fat: Ticket to Weight Gain. The Daily Value for Total Fat is the only one that is truly off the mark. The others are reasonable. At 180 sat-fat calories, the DV for saturated fat is actually low for most people and that's healthy. But at 30 percent of 2000 total calories, the DV for total fat is so huge (65 grams or 585 fat calories) that anyone using it is bound to gain weight. This DV of 585 fat calories is supposed to fit your needs whether you are a man or a woman, 4 feet 10 inches or 6 feet 7 inches tall, and your goal weight is 105 pounds or 195. How does 585 fat calories compare with your Fat Budget?

% Daily Value: Complicating a Simple System

Choosers to Lose have always found reading food labels a snap because they have their own personal Fat Budget by which to judge the fat content of any food. But because everyone else had no way of judging if 90 fat calories is a lot of fat or a little, the FDA invented % Daily Value (% DV). This turns out to be the most confusing part of the nutrition label.

Wheat Snack Crackers

	% Daily Value*
Total Fat 6g	9%
Saturated Fat 1.5g	8%
Cholesterol 0mg	0%
Sodium 320mg	13%
Total Carbohydrate 16g	5%
Dietary Fiber 2g	6%

In case you are curious, this is how it works. Look at the Daily Value section of the Wheat Snack Crackers label. Nutrients and the amount of each nutrient contained in the product are listed on the left. On the right is a list of percentages. Each of these percentages represents the percentage of the Daily Value for that nutrient which is contained in 1 serving of this product. Clear as mud?

Total Fat 6g__________9%. A specific example will make it clearer. Let's take total fat. The label lists Total Fat 6g__________9%. This means that 1 serving of these snack crackers contains 6 grams of fat (really 50 fat calories*), and 50 fat calories represents 9 percent of the DV for fat (9 percent of 585 fat calories). What are you supposed to do with this information? If you didn't have a Fat Budget, you might say, "9 percent? That's not much," and finish off the whole box, which contains 7 servings for a total of 350 fat calories. Would you realize that *each* serving was 9 percent of your Daily Value and that if you eat 2 servings, you would be eating 18 percent of your DV? Would you keep track each day of all the percentages of all the commercial food you eat until you reach 100 percent and then stop eating? Most likely, you would mistakenly think that the 9 percent represents the percentage of fat** in the food and eagerly gobble up the entire box of crackers.

Choose to Lose: Getting Straight to the Facts

Not only is the % DV unnecessarily complicated and thus extremely confusing, it is misleading, too. How should an educated Chooser to Lose analyze this cracker label? First, look at the calories from fat: 50 fat calories per serving. Next, look at the serving size and figure out how much you are going to eat.

The cracker label tells you that the serving size is 15 crackers (30 g). Unless you open the package, you really don't know how filling 15 crackers will be, but you do know that 1 ounce is a picayune amount of food. Eyeballing the small box, you might judge that you could polish off the entire contents (350 fat calories) in an evening. Assume you used restraint and ate only 2 servings (30 tiny crackers). At 50 fat calories a serving, you would be eating 2 × 50 fat calories = 100 fat calories. Compared to a DV of 585, even 100 fat calories doesn't seem so high, but compare it to your Fat Budget and you'll appreciate the cost.

---

*The 6 g comes from the following: One serving of the crackers has 50 fat calories. Since there are 9 calories per gram of fat, to determine the number of grams, the manufacturer divides 50 by 9 = 5.5. He rounds up to 6 grams.

**The percentage of fat in individual foods is irrelevant. See explanation on page 36.

TRENDS

Bigger Is Better

When we wrote the 1995 edition of *Choose to Lose,* we were aware of the many ways the food industry was conning the American public into thinking it was eating low-fat. Foods that claimed they were lower in fat were still mighty high. Foods that claimed to be healthy might be filled with coconut oil.

Although there are still hundreds of products out there appealing to your quest for a perfect body, the tendency these days is for packages to scream, We give you MORE. We are BIGGER!!

Banquet has "The Hearty One — EXTRA helpings of Salisbury steak"; Sara Lee's apple pie is "40% more . . . Large 9″ slice"; Mrs. Smith's pumpkin pie is "Hearty — 40% larger." Prego Lasagna is "made with LOTS of real cheese." Even Lean Cuisine now has "Hearty Portions." Every brand of frozen dinners has a new line of larger meals. For example, the Lean Cuisine normal Lasagna weighs 326 grams and has 35 fat calories; the Lean Cuisine Hearty Portions Lasagna weighs 425 grams and has 80 fat calories. It is as if the food industry has decided that people no longer care about healthy eating; they just want a lot to eat.

Fat Is Beautiful

Some of the foods are getting higher and higher in fat. Just look at the Marie Callender's Chicken Pot Pie (880 fat calories a pie) we discussed earlier. It is not unique. Look at Ben & Jerry's Ice Cream if you want to see big numbers. A dinky *half* cup of butter pecan ice cream is 220 fat calories — 220 fat calories! No wonder they call one flavor "chubby hubby."

Some manufacturers are even putting the fat back. SnackWell's is now making only one type of fat-free cookies: devil's food cookie cakes. Some of the formerly fat-free SnackWell's crackers now have 14 fat calories each; some of the cookies have 35 fat calories each. 35 + 35 + 35 + 35 = 140 fat calories. The newly designed boxes make no claims. SnackWell's business director explained that these claims are unnecessary because "Consumers already identify SnackWell's as being low fat or reduced fat. There's no longer any need to scream it on the package." Sounds a bit misleading to us since many of the products are no longer low fat.

As the portions get larger and the fat calories grow, it becomes more and more important that you read the label because who knows how much fat you might be consuming?

Fat-Free May Not Be

Now when you grab a food that has always been nonfat, you need to look at the label. SnackWell's is not the only brand that is adding back the fat. Entenmann's "Light" desserts now include both fat-free and reduced fat. As always, you need to **READ THE NUTRITION LABEL.**

"Fat Free" Doesn't Make *You* Fat Free

You also have to be wary of truly fat-free. Don't assume that fat-free or nonfat means healthy. Fat-free often means nutrition-free and fiber-free. It also may mean high sugar and empty calories. The "fat-free" label makes people feel they have license to eat unlimited quantities. Eating boxes of nonfat cookies and crackers adds nothing to your health. You could be eating nutritious food instead. Eating nonfat foods will *not* help you lose weight. In fact, your weight loss will be stalled as your body burns all those empty calories instead of stored fat.

If you are eating fat-free made with olestra, you might even suffer immediate physical discomfort as well as long-term health problems. (See page 124.)

CAVEAT EMPTOR

Deceptive Labeling

The FDA regulations on food labeling have greatly restricted the misleading and deceptive practices so common in the past. But the rules still allow a considerable amount of duplicity and confusion to slip through.

What's in a Name? The FDA has not cracked down on established brand names that imply good health and the key to weight loss when they may be neither. When you see the names Weight Watchers, Lean Cuisine, and Healthy Choice, you naturally assume that eating these products will help you become lean and healthy. Not true. Never assume anything. Read the label: Weight Watchers Smart Ones Swedish Meatballs — 90 fat calories; Lean Cuisine Chicken Florentine — 80 fat calories; Healthy Choice Breaded Country Chicken — 80 fat calories. Eating this much fat for an entrée day after day will make you neither slim nor healthy.

Eat Your Veggies. You need to be ever-vigilant. Always be suspicious of hype implying that a product is healthy or weight-reducing. Never devour the contents of a package that screams, "I am healthy," "I will help you lose weight," until you read the nutrition label. For a real-life example, take a look at the Veggie Stix packaging and label on page 152.

All Natural Fat. Although the food processor does not claim outright that this snack is healthy and a good choice for weight loss, he has designed his package so that you will leap to this conclusion on your own. The brand name is GOOD HEALTH. Translate the words "100% Natural Snack" and "Mixed Vegetables" into "healthy." Read "40% less FAT than Potato Chips" as "low in fat." This is the hype on the package. But, if like Sergeant Friday, you just want the facts, ma'am, just the facts, read the label. The food label tells the true story.

Calories from Fat: 60; Serving Size: 1 oz. Wow! 60 fat calories for 1⅓ cups (we measured it) of fried potato flour. While 60 fat calories per ounce is less than the average 90 fat calories per ounce for potato chips, it would be a stretch to consider it low in fat. Even the "40% less FAT than potato chips" claim is an exaggeration; 60 fat calories is actually "33% less FAT than potato chips."

This snack may contain no unhealthy additives, but the fat content negates any chance of its being a healthy choice.

"Lean" Meat. Consumers associate the word "lean" with low in fat.

GOOD HEALTH

VEGGIE
100% Natural SNACK
Irresistibly Delicious
40% less FAT than Potato Chips
STIX
Mixed Vegetables
NET WT. 8 OZ. (227g)

Nutrition Facts

Serv. Size 1oz (28g 1/8 pkg)
Servings 8
Calories 140
Fat from Calories 60

* Percent Daily Values (DV) are based on a 2,000 calorie diet.

Amount / Serving	%DV*	Amount / Serving	%DV*
Total Fat 6g	10%	**Total Carb.** 19g	6%
Saturated Fat 0.5g	4%	Dietary Fiber 1g	2%
Cholesterol 0mg	0%	Sugars 0g	
Sodium 310mg	13%	**Protein** 1g	

Vitamin A 0% • Vitamin C 8% • Calcium 0% • Iron 6%

INGREDIENTS: WHOLE POTATO FLOUR, EXPELLER PRESSED NON-HYDROGENATED SUNFLOWER AND/OR CANOLA OIL, NATURAL WHEAT STARCH, RICE FLOUR, TOMATO AND SPINACH PUREE, SALT.
Distributed by: Good Health Natural Foods Inc., Northport NY 11768

However, in the FDA rule book, to qualify as lean, a meat must contain less than 10 grams of fat (90 fat calories) per 100 grams (3½ ounces) of food. You are not going to remain lean long if you spend 90 fat calories on a mere 3½ ounces of meat, or, more likely, 128 fat calories for 5 ounces of "lean" meat.

Meat processors are now taking advantage of the weakness in the "lean" claim regulation to promote meat that is quite high in fat as a healthy choice.

***Lean* Ground Turkey?** Look at the label below. First, your eye is attracted to the words *Lean* Ground Turkey. You know turkey is one of the leanest meats, so you aren't surprised to see it linked with the word "lean."

You might wonder at the claim "7% FAT." Whole milk, a high-fat food, is 4 percent fat. Why would the turkey processor emphasize such a high percentage of fat? It is obvious that they count on the public's ignorance. What does the 7 percent fat (or 93 percent fat-free) mean anyway?

Nutrition Facts

Serving Size 4 oz. (112g)
Servings Per Container Varied

Amount Per Serving	
Calories 160 Calories from Fat 70	
	% Daily Value*
Total Fat 8g	**12%**
Saturated Fat 2g	**10%**
Cholesterol 85mg	**28%**
Sodium 85mg	**4%**
Total Carbohydrate 0g	**0%**
Protein 22g	
Iron 10%	

Not a significant source of dietary fiber, sugars, vitamin A, vitamin C, and calcium.

*Percent Daily Values are based on a 2000 calorie diet.

The 7 percent fat claim means that fat accounts for 7 percent of the *weight* of the meat. (The meat weighs 112 grams and the fat weighs about 8 grams.) Who cares? Actually the label could just as easily say 44 percent fat, because fat represents 44 percent of the total calories. Of course, that information is also irrelevant (see page 36). All you really need to know is that there are 70 fat calories for 4 ounces of meat (plus fat and skin), or almost 18 fat calories an ounce. This is a far cry from 2 fat calories an ounce for ground turkey you grind yourself from white-meat turkey breast.

Zero Fat? Many products that claim to be 0 fat have no fat, but not *all* products that claim to be fat-free really are. A good clue can be found in the ingredient list. If the first ingredient is olive oil, you have a right to be suspicious.

Take, for example, cooking sprays. Pam boasts ". . . for Fat Free cooking," and I Can't Believe It's Not Butter! prints "Zero Calories" in bright red letters on the front label. Zeros have taken over the Nutrition Facts box: Calories 0, Total Fat 0 g, Calories from Fat 0, Saturated fat 0 g, Cholesterol 0 mg, etc. If you weren't a cagey Chooser to Lose, you might be convinced that cooking sprays are fat-free.

Take a look at the ingredient list on the Pam can. Ingredient number one: 100% imported olive oil. Fat-free olive oil? The I Can't Believe It's Not Butter! bottle lists water first and then liquid soybean oil followed by sweet cream buttermilk. Fat-free soybean oil?

How do they get away with it? The Pam serving size — a spray lasting 1/3 of a second, or .266 gram — is so minute that they can claim fat-free. Look again at the I Can't Believe It's Not Butter! bottle. It says "Zero Calories" but in tiny, tiny print adds "per serving."

A tablespoon of Pam actually contains 119 fat calories. A tablespoon of I Can't Believe It's Not Butter! contains 52 fat calories. Of course, because you are spraying, you don't use as much fat. But you do use some. You probably spray about half a teaspoon (20 fat calories if you use the olive oil spray) to grease a 9-inch cake pan. That's no saving. It takes one-half teaspoon of margarine to grease a 9-inch cake pan.

If you don't mind the added cost, there is nothing wrong with using

I Can't Believe It's Not Butter!

Nutrition Facts
Serving Size 1.25 Sprays (0.25g) For Cooking Spray
5 Sprays (1g) For Topping
Servings Per Container 904 For Cooking Spray
226 For Topping
Calories 0

Amount/ serving		Cooking Spray	Topping	Amount/ serving		Cooking Spray	Topping
		% DV*	% DV*			% DV*	% DV*
Total Fat	0g	0%	0%	Cholesterol	0mg	0%	0%
Sat. Fat	0g	0%	0%	Sodium	0mg	0%	15mg/1%
Polyunsat. Fat	0g			Total Carb.	0g	0%	0%
Monounsat. Fat	0g			Protein	0g		

Not a significant source of dietary fiber, sugars, vitamin A, vitamin C, calcium and iron.
*Percent Daily Values(DV) are based on a 2,000 calorie diet.

MANUFACTURED AND UNCONDITIONALLY GUARANTEED BY ©1997 LIPTON, ENGLEWOOD CLIFFS, NJ 07632 QUESTIONS OR COMMENTS? CALL TOLL-FREE 1-800-735-3554 Ⓤ D 27895-1

INGREDIENTS: WATER, LIQUID SOYBEAN OIL, SALT, SWEET CREAM BUTTERMILK, XANTHAN GUM, SOY LECITHIN, POLYSORBATE 60, LACTIC ACID, POTASSIUM SORBATE AND CALCIUM DISODIUM EDTA ADDED AS PRESERVATIVES, ARTIFICIAL FLAVOR, COLORED WITH BETA CAROTENE, VITAMIN A (PALMITATE) ADDED. BEST WHEN PURCHASED BY DATE PRINTED ON CONTAINER REFRIGERATION RECOMMENDED

0 40600 34122 6

cooking sprays. But you also have to keep in mind that you are using fat and it adds up. To determine more accurately how much you use, spray your cake pan, waffle iron, or other cooking implement as you normally do. Take a rubber spatula and collect the fat and pour it into the appropriate measuring spoon to measure it. If you spray popcorn or chicken with a cooking spray, spray into a container for the same length of time you would if you were spraying the food, then measure the amount of fat you used.

IgNobel Prize for Fiction. The marketing copy that you read on some packages will make you shake your head in wonder. Read this text on a Kid Cuisine frozen dinner for a superb example of creative writing: "Dear parents, We know how important nutrition is to you and your family. Look below to see how this meal fits into your child's balanced diet." If you "look below" you will find that the Cosmic Chicken Nuggets meal contains 230 fat calories and 90 saturated fat calories. How does this meal fit into your child's balanced diet? NO WAY. Conagra doesn't exactly *say* the meal is healthy, but the implication is certainly strong and if you didn't understand food labels, you would be taken in.

~~Smart~~ Choice. This peanut butter label is practically jumping off the jar it is so excited: "NEW 25% LESS FAT." This must be a SMART CHOICE. You must be smart to be choosing such a healthy, "low-fat" product. Really? Check out the label.

Peanut Butter

Nutrition Facts
Serv. Size 2 Tbsp (36g), Servings about 14

Amount/serving – % Daily Value**

Calories 190 Calories from Fat 110

Total Fat 12g - **18%**, (Sat. Fat 2g - **10%**), **Cholest.** 0mg - **0%**, **Sodium** 140mg - **6%**, **Total Carb.** 15g - **5%**, (Fiber 1g - **5%**), (Sugars 6g), **Protein** 8g

Vit. A <2% • Vit. C <2% • Calcium <2% • Iron 4%

**Percent Daily Values are based on a 2,000 calorie diet.

*THAN REGULAR PEANUT BUTTER WHICH CONTAINS 16g FAT PER SERVING. SMART CHOICE CONTAINS 12g FAT PER SERVING.

Calories from Fat: 110; Serving Size: 2 Tbsp (36g). Why don't we think 55 fat calories a tablespoon is such a smart choice? Yes, it is less fat than the 75 fat calories for a tablespoon of regular peanut butter, but 55 fat calories per tablespoon does not exactly make it diet food.

% Fat Free: Noninformation. 95% fat-free! 80% fat-free! 80% lean! The food manufacturers who use % fat-free claims are not lying. They are just telling you an irrelevant fact — that fat accounts for a small percentage of the *weight* of a product. (In most foods, much of the weight is water, so even a glass of whole milk, which has 73 fat calories, is only 4 percent fat by weight.) You assume they are telling you the percentage of calories coming from fat and conclude the product is low-fat and you can eat it in unlimited quantities. Wrong.

For example, take the Weight Watchers Smart Ones Pasta with Tomato Basil Sauce. Weight Watchers dubbed these frozen entrées Smart

Ones because originally each entrée contained only 1 gram of fat. They are still called Smart Ones even though they now contain many grams of fat.

The Smart Ones package pictured on page 158 screams: 97% FAT FREE. That claim is based on the fact that the weight of the fat (7 grams) is 3 percent of the weight of the entrée (272 grams). Who cares? If the meal had weighed more, the percent would be less. The only important number you need to know is that this dinky little dinner contains 60 fat calories, and that is a lot of fat for very little food.

More Claims and What to Make of Them

1. **Claim:** Cholesterol-free

   Truth: The average consumer translates "cholesterol-free" to mean "heart-healthy" or "healthy." Cholesterol-free or low cholesterol are claims that should be totally ignored.

   First, since dietary cholesterol has a minor effect on raising blood cholesterol, the cholesterol content of a food is of little importance. What is significant for your heart health is the amount of saturated fat and fat the food contains. Second, cholesterol-free just means that a foodstuff contains no animal products, because cholesterol is found only in animal products. Every plant product, no matter how high in saturated fat, is cholesterol-free. Coconut, palm, and hydrogenated plant oils, which are among the most heart-risky foods around, are cholesterol-free because they are plant products.

   According to the FDA regulations, if a product claims it is cholesterol-free, it cannot be loaded with saturated fat and thus be heart-risky. Specifically it must contain no more than 18 calories of saturated fat per serving and 27 calories of fat. The regulation limiting total fat was created to prevent high-fat and thus unhealthy products being chosen because they were cholesterol-free.

   The food manufacturer who made the wheat snack crackers above obviously didn't take that regulation seriously. The box

Nutrition Facts

Serving Size: 1 Entree (272g)
Servings Per Container: 1

Amount Per Serving	
Calories 260	Calories from Fat 60
	% Daily Value*
Total Fat 7g	**11%**
Saturated Fat 2.5g	**13%**
Polyunsaturated Fat 1g	
Monounsaturated Fat 3.5g	
Cholesterol 10mg	**3%**
Sodium 360mg	**15%**
Total Carbohydrate 40g	**13%**
Dietary Fiber 3g	**12%**
Sugars 4g	
Protein 10g	

Vitamin A	15% •	Vitamin C	15%
Calcium	15% •	Iron	10%

says "no cholesterol," but the total fat per serving is 50 fat calories — an unhealthy amount of fat. In addition, the fat that is used is "partially hydrogenated vegetable shortening," a heart-risky fat. The label lists saturated fat as 1.5 grams (13.5 sat-fat calories) but doesn't account for the amount of heart-risky trans fats it contains.

2. **Claim:** Low cholesterol
   Truth: A product can claim it is "low cholesterol" even if it is horrendously high in total fat. Read the information above on cholesterol-free and ignore both of these claims.

3. **Claim:** Light
   Truth: Food producers bank on the fact that we associate the term "light" with low-fat. However, although a product that advertises itself as "light" must contain one-third fewer calories or half as much fat as the "regular" product, it may still not qualify as low-fat in our book. A "lite" brand of potato chips that contains 60 fat calories per ounce rather than 90 is definitely an improvement, but 60 fat calories is still a whopping amount of fat for such a small amount of food.

4. **Claim:** 2% milkfat reduced-fat milk
   Truth: Because of FDA regulations, what used to be labeled 2% low-fat milk is now labeled 2% milkfat reduced-fat milk. Somehow reduced fat doesn't seem like much of an improvement on low-fat. People still think they are getting a low-fat milk; after all, it's only 2% fat. But what does 2% fat mean? The 2% refers to the fact that fat accounts for 2% of the *weight* of the milk. Milk is mostly water, so the percentage contributed by fat to the total weight is small. It would be more accurate to advertise this milk as 35% high-fat milk since fat contributes 45 of the total 130 calories (45 ÷ 130 = 35%). However, neither the claim 2% fat nor 35% fat tells us what we need to know — that one 8-ounce glass contains 45 fat calories, two glasses contain 90 fat calories, three glasses contain 135 fat calories, and that this is a drink your body can ill afford.

5. **Claim:** Extra Lean
   Truth: Under the FDA regulations, a meat can qualify as "extra lean" if it contains less than 45 fat calories for 3½ ounces of food. This amount hardly seems extra lean when you compare it with turkey, cod, crab, haddock, lobster, pike, pollock, and scallops that contain only 7 fat calories for 3½ ounces, or skinless chicken breast, red snapper, or sole at only 11 fat calories for 3½ ounces.
6. **Claim:** 60% less fat
   Truth: Less fat than what? The reduced fat description tells you nothing you need to know. A Weight Watchers Smart One Swedish Meatballs entrée claims 60% less fat. It contains 90 fat calories, which is a lot of fat. Perhaps the original recipe contained 150 fat calories, but who cares? Ignore the reduced fat claims and look at the fat calories and the serving size on the label to determine if you want to eat this product.

Words of wisdom: only an educated consumer can remain healthy and lean.

Remember:

1. Read food labels to make wise selections.
2. The ingredient list will give you insight into what the product contains. Ingredients are listed in descending order of weight.
3. The only two numbers on the food label that you need to know are:
   a. Calories from fat per serving
   b. Serving size
4. Be sure to adjust the fat calories for the number of servings you eat.
5. Products are getting larger. What are the fat calories per serving? How many servings are you eating?
6. Fat-free does not mean healthy. Treat fat-free empty calories as splurges.
7. Beware of misleading advertising claims such as No fat! Lite! Lean! 95% fat free! Low cholesterol! Lower in fat! All Natural!

8

Taking Control: Eating In

"I feel very confident that the pounds lost will not come back since I don't feel I 'dieted' but rather 'ate healthier.'"

— Mark Mandel, Erie, Pennsylvania

YOU ARE #1. That means you need to take time for yourself. You need to set priorities, and eating right is up there at the top. Eating healthfully is a top priority for keeping you fit and feeling great. Sure, your life is hectic, but you have to make time for the important things. No longer can your mantra be: "Busy, busy, busy. Too busy to cook. Too busy to eat. Too busy to exercise."

Are you too busy to be slender and healthy? If so, will you have time to be sick later? If you want any weight-loss plan to work, and work forever, you have to make time. You have to take time to enjoy lots of delicious low-fat, nutritious food. Read Chapter 12, "It's Up to You," about setting priorities so that you will have time to devote to *Choose to Lose.*

MAKING TIME

It Takes Less Time Than You Think

It just takes a little thinking and planning. Preparing food does not have to take a lot of time. Lynn Karnitz, a Choose to Lose Leader in Two Rivers, Wisconsin, had a participant who complained that she didn't have time to make lunch. Lynn brought bread, sliced turkey, hot peppers, tomatoes, mustard, carrots, and an apple to the class. In 45 seconds she prepared and packed a bag lunch. You may think you don't have 45 seconds to make lunch, but you do.

You need to make time to shop for food (or make a list so someone else can shop for you) and time to prepare it (or encourage some family member or friend to be the cook). It doesn't have to be fancy unless you want it to be.

Taking Time Is Essential for Success

Why do you need to make time? Because if you are eating full meals that are low, low, low fat and taste delicious, you can follow *Choose to Lose* forever. We do.

This doesn't mean you can never eat out. Of course you can. Chapter 10, "Taking Control: Eating Out," discusses ways to make eating out as low-fat as possible. However, if you eat most of your meals out (that includes carry-in), you will find it difficult to really eat low-fat and eat enough healthy food.

A CLEAN SLATE

Out with the Old

You need to start *Choose to Lose* with a low-fat kitchen. Crackers in the cupboards, pizza and ice cream in the freezer, cookies in the cookie jar, cheese dips in the refrigerator — all these foods have to go. Otherwise, they may cry "Eat me! Eat me!" and weaken your resolve. Don't fret about what they may have cost you in dollars. In the long run you'll be saving in both health and emotional costs.

In with the New

Next step is to fill up the kitchen with delicious, nutritious low-fat food. Even if you are not the cook, you should help with the grocery list. Make sure your fruit bowl is filled, a jar of popcorn kernels awaits air-popping in your hot-air popper, and a variety of vegetables are at the ready for lunch and dinner.

This is a great opportunity to enlarge your life. Experiment. Try new foods, new preparations, new dishes. Do you think you hate spinach? Buy it and try it. You might be surprised. Is your diet bland and dull? Add spices. Use *Choose to Lose* as an excuse to lift yourself out of your rut. Who knows where it will lead?!

You need an **ATTACK PLAN.** Planning spells success.

REALLY SMART SHOPPING

Fresh Ingredients: An Unlimited Source of Riches

Tomatoes, potatoes, green peppers, onions, garlic cloves, broccoli, carrots, mushrooms, apples, lemons, limes, salmon steaks, shrimp, boneless chicken breasts, turkey cutlets, pasta, rice, chicken broth, and on and on. Think of all the wonderful dishes you can make — and eat! If your approach to grocery shopping has been hit-and-run forays into the frozen dinner freezers, throw the microwave boxes out of your cart and take time to explore the natural riches along the perimeter of the store.

Get to know the produce section. Just walking through the aisles and looking at the colors, shapes, and textures that make each fruit or vegetable unique and beautiful is a pleasure. Fill up your cart with produce. It's a capital investment. You can't take an orange for a snack if you have no oranges in the refrigerator. You can't make *Sweet Potatoes à L'Orange* (page 284) if you don't have sweet potatoes. You can't make *Zucchini Soup* (page 268) if you don't have zucchini.

Be adventuresome. If your supermarket stocks a variety of produce, experiment. Try cilantro and fennel. Try bok choy and kale. Buy peppers of all colors and types — red, yellow, green, jalapeño. Even if your grocery store is not so well endowed, each week choose a vegetable and fruit you never brought home before. On page 164 is a check list to take with you to the store.

Check out the fish counter. Try something new each week.

Make sure you have the basics in stock at home (see Stocking Up, page 177). This means foods such as flour, sugar, rice, canned tomatoes, chicken broth, molasses, onions, and potatoes in the cupboards; lemons, limes, garlic, and ginger in the refrigerator; chicken breasts, turkey cutlets, corn, and peas in the freezer. Have a variety of spices on hand to enhance your dishes instead of adding fat.

FRUIT AND VEGETABLE CHECKLIST

FRUITS

❑ Apples	❑ Apricots	❑ Bananas
❑ Blackberries	❑ Blueberries	❑ Boysenberries
❑ Cantaloupe	❑ Carambola	❑ Cherimoya
❑ Cherries	❑ Currants	❑ Dates
❑ Figs	❑ Grapefruit	❑ Grapes
❑ Guavas	❑ Honeydew	❑ Kiwi fruit
❑ Kumquats	❑ Lychees	❑ Mangos
❑ Nectarines	❑ Oranges	❑ Papayas
❑ Passion fruit	❑ Peaches	❑ Pears
❑ Persimmons	❑ Pineapple	❑ Plantains
❑ Plums	❑ Pomegranates	❑ Prickly pears
❑ Prunes	❑ Quince	❑ Raspberries
❑ Rhubarb	❑ Strawberries	❑ Tangerines
❑ Watermelon		

VEGETABLES

❑ Alfalfa sprouts	❑ Artichokes	❑ Arugula
❑ Asparagus	❑ Beets	❑ Bok choy
❑ Broccoli	❑ Celery	❑ Cabbage
❑ Carrots	❑ Cauliflower	❑ Collards
❑ Cucumbers	❑ Corn	❑ Eggplant
❑ Endive	❑ Green Beans	❑ Kale
❑ Leeks	❑ Mushrooms	❑ Okra
❑ Parsnips	❑ Peas	❑ Peppers
❑ Pumpkin	❑ Potatoes	❑ Radicchio
❑ Rutabaga	❑ Squash	❑ Spinach
❑ Turnips	❑ Watercress	❑ Zucchini

The message! If you want to become lean and stay that way for a lifetime, you need to keep your refrigerator and cupboards jam-packed with the ingredients needed to create great low-fat, healthy dishes.

Here are some tips that will make eating in as easy and delicious as possible.

HEARTY, HEALTHY BREAKFAST

First, some advice: do NOT skip breakfast!

Quick and Easy

Here are some ideas for delicious, low-fat breakfasts that you can whip together in seconds.

- Whole-grain toast (whole-wheat bread has more fiber and nutrients), English muffins, or bagels instead of sweet rolls, biscuits, muffins, Danish pastries, doughnuts, or croissants.
- Jelly, honey, apple butter, or nonfat cream cheese instead of butter, margarine, or cream cheese.
- Hot cereals such as oatmeal, wheatena, cream of wheat, farina, oat bran plain or with skim milk (*not* 2% or whole).
- Cold cereals (Check the nutrition label for fat and dietary fiber. Find cereals you like with at least 4 or 5 grams of dietary fiber.) with skim milk (*not* 2% or whole).
- Half a cantaloupe filled with 1% or fat-free cottage cheese.
- Fruit-flavored or vanilla nonfat yogurt plus cut-up fruit.
- An omelet made with egg whites and one egg yolk or egg substitute instead of whole eggs. Remember — an egg yolk contains 50 fat calories.
- Ron's favorite: 1% cottage cheese spread over a piece of whole-wheat bread toasted twice. If you're using commercial bread, put toaster on low so it won't burn.
- Nancy's favorite: whole-wheat toast covered with banana slices, topped with 1% cottage cheese.
- Fruit: banana, orange, a bowl of strawberries, blueberries, raspberries, half a cantaloupe, melon slices.

> Always read the nutrition label on the cereal box to make sure you have made a good choice. For example, Post Crunch Pecan cereal contains 60 fat calories per ⅔ cup serving. It also contains dried coconut and partially hydrogenated cotton oil. Nasty.

- A weekend treat: waffles or pancakes using buttermilk and skim milk, nonfat yogurt and skim milk, or orange juice and skim milk instead of whole milk, sour cream, or sweet cream. Substitute a small amount of olive oil in the batter for butter or margarine. Spray a nonstick waffle iron with canola or olive oil spray before you plug it in. Use a cast-iron grill with a light application of margarine or olive oil for the pancakes. For a divine experience, use real maple syrup not cane syrup or maple-flavored syrup. Ron eats maple syrup with waffles on the side.
- Beware convenience breakfast foods such as Great Starts breakfasts, ranging from Breakfast Burritos at 100 fat calories a serving to Sausage, Egg, and Cheese on a Biscuit at 250 fat calories a serving and Aunt Jemima Silver Dollar Pancakes with Sausage at 160 fat calories a serving.
- Be careful of Pop Tarts and Granola Bars. Some are candy bars in disguise. A brown sugar cinnamon Pop Tart has 60 fat calories.
- Avoid egg dishes (made with whole eggs), sausage, bacon, and home fries.

HEARTY, HEALTHY LUNCH

- The best advice is to bring lunch from home. To save time in the morning, make and refrigerate it the night before. Invite your coworkers to bring their lunch and join you. Have picnics if the weather permits.
- Make a sandwich of last night's *Bar-B-Que Chicken* (see page 270). Fill a whole-wheat pita with *Ted Mummery's Turkey Barbecue* (see page 292) and heat it in your office microwave.
- Cook up a large batch of *Chili Non Carne* (page 287) or *Pawtucket Chili* (in *Eater's Choice Low-Fat Cookbook*) and prepare 1 or 2 cups of rice. Keep it all in the refrigerator and bring a combination of chili and rice to work and heat it in a microwave. You can also put it in a whole-wheat pita pocket, heat it up and add lettuce, tomatoes, and onion.
- Spruce up your turkey breast sandwich with tomato slices, mustard, and hot peppers. Avoid mayonnaise, butter, or margarine.

- Use poultry, fish, or lean meat in place of processed luncheon meats, such as salami, bologna, or frankfurters.
- Beware of any luncheon meats, even, or especially, innocent-looking franks made with chicken or turkey; read labels for fat content, and be sure to adjust fat calories for amount eaten.
- Beware of low-fat cheeses or meats. How much do you eat? Fat calories add up quickly.
- Eat nonfat cheeses with care. Eating fatty-tasting (although nonfat) food keeps up your high-fat taste.
- Make your tuna salad sandwich with low- or nonfat mayonnaise and nonfat yogurt. Again, don't overdo the high-fat-tasting mayonnaise if you want to lose your high-fat tooth.
- Keep plenty of fruit available.
- Pack carrot or celery sticks.
- Pack a nonfat flavored yogurt.
- Bring air-popped popcorn (take it with you to the movies, too). Or, better still, keep a hot-air popper in your office and make a fresh batch each day.

HEARTY, HEALTHY DINNER

Let's Hear It for Eating Home Cooking at Home

Don't let your schedule, job, hobbies, or activities get in the way of eating dinner at home. It's a MUST. The emotional benefits of sitting together at a table every evening and sharing a meal and conversation with spouse and children are immeasurable. For your own mental health, sitting down to a meal, taking time to eat, to digest, to think, to talk, to unwind can have long-term psychological rewards.

Of course, the health and dietary benefits are substantial. When you choose the food you eat, you know you are getting enough vegetables, fruit, dairy, and grain. You know that the meat you buy is low-fat, because you have purchased it. You know the preparation is low-fat, because you prepared it. You are not tempted by high-fat no-no's, because you don't have them in the house. Dinner can always taste divine, because you are in control of the cooking.

Spice up the Main Course

Delectable dishes can be made without cream, butter, or cheese. Beef doesn't have to be eaten in a slab. Foods can taste good without being drowned in oil.

Rethink your preferences and retrain your palate. Modify favorite recipes by reducing the fat. Make "cream" sauces with skim milk, non-fat yogurt, or buttermilk.

To reduce fat and add texture and flavor, cut a small piece of meat into bite-size pieces and mix it with masses of vegetables. Make your dish more rice and less meat.

Give up the rich, bland, fattening dishes you eat too much of because they are so unfulfilling. Use spices instead of fat to create savory dishes. Spicy doesn't necessarily mean "hot." Spicy means tasty. Spices satisfy your palate and you. And, speaking of spices, don't blame spices for your indigestion. It's the fat in greasy dishes that give you pain. Spices give you only pleasure.

It's easy to eat well on a low-fat diet. Does *Spicy Shrimp Louisiana* (page 279) or *Sesame Chicken Brochettes* (page 275) sound like diet food? With 28 or fewer fat calories each, they slip easily into anyone's Fat Budget.

Eat, Eat, My Darling

You need to eat real, full meals. If you are eating a baked potato and broccoli and calling it dinner, your days as a Chooser to Lose are numbered. You need to be satisfied. Eat the baked potato and broccoli, but eat it with *Broiled Ginger Fish* (page 278), steamed rice, *Cucumber Salad* (page 290), and *Tortilla Soup* (page 267).

Believe it or not, eating nutritious, low-fat foods will make you thin.

A Chooser to Lose told us that, without telling her husband, she started cooking *Eater's Choice* recipes to reduce fat in their diet. He was delighted with the new dishes and ate with gusto. After several months he came to her visibly distressed. "Dear," he began, "I think I have cancer. I have never eaten so much, and I keep losing weight." Can you imagine the anguish he felt? She calmly explained that he was losing weight because he was eating a low-fat, nutritious diet. He had become an unwitting Chooser to Lose. Don't be distressed if *you* are eating more than ever and losing weight. Celebrate.

Read Chapter 9, "Cooking Low-Fat and Delicious," for some helpful tips that make creating scrumptious dinners easy and fun.

Remember:

1. If you are too busy to live a healthy lifestyle now, will you have time to be sick later?
2. Preparing healthy, nutritious food doesn't have to take a lot of time.
3. Start with a clean slate: replace high-fat stores with nutritious, low-fat foods.
4. Stock your house with fruits, vegetables, and low-fat meats so that you can have delicious, healthy snacks and delicious, healthy meals.
5. Skipping meals does not make you thin. It makes you hungry, tired, irritable, and out of control. Eating delicious, low-fat, nutrient-dense/fiber-rich meals makes you happy, full, in control, and helps you lose weight.

9

Cooking Low-Fat and Delicious

"A friend recently recommended *Choose to Lose* to me. I am absolutely delighted with it. I feel there is hope! I don't feel deprived! Ingredients are usually ones I have on hand. Most of the recipes are able to be made quickly, which fits into my busy schedule."

— Barbara Karpinski, Rockville, Maryland

THE WAY TO HAVE the greatest control over what you eat is to prepare it yourself. Cooking doesn't have to be complicated or time-consuming. Preparing food often takes much less time than eating out. "But," you argue, "when you eat out, you aren't doing the work." True, but for a little effort the benefits are magnificent. Enlist your family to help you. Why should you be the only cook? Be experimental. Have fun.

HINTS FOR QUICK AND EASY MEAL MAKING

Preparing dinner need not be a three-hour endeavor. If you plan ahead, meal-making can be quick and easy. Here are some hints.

- At the beginning of the week, glance through *Choose to Lose, Eater's Choice Low-Fat Cookbook,* or other low-fat cookbooks and choose recipes for the week. Jot down the ingredients you will need so that you won't waste time when you shop. Be sure to add foods such as fruits, vegetables, and bread, which you will need for snacks, side dishes, and desserts.
- When you go to the supermarket, follow your shopping list, but if you see a food that entices you, buy it. For example, perhaps you

hadn't even thought of making Shrimp Creole, but the shrimp look so large and appetizing, you decide to buy them for that night's dinner.

- Buy enough food so you have plenty of material to work with. Buy a variety of vegetables so that you always have a vegetable or two with dinner. Make sure you keep your larder (excuse the expression) well stocked with the basics (see Stocking Up, page 177).
- Cook up several soups (double or triple the recipes) over the weekend. Freeze them in small containers (enough for one meal) to reheat and eat all through the week.
- Make enough of an entrée so that you can eat it for two days. Make even more so that you can also freeze some to reheat later on.
- To save preparation time, flour chicken breasts and bake them (with no fat) the night before so that you can use them in a recipe the next day.
- Make enough rice for several days. Each day, add a little water and heat up enough for that meal.
- Try basmati rice. It cooks in 10 minutes and has a delicious flavor.
- Store canned chicken soup in the refrigerator. The cold hardens the fat so you can remove it easily.

COOKING LOW-FAT

You can't help but notice while reading this book how many times we have mentioned the *fantastic* recipes in the *Eater's Choice Low-Fat Cookbook*. We are not trying to strong-arm you into buying another book, but as you can see from our hearty endorsement, we believe you would greatly enjoy the 240 additional recipes in the *Eater's Choice Low-Fat Cookbook*. They are generally quite low in fat, delectable, and helpful for losing weight. In fact, one of the reasons we wrote *Choose to Lose* was that so many people told us they had lost weight using *Eater's Choice* recipes.

You probably also have your own favorite recipes that are too high-fat to keep your Fat Budget in the black. By making a few substitutions and modifications, you may be able to fit these favorites into your new eating plan.

Silent Substitutions

In most cases, the following low-fat or nonfat substitutions will greatly reduce the fat content of your recipes without diminishing the taste or texture.

When the recipe calls for:	Substitute:
cream	nonfat yogurt, buttermilk, skim milk, evaporated skim milk
whole milk	skim milk
sour cream	nonfat yogurt, buttermilk
veal cutlets	turkey cutlets
pork	white-meat chicken
ground beef	ground chicken or ground turkey (without skin and fat) that you grind yourself
3 egg yolks*	1 egg yolk
2 whole eggs	1 egg + 1 egg white or 3 egg whites
margarine	try apple sauce

Warning

If all your recipes continue to taste cheesy or creamy or fatty, even though they are made with nonfat substitutes, you will never lose your taste for fat. Try moving toward tasty recipes instead, and your weight loss success will be greater.

Reducing the Fat

Recipes often recommend more fat than is necessary. Another way to modify a recipe is to reduce the amount of fat you use.

- If the recipe calls for sautéing in ¼ cup oil or butter, use 2 tablespoons of olive oil instead. If the reduced amount works, try 1 tablespoon or even 1 teaspoon. If a recipe calls for sautéing in 2 tablespoons of fat, use 1 teaspoon of olive oil or less plus a splash of water or sauté in strained chicken broth.

---

*The egg yolk contains all the fat (50 calories) as well as all the cholesterol in an egg; the egg white has neither fat nor cholesterol.

- DON'T dot the top of your fish or pie with margarine or butter. The dish probably doesn't need the extra fat.
- Instead of sautéing chicken breasts in a few tablespoons of olive oil, dredge them in a mixture of flour, salt, and pepper, and then bake them in a single layer in a shallow baking pan. Add no fat. When the breasts are cooked, refrigerate or freeze them for later use or cut them up and add them to the sauce you have just prepared.
- Try substituting apple sauce for margarine in a recipe. Don't experiment on guests! Sometimes (as in *The York Blueberry Cake* recipe, page 297) the results are great. Sometimes the results land in the garbage can. It is certainly worth a try. York Blueberry Cake originally contained 46 fat calories a slice. After the applesauce substitution, the cake tastes just as good but it contains 3 fat calories per square.

Modifying a Recipe

We easily modified the high-fat recipe for Curry Soup into a tasty low-fat treat.

A Recipe: Before and After

HIGH-FAT CURRY SOUP		LOW-FAT CURRY SOUP	
INGREDIENTS	FAT CALORIES	INGREDIENTS	FAT CALORIES
¼ cup butter	400	1 tsp olive oil	40
2 medium onions		2 medium onions	
3 stalks celery		3 stalks celery	
2 tbsp flour		2 tbsp flour	
1 tbsp curry powder		1 tbsp curry powder	
2 apples		2 apples	
1 cup diced chicken (light and dark meat)	31	1 cup diced chicken (white meat)	13
2 qts chicken broth	184	2 qts chicken broth, defatted	5
1 cup light cream	417	1 cup nonfat yogurt	0
Total fat calories	1032	Total fat calories	58
Makes 10 cups		*Makes 10 cups*	
Per cup of soup	**103**	Per cup of soup	**6**

This is how we did it:

- Four tablespoons of butter was more fat than necessary to sauté the onions. In addition, butter is a heart-risky fat. We reduced the fat to 1 teaspoon of olive oil and a splash of water and saved 360 fat calories. (We chose olive oil because it is a heart-healthy oil; but being heart-healthy does not make it any less fattening.)
- Instead of dark and light chicken meat, we used only white breast meat and saved 18 fat calories.
- Defatting the chicken broth saved 179 fat calories.
- Replacing the cream with nonfat yogurt saved 417 fat calories.

The resulting soup tastes fine, and at 6 fat calories per cup you could easily eat several cups.

Determining Fat Calories per Serving. If you want to determine the fat calories per serving for a recipe, add up the fat calories for all the fat-containing ingredients and then divide by the number of servings. For example, in the high-fat recipe above, the fat calories for all the fat-containing ingredients total 1032. To determine the fat calories per serving, you divide 1032 by the number of servings, which is 10. 1032 ÷ 10 = 103 fat calories per serving.

Cooking Tips

The following cooking tips will help you stay within your Fat Budget.

- Cook vegetables with an inexpensive metal steamer. Put the steamer in a saucepan filled with about an inch of water. Place cut-up fresh vegetables on the open "leaves," cover the saucepan, and steam the vegetables until they are tender. For added flavor, mix in some pressed garlic, vinegar, or herbs (thyme, basil, etc.) or top cooked vegetables with a dollop of nonfat yogurt mixed with Dijon mustard.
- Sauté vegetables in 1 teaspoon of olive oil and add splashes of water to steam-cook them.
- Sauté vegetables in broth.
- Always steam or bake eggplant. When it is sautéed, it absorbs gobs of oil.

- Always remove skin *before* cooking chicken to keep the chicken from absorbing the fat from the skin. A skinless chicken breast has 13 fat calories. A chicken breast that was cooked with the skin has 28 fat calories after the skin is removed. Eating the skin more than quadruples the fat calories in a chicken breast. A chicken breast with skin has 69 fat calories.
- NEVER deep-fry fish or chicken. NEVER flour or bread and fry fish or chicken. Breading absorbs fat like a sponge.
- Bake, broil, grill, or poach chicken or fish. Limit the amount of fat you use. Use low-fat sauces for added taste.
- If you cook beef, lamb, pork, or veal, trim off all visible fat. Broil or bake on a rack to drain the fat. Wrap cooked meat in paper towels to remove more fat. Avoid pan-frying.
- In any recipes that call for canned chicken broth, either strain the broth or if the fat is hardened, lift it off the soup with a spoon, or use a gravy skimmer. Keep cans of broth in the refrigerator so that the fat will be hardened and easy to remove.
- Use *unbleached* white flour. When flour is bleached, it loses many of its important nutritional qualities.

For Superior Taste

- Use freshly ground black pepper. There really is a difference between the stale, tasteless pepper that comes in a can and the peppery taste of pepper that has just been ground. It is definitely worth buying a peppermill (usually not expensive) and whole peppercorns (available in the spice section of your local grocery store).
- Grate a whole nutmeg (available in the spice section of your local grocery store) with a grater for a fresher, nuttier taste.
- Fresh garlic is available in the produce section of your grocery store and is far superior to garlic powder or garlic salt. It will stay fresh for weeks in the refrigerator.
- Fresh ginger root is available at many grocery stores and is easily kept for months in the refrigerator in a jar filled with sherry. Slice off the outer covering and use ginger root with superior results in recipes that call for ginger (but not in cake or cookie recipes).

Fat Reducing Cooking Equipment

- Cast-iron makes a great nontoxic, nonstick cooking surface. You can purchase cast-iron griddles that fit over two stove burners and quickly cook chicken breasts, turkey cutlets, French toast, and pancakes with little or no added fat. Small cast-iron griddles that fit over one stove burner are also available.
- Steamer: See the preceding list of Cooking Tips.
- Clay cooker: Keeps chicken moist and tender without adding fat.
- Gravy skimmer: removes fat from canned chicken broth.

Our Favorite Cooking Gadgets

- Defrosting tray: if you are like Nancy Goor and forget to thaw chicken breasts in time for preparing dinner, the defrosting tray (one brand is called Miracle Thaw) is a lifesaver. Just run the black metal sheet under hot water until it becomes hot, place the frozen breasts across it, and 15 minutes later you will have either almost or completely thawed meat. If the meat is not completely thawed, repeat the process. Buying the defrosting tray was Ron's idea. Nancy was skeptical until she tried it.
- The best vegetable peeler in the world: Good Grips makes a vegetable peeler that works like a dream.
- Lemon zester: what a neat gadget! Just press as you pull the zester down a lemon and you create skinny, fragrant pieces of lemon zest.
- Garlic press: the easiest way to mince garlic for most recipes is just to squeeze it through a garlic press.
- Mini-chopper: you may have noticed how many of our recipes use garlic and ginger. A mini-chopper makes mincing cloves of garlic and ginger root effortless. If you have no lemon zester, you can chop strips of lemon peel (just the yellow) and create grated lemon zest.

Is Low-Fat Cooking and Eating Expensive?

No way. Replacing foods high in fat with vegetables, fruits, and whole grains will keep you thin and your wallet fat. Convenience foods and snacks are high in fat and expensive. Compare the cost of a pound of potatoes with a pound of potato chips. Frozen dinners are about five times as expensive as the raw ingredients used to prepare them. Beef costs more than chicken or turkey. Sour cream, sweet cream, and cheeses are rich foods that impoverish both your pocketbook and your Fat Budget.

In contrast, a variety of vegetables, rice or pasta with small amounts of chicken or beef keep both budget and body trim. Even if low-fat eating were more expensive — which it definitely is not — the saving in health costs and the psychological benefits of not being overweight would balance the expense.

STOCKING UP

We will say it again. You need to have lots of fresh vegetables and fruit in stock so that you will have plenty of material with which to create delectable dishes. You can't make sweet potatoes whipped with orange juice if you have no sweet potatoes or orange juice. But if you have them on hand, in a short, short time, with very little effort, you can create a luscious vegetable side dish.

For some of you, reducing fat will mean a big change in the foods you cook. You may wonder what foods you should have on hand to make cooking a snap. Here is a list of foods and spices to keep in your cupboard, your refrigerator, and your freezer.

FOODS TO KEEP ON YOUR SHELVES

Almonds, slivered
Anchovies
Apricots, dried
Apricot jelly
Artichoke hearts in water
Baking powder
Baking soda
Barley
Beans, dried (black, pinto, etc.)
Beans, kidney, canned
Chicken broth
Chickpeas
Cocoa, unsweetened
Cornstarch
Cream of tartar
Flour, unbleached white
Flour, whole-wheat
Honey
Lentils
Molasses
Oat bran
Oatmeal (not quick cooking)
Olive oil
Olives, black and green
Pastas (spaghetti, noodles, etc.)

Pineapple chunks, canned
Peaches, canned
Peanuts, unsalted
Prunes
Raisins
Salt
Sherry, dry
Sugar, brown
Sugar, white
Sugar, confectioners'
Tomatoes, canned
Tomato juice
Tomato paste
Tomato sauce
Vanilla
Vermouth
Vinegar
Yeast (if you plan to make bread)

ASIAN FOODS TO KEEP ON YOUR SHELVES

Bamboo shoots, canned
Bean sauce
Black beans (fermented)
Chili paste with garlic
Hoisin sauce
Mushrooms, dried black
Sesame chili oil
Sesame oil
Soy sauce and double black soy sauce
Water chestnuts, canned

SPICES AND HERBS

Basil leaves
Caraway seeds
Cardamom, ground
Cayenne pepper
Chili powder, hot (if you like)
Chili powder, mild
Cinnamon, ground
Cinnamon sticks
Cloves, ground and whole
Coriander
Cumin, ground and seeds
Curry powder
Dillweed
Ginger, ground
Mustard seed, black
Nutmeg, whole
Oregano
Paprika
Peppercorns, black
Rosemary leaves
Sesame seeds
Tarragon
Thyme leaves
Turmeric

FOODS TO KEEP IN YOUR FREEZER

Chicken breasts, boneless, skinless
Corn, frozen
Margarine
Peas, frozen
Turkey cutlets
Whole-wheat bread
Whole-wheat pita bread

FOODS TO KEEP IN YOUR REFRIGERATOR

Buttermilk
Ginger root (store in jar with sherry)
Mayonnaise or salad dressing
Mustard, Dijon
Yogurt, nonfat plain

ALL-SEASON FOODS TO KEEP ON HAND

Apples	Limes	Snow peas (Chinese peapods)
Bananas	Onions	Spinach
Broccoli	Oranges	Squash (acorn, butternut, etc.)
Carrots	Pears	Tomatoes
Cauliflower	Peppers, green	Zucchini
Celery	Potatoes, sweet	
Garlic cloves	Potatoes, white	
Lemons		

And Coming Next . . .

Learn how you can become the master of your fate in every eating situation away from home.

Remember:

1. The most delicious and effective way to lose and maintain weight is to prepare your own food from scratch.
2. It is not difficult to cook.
3. To reduce fat in recipes:
   a. Use low- or nonfat substitutions.
   b. Use less fat in cooking.
   c. Steam, broil, grill, or bake with little or no fat instead of frying.
   d. Use cast iron, steamers, clay cookers to reduce fat use.
4. Keep your larder well stocked so that you have lots of food with which to create delicious recipes.

10

Taking Control: Eating Out

"I look back and I consider myself a convenience food store junkie. I would go in and buy a pack of salted peanuts, and three hours later I would go back to get something else because I was hungry again. Now I take something with me. And when I eat out, the restaurants we usually go to are really good about making sure they cook something just the way I like it. I'm not saying I'm a health nut. When I want a milkshake, I have one."

— Boonie Mitchell, Florida

YOUR GOALS are high. You are making all the right changes — and loving it. Soon you will find that your clothes are slipping off your bony body. You'll be so perky and alert, your boss will give you a raise, or if you're the boss you'll give yourself a bonus. You are beginning to prefer plain air-popped popcorn to buttered. Your life is shaping up. And so are you.

We hope this first paragraph describes you. However, the road to thinness may not be quite so smooth. Temptation abounds. Television ads assault you with visions of rich, dark chocolate flowing smoothly over clusters of peanuts, thick, juicy hamburgers sparkling with fat, breaded shrimp floating in pools of butter. Birthday parties, office parties, dinner parties threaten your resolve. Here are some tips to get you through the hard times.

SPLURGING

Don't Act Natural

There are two natural reactions to gluttonous behavior.

1. Wringing the hands and saying, "Since I have been bad, I might as well just keep stuffing myself."
2. Wringing the hands and saying, "I have been bad. Tomorrow I will starve myself to make up for today."

Reject your normal urges. Make the following be your reaction.

1. I lost control, but I am not bad. I will stop stuffing myself right now. I will get back on the CTL track right now.
2. I lost control, but I am not bad. Starting now I have to eat a lot of nutrient-dense/fiber-rich foods in the form of real, full meals. I need to include an occasional high-fat splurge so I don't feel like a martyr and this behavior won't occur again.

The beauty of *Choose to Lose* is that you can fit in high-fat favorites. You just have to plan ahead.

Game Plan

Plan in advance exactly how you are going to splurge. Is your hostess's lemon meringue pie close to divine? Take one slice (100 fat calories) and leave the dessert table. Do you love sour cream onion dip? Spoon 1 or 2 tablespoons (50 fat calories) on a plate and then move away to dip and enjoy. Determine your splurge *before* you arrive at the affair and then limit yourself to that high-fat treat.

Plan ahead for bigger splurges. Say your birthday is Saturday and you want to celebrate with a special splurge dinner: a 6-ounce steak (276 fat calories), a baked potato with a teaspoon of butter (34 fat calories), and a piece of apple tart (100 fat calories) for a total of 410 fat calories. WOW! Even if your budget is 315, you can afford the splurge.

Here's how: If you eat 220 fat calories each day from Sunday to Saturday, you will have accumulated 315 − 220 = 95 fat calories per

Important: *Choose to Lose* is quantitative. You have a Fat Budget. If you weaken and eat two large slices of Aunt Miranda's famous whipped cream pie, just compensate by making low-fat choices for the next few days — and you know exactly how many fat calories that is. Don't say, "I've been bad," give up, and pig out for the rest of the day because you splurged. You can fit it in. You are in control.

You should also ask yourself, "Why did I eat two slices, when I only needed one? Was I hungry because I hadn't eaten enough before dessert?" Try to figure out why so it won't happen next time.

day; 95 fat calories × 6 days = 570 extra fat calories to spend on Saturday. You can enjoy your splurge. Just plan ahead.

A warning: ("Ah, always a warning," you sigh. Yes, we always give a little warning. Nothing in life is free.) If you take a total meltdown splurge every weekend (which may be even more fat calories than you anticipated), you may cancel your fat-loss progress for the week. You probably won't gain weight but your weight loss may slow down. Instead of a wild all-weekend binge, satisfy your id with an *occasional* high-fat splurge. This will keep you progressing at a steady pace.

TAKE CONTROL IN EVERY SITUATION

Here are some hints for the many occasions that will arise when you haven't saved for a splurge but don't want to destroy your budget.

Important Advice to Chronic Dieters

Don't starve all day so that you go wild when you arrive at the table. You'll break your Fat Budget in a millisecond or two. If you want to save fat calories for the affair, just eat extra low-fat all day (but eat a lot so you won't be hungry) or for several previous days. (See Splurging on page 181.)

At Parties

It's guaranteed — any party you attend will be overloaded with high-fat foods. In fact, at some parties the only low-fat choice will be soda water.

Before the Party. Ask your host or hostess if you can bring something — hors d'oeuvres, bread, salad, vegetable, main course, or dessert. Of course, you'll make it low-fat. Most hostesses will be delighted, but if one should resist, insist. Your contribution may be all you get to eat.

Eat a light snack before the party so that blinding hunger won't cause you temporary failure of will power and result in overeating.

Plan ahead and know exactly how you are going to splurge.

At the Party. The first glance may make your heart sink. How can you resist the sumptuous array of the most delectable foods you have craved for your entire life? But look again. Does the breaded shrimp look luscious or just greasy? Does the twelve-layer rum cake taste good or just look beautiful? Granted, some of the food will be delicious, but not all. Pick and choose.

Leave the Scene of the Crime Before It Becomes a Crime. When you have finished savoring a few offerings, leave the area. Don't stand around listening to the food cry, "Eat me! Eat me!" Go to a room where there is no food. Out of sight, out of mind; in sight, in mouth. (Eventually, you may not have to avoid rooms with food. A Chooser to Lose told me that this is the first holiday season she was able to pass by doughnuts, cookies, and cakes rather than frantically scarfing them down. She knew she could have them, so she could ignore them. She ate the one piece of cake she had chosen for herself. This may even happen to you.)

Regrets to the Hostess. Before you arrive at a dinner party devise a plan of action. Be ready with an appeasing comment, "You know I adore your desserts, but I'm off sweets until I can button my slacks." Or privately before dinner (if you know the food will be full of fat), "Please give me tiny portions. I'm trying so hard to lose weight."

At Your Own Party. This is a perfect opportunity to show your friends how well they can eat low-fat. Give a dinner party and start with *Asparagus Soup Chez Goor* (page 264) at 0 fat calories per cup. Then serve *Lemon Chicken* (page 273) over rice at 13 fat calories, steamed carrots and zucchini (0 fat calories), steamed cauliflower with *Dijon-Yogurt Sauce* (page 281) for 0 fat calories, *Onion Flat Bread* (page 294) for 7 fat calories a slice or made with no fat (0 fat calories per slice), and, for dessert, *Key Lime Pie* (page 300) at 36 fat calories. When is the party? We'll be there.

If time limitations force you to weaken and buy a high-fat dessert (better to serve fruit), be sure no leftovers remain. Stuff them into your guests' pockets and purses if necessary. Don't leave a trace even in your garbage can because we all know that food in garbage cans can be retrieved.

At Restaurants

Eating at home makes keeping to your Fat Budget a breeze. Although you can't always be expected to eat at home, keep eating out to a minimum. Take a look at the fat calories for **RESTAURANT FOODS** and **FAST FOODS** in the Food Tables if you need a reason to dine at home. How does a serving of fettuccine Alfredo at 873 fat calories fit into your Fat Budget? Or lasagna at 477 fat calories per serving? Or a bowl of wonton soup at 108 fat calories, or a dish of kung pao chicken at 1134 fat calories? Can your Fat Budget be stretched to accommodate that amount of fat every couple of days and bounce back?

> **Remember:** If you eat out once or more a week, it is no longer a special treat that justifies splurging. Eat as you do at home — with care.

Here are some tips for eating out.

Be Choosy. If possible, *you* pick the restaurant. If others choose, you may have no options. Select a restaurant that you know serves food you can eat. Want to try some place new? Call early in the day and ask what's on the menu. If nothing suits your Fat Budget, ask if the chef can prepare something with little fat or with a low-fat sauce. Be specific.

Be Careful. Don't choose restaurants with unlimited portions. But if you find yourself at a smorgasbord through no fault of your own, don't pile up your plate with second and third helpings. Take a modest amount. It's hard to resist being a glutton because you feel you are getting lots of extra food for free. But bargains like this will cost you dearly in fat calories and health. In addition, the food is generally not worth the damage to your Fat Budget.

Research First. Check the **RESTAURANT FOODS** section of the Food Tables and Chapter 4, "Where's the Fat?" to help you choose a restaurant. You may find that your "healthy choice" is really a den of high-fat iniquity.

Beforehand. Don't starve all day in preparation for dinner, or you'll frantically gobble up your Fat Budget before you get through the soup and salad. Eat low-fat all day, but eat. In fact, you might want to munch on a carrot or two or even snack on a nonfat flavored yogurt an hour before you depart for the restaurant.

Take Control. The aim of most restaurants is to offer food they think customers will enjoy — that is, food oozing fat. They are proud of their cheese soufflés and their Death By Chocolate desserts. Although some offer a few low-fat dishes (be suspicious), most don't even pretend to be healthy. It is up to you to take control and make sure your meal is as low-fat as possible. Remember: you're the boss. You have no excuse if you give into temptation and throw your Fat Budget to the winds.

Don't Be Intimidated. Don't let the restaurant staff intimidate you and don't be cowed by your dining companions, either. If they roll their eyes when you order a potato instead of French fries or ask for salad dressing on the side, ignore them. If they razz you because you want to know how the chicken is prepared, roll your eyes at them. Explain to them that you are trying to eat healthfully. Do you wonder why they are threatened by your behavior? Could it be that they know they should be following your example? If eating with them becomes too uncomfortable, choose more caring dining partners.

Ask Questions. Don't be shy. Always ask the waiter what you want to know. It's your body, not his. Ask how the Sole Amandine or

Chicken Dijon is prepared. Is it fried? Baked? Broiled? Is it made with butter? Oil? What kind of oil? What's in the soup or the sauce? Cream? Milk? Sour cream? Ask enough questions to make educated choices.

Since When Is Coconut Heart-Healthy? Even (or especially) ask questions if you see a heart or symbol on a menu that indicates a dish is either heart-healthy or a low-fat choice. In fact, be especially careful with "healthy" options. We have seen "heart-healthy" choices such as fruit salad topped with shredded coconut (coconut is what they feed rats to give them heart attacks) and hamburgers without buns. We just saw a menu with a bright red heart ("a healthy choice approved by Suburban Hospital") in front of a steak and pork dinner entrée.

Make Requests. Ask the waiter if the chef will broil, bake, poach, or grill the chicken or fish without fat. Request that the sauce be served on the side so that you can decide how much you want to use. Ask to have the skin removed from the chicken before it is cooked. Ask if the nonfat mustard sauce that spices up the steak dish can be used on a grilled fish instead of fat. Be creative.

Send It Back. If the toast comes buttered, send it back. If the chicken comes with skin, send it back. If your order is not served as you requested, send it back. You wouldn't accept a purple couch if you had ordered a beige one.

On the Side. Always ask to have butter, salad dressings, and sauces served on the side so that you can regulate how much you use. You'll see how quickly a few tablespoons add up to more than you need to eat.

Dip in or BYO. When you order your salad dressing on the side, keep it in the serving dish instead of pouring it over your salad. Use the Fork Method: dip your fork into the dressing, then spear your salad greens or vegetables. You will get the salad dressing taste with many fewer fat calories. You might want to order your salad plain and enhance it with packets of nonfat or low-fat salad dressing you bring from home. Or you could just not order salad at all. It has little to recommend it except as a vehicle for salad dressing.

Bar Salad Bars. As we have mentioned more than once, salad bars are to be avoided. (To see why in specific detail, check out the **SALAD**

Fat Attack at the Salad Bar		
SALAD BAR OFFERING	**AMOUNT**	**FAT CALORIES**
Ranch dressing	1/3 cup	453
Carrot waldorf salad	1/2 cup	110
Creamy cole slaw	1/2 cup	160
Macaroni salad	1/2 cup	160
Pasta fiesta	2/3 cup	80
Sunflower seeds	2 tablespoons	55

BAR FOODS section of the Food Tables.) Unless you can restrict yourself to the fresh fruit, unadulterated carrots, tomatoes, peppers, and cauliflower, stay away. You don't need all those mixed fat salads. Your delicious, full meal is on the way.

Vegetables, Plain. Order your vegetables steamed with no butter. They will taste delicious that way. (Your request may be ignored. Be suspicious if your vegetables come swimming in a glistening yellow pool.)

Substitutions. Order a baked potato, plain instead of fries; a fruit salad instead of coleslaw. Not only will you save hundreds of fat calories, you will be getting more delicious, more healthy, and more filling choices.

Give It Away. If a portion served is more than you should eat, place the extra on a salad plate and ask the waiter to take it away. Send back the butter served with the bread. Ask for the potato plain. You can't eat what's not there.

You may feel compelled to lick your plate clean because you paid for your meal, but resist the urge. The cost to your Fat Budget is not worth the salve to your conscience.

Share. If you are dying for a taste of that sinful-looking Black Forest cake that keeps haunting you from the dessert card on your table, share it with a friend, or better yet, several friends. Choose your friends wisely and you might find one who will give you a bite of everything she eats. Main dishes and appetizers can also be shared without any-

Taste

Don't just shovel in the food, particularly a high-fat splurge. Concentrate when you eat. Taste every bite. Don't just grab some vegetables at a salad bar and call it dinner. Make a quick but glorious *Eater's Choice* meal and savor every morsel. Eating is one of life's great pleasures.

one starving. Restaurants often serve single portions large enough for two.

If you do eat restaurant food without trying to modify your choices, be sure to triple the amount of fat and calories you think you are eating. Even that may be an underestimate. The *New York Times* reported a study in which experienced nutritionists were asked to estimate the fat calories for restaurant dishes. Even they were way under on the numbers. For example, nutritionists judged that a lasagna containing 477 fat calories contained only 315 fat calories. They estimated that a porterhouse steak dinner contained 576 fat calories when the amount was actually 1125!

If you find that you are not losing weight following *Choose to Lose,* ask yourself how many times a week you eat out (including carry-in). You will find it extremely difficult and nearly impossible to reach and maintain your weight loss goals if you eat out all the time.

Traveling

Flying? Plan Ahead. Why waste hundreds of fat calories on mediocre airline food? If you plan ahead, you won't be forced to choose between two greasy sausages (140 fat calories), a greasy cheese omelette (125 fat calories), and a greasy muffin (40 fat calories) for breakfast or nothing to eat for three hours. Just plan ahead. Call the airline at least 24 hours before your flight and order a special meal. Explain that you are on a low-fat diet. Be as specific as they allow in making your request.

Ask for a fruit plate. Ask for chicken breast without skin. You'll find that the meals you order will not only be lower in fat, they will also taste better. If you fear the meal will be mediocre or minuscule, eat in the airport before you depart or pack a brown bag lunch to eat on the plane.

Taking the Train? Plan Ahead. Train food has expanded to include a few low-fat offerings. On some trains you can buy nonfat pretzels and low-fat yogurt. You may even find a chicken or turkey sandwich. However, to ensure that you will be eating a full healthy meal that you will really enjoy, prepare or buy sandwiches, nonfat yogurt, and fruit before you board the train. The food will be better and probably less expensive than offerings at the cafe car.

Driving? Plan Ahead. You probably assume that plenty of low-fat options are available if you drive, but often it isn't so. The only restaurants available along many turnpikes and highways are fast-food restaurants. The occasional so-called family restaurant may be abysmally high-fat. Use the advice above to pick the best of the food choices available or stop at a local grocery store and concoct your own healthy, low-fat meal.

To keep from starving between meals, find a supermarket and stock up on fruit. Carry along a cooler for fruit juice and nonfat fruit yogurts. Buy a box of nonfat pretzels (to be eaten judiciously). Have a deli make you turkey sandwiches with mustard, tomatoes, and hot peppers for lunch.

Bring your snack bag into the hotel or motel so that when your stomach grumbles at night, you have low-fat snacks to satisfy it.

At Work

A Fat Minefield. In the abstract, one might think that a worksite would be a safe haven from temptation, but no such luck. Boxes of big, shiny, glazed doughnuts arrive every morning like clockwork. Vending machines full of little bags of fat are ever ready in the lounge. Is there ever a day without a cake to celebrate a birthday, anniversary, team victory, big contract? Then there are the meetings, office lunches, and office parties. Although it seems an impossible job to avoid all the booby traps, you can take control of each fat mine and disarm it.

Remove the Lure

- Move the doughnuts or birthday cake to another office so they won't keep yelling, "I'm here!" in your ear.
- Bring a large bag of fruit, bagels, and nonfat pretzels for morning and afternoon snacks so that you never need to invade the vending machine.
- Keep an air popper in your office so that you can enjoy a snack of fresh air-popped popcorn every day.
- Help plan the refreshments for a meeting so that there will be food for you to enjoy.
- Instead of the typical Danish, doughnuts, cookies, and sweet cakes for the morning break, make sure fruit (how about a beautiful arrangement of cut-up fruit?), English muffins, and bagels are offered.
- For lunch, insist that the cold cut platter include turkey breast, the bread tray include French bread, whole wheat, or rye, and that mustard, pickles, and hot peppers are available.
- If a birthday is going to be celebrated at a restaurant, make certain that it is a restaurant with low-fat choices.
- For the office party, join the planning committee to make sure a table of low-fat offerings (fruit, vegetables, low-fat dip, low-fat dishes) is included.
- Bring a favorite *Choose to Lose* or *Eater's Choice Low-Fat Cookbook* recipe as a treat for yourself and the crowd.

You can take control of the fat aspects of your working conditions if you make an effort — and it's worth it.

Don't Skip Meals

Don't allow work to interfere with eating. *Never go without lunch!* If you know you are going to be chained to your desk through your lunch hour, be sure you have brought lunch so that you have something to eat. Keep a few flavored nonfat yogurts in the office refrigerator so that you can take one for your afternoon break. Remember: If you do not eat lunch or an afternoon snack, you will be hungry and tired all day.

The fatigue will follow you home and wipe you out for the entire evening. The hunger-fatigue combo will make you fall off the *Choose to Lose* wagon, go berserk, and eat margarine sandwiches.

Plan ahead so that you can Eat! Eat! Eat! low-fat, vitamin-mineral-fiber-packed foods when you are hungry during the day. Remember: bingeing doesn't happen unless you are starving.

Remember:

1. The key to surviving every social situation is planning ahead.
2. Splurging: *Choose to Lose* allows you to splurge without guilt. By eating fewer fat calories than allowed on your Fat Budget, you can save up for special occasions. Plan for specific splurges.
3. Party Survival
   a. Eat a light snack before the party.
   b. Bring a low-fat dish.
   c. Station yourself AWAY from the food.
   d. Have a party and serve low-fat dishes.
4. Restaurant Survival
   a. Choose a restaurant with low-fat options.
   b. Ask questions about preparation and ingredients.
   c. Specify exactly how you want your food prepared; be creative.
   d. Have sauces, dressings, gravies, toppings, and butter served on the side so that you can regulate the amount you use.
   e. Eat with care (when you dine out once a week or more, it is no longer a special occasion justifying high-fat gluttony).
5. Traveling
   a. Order meals ahead when flying.
   b. Buy or prepare food before train and car trips.
6. Work Survival
   a. Remove high-fat temptation from your work area.
   b. Take control in choosing restaurants, planning menus, or bringing food.
   c. Always eat lunch.
   d. Have fruit, bagels, air-popped popcorn, pretzels, and other low-fat food around to snack on.

11

Exercise: Is It Necessary?

"I have recently passed the 100-pounds milestone. I have lost 12 inches in my waist and have gone from 54″ to 42″. I have stamina and endurance at physical activity I would never have believed. I now get out of bed one hour earlier than I used to, and commit that hour to my exercise program of walking, leg lifts, and stomach crunches. When I started, I could barely make a half-mile walk. I gradually have increased my walking distance and now I am walking 3 miles in 45 minutes each morning."

— Don King, California

EATING RIGHT + EXERCISE = PERFECTION

You're almost there. You now know how to modify your fat intake to lose weight. Soon you will be watching your clothes get baggier and baggier as you become slimmer and slimmer. To keep *you* from getting baggier and baggier as you change your eating habits, you will need to indulge in an activity you may have tried to avoid — *exercise.*

Relax. We are not proposing marathon running, Olympic swimming, 2000 push-ups six times a day. We are suggesting that you walk at least 30 minutes every day. Is that so bad?

LET'S HEAR IT FOR EXERCISE

Please yell the following line as loud as you can: EXERCISE IS ESSENTIAL! You gotta do it! We cannot say it often or loud enough — exercise is essential for weight loss and long-term health. Daily aerobic exercise forms the third and equally important side of the *Choose to Lose* Success Triangle.

Exercise Builds and Preserves Muscle

Exercise will help you lose weight as well as benefit your health and make you feel good.

As we discussed in Chapter 1, "Fat's the One," exercise helps you lose weight because exercise builds muscle and muscle burns fat. The more muscle you have, the more fat-burning capacity you have.

If you don't exercise, the muscle is broken down and not rebuilt. The protein released from the muscle is burned instead of fat stored in the adipose tissue. Exercise protects your muscle from being broken down and burned for energy. By exercising you force your body to burn fat from the fat stores instead of protein from the muscle tissue.

Exercise Raises Your Metabolic Rate

Exercise increases your metabolic rate, the rate at which you burn calories. Not only does your metabolic rate increase *while* you are exercising, there is evidence (though it is somewhat controversial) that your metabolic rate *remains elevated* for a number of hours *after* you finish exercising. If this is true, then you continue to burn off calories at an accelerated rate even after you stop exercising.

Fat but Not Overweight

If you do not exercise when you reduce your fat intake, you will lose muscle tissue instead of fat tissue. You can end up weighing less but being fatter; that is, a higher percentage of your weight is fat. "So what?" you say. "As long as I weigh less, who cares what percentage is what?" You *should* care. Being fatter and having less lean body mass puts you at greater risk of regaining weight because you have less metabolically active muscle tissue to burn off the fat you eat.

Losing Pounds Doesn't Always Make You Thin

To make this concept clearer, let's look at Kim and Carol. Initially, Kim and Carol both weighed 180 pounds and both were 30 percent fat (54 pounds of body fat). They both lost 20 pounds, but because Kim exercised, she lost 25 pounds of fat and gained 5 pounds of muscle. Sedentary Carol lost 4 pounds of fat and 16 pounds of muscle. If you subtract Kim's fat loss from her original body fat (54 pounds − 25 pounds = 29 pounds of fat), you will find that Kim is now 18 percent fat (29 pounds body fat ÷ 160 pounds total weight = 18 percent body fat) and considered lean. Although Carol also lost 20 pounds, her fat loss (4 pounds) is much less than Kim's, making her percentage of body fat considerable (54 pounds of total body fat − 4 pounds = 50 pounds of body fat ÷ 160 pounds total weight = 31 percent body fat). Carol is fatter than when she started even though she weighs less than she did.

	KIM (SEDENTARY)	CAROL (SEDENTARY)
Before weight loss		
Total weight	180 lbs	180 lbs
Body fat	54 lbs	54 lbs
Percent body fat	30%	30%

	KIM (EXERCISING)	CAROL (SEDENTARY)
After weight loss		
Total weight	160 lbs	160 lbs
Body fat lost	25 lbs	4 lbs
Body fat remaining	29 lbs	50 lbs
Percent body fat	18%	31%

Kim and Carol may weigh the same, but Kim has much more metabolically active muscle and much less metabolically inactive fat. Kim will have an easier time maintaining her new weight.

This graphic illustration shows why the scale gives no meaningful information and may even mislead you into thinking you are making no progress. If you change your sedentary ways and begin to exercise, you will build new muscle. Muscle has weight. At first, the scale may show you are actually gaining weight. But this weight (new muscle) is fat-burning tissue and soon it will burn the fat away. People who weigh in daily may get discouraged and stop just as everything is beginning to work. Once again, do yourself a favor. Throw away your scale or give it to someone you wish to torture. You'll know you are losing fat when your clothes get looser and you fit into smaller sizes.

Give a Cheer for Exercise

In addition to helping you become lean, regular aerobic exercise provides the following important benefits:

- Cardiovascular fitness — helps protect you against fatal heart attacks
- Lowers blood pressure
- Helps protect you against osteoporosis (if weight-bearing)
- Helps protect you against some forms of cancer
- Helps you cope with anxiety, stress, and tension
- Relieves insomnia
- Enhances mood — makes you feel good
- Strengthens, tones, and shapes muscles
- Increases stamina

> **Exercise: The Best Sleeping Pill**
>
> Insomnia is a common complaint. People toss and turn all night and rip their sheets apart in frustration. One of the reasons they can't sleep is that they have been inactive all day, and they aren't tired enough to sleep. If insomnia is your problem, the best solution is to exercise during the day. Take a 30-minute walk around the neighborhood before it gets dark. Move that body. Give your body a reason to sleep.

WHAT IS THIS THING CALLED EXERCISE?

There are three types of exercise that are crucial for long-term health and peak performance. These are aerobic exercise, resistance or strength training, and stretching. We will discuss resistance training and stretching at the end of the chapter, but for fat loss, you need to focus your attention on aerobic exercise.

Consult your physician before you begin an exercise program if you are over forty, have heart trouble, experience pain during or after exercise, or suffer from arthritis, dizzy spells, or any condition that would lead you to believe you should consult your physician.

AEROBIC EXERCISE

Aerobic exercise is repetitive and rhythmic. It includes walking, jogging, continuous aerobic dancing, biking, and swimming. It does not include lifting weights, bowling, golf, tennis, or most other competitive sports because they are start-and-stop activities.

Aerobic exercise need not be strenuous, nor does it require any athletic ability. But to realize the maximum weight loss benefits, you should do at least 30 minutes of non-stop aerobic exercise daily.

Why Aerobic Exercise?

Back to the science. When you eat carbohydrates your body breaks them down into glucose, a simple sugar. Some of that glucose is burned immediately, some is stored as glycogen. When you exercise in spurts and bursts, as when you play tennis, your body burns mainly glucose for quick energy. Burning glucose has no effect on removing fat from your fat stores. In contrast, when you exercise aerobically for more than 20 minutes, you burn mostly fat.

How Long Should I Exercise? When you exercise you burn a mix of fuels. During the first 20 minutes you burn mostly carbohydrates. After 20 minutes you burn more and more fat. The longer you exercise nonstop, the more fat you burn. For fat loss, it is more effective to walk for 40 minutes than to run for 15.

How Often? If you exercise every day, you can't put it off until tomorrow because you are already exercising tomorrow. Just make exercise a normal part of your daily routine, like brushing your teeth.

How Intensely? For fat loss, it is better to exercise longer and less intensely than intensely for a short time. When you exercise intensely you burn more glucose than fat. You can tell if you are exercising at the

> If you are overweight, you may find that walking literally takes your breath away. Don't give up. Begin by walking for 5 minutes, slowly at first. Stop for a few moments when you get out of breath. But keep walking. And continue to walk — a little faster and a little longer each day. As you get thinner, walking will become easier.

correct intensity if you can carry on a normal conversation without getting out of breath.

Aerobic Exercise Choices

The Number-One Choice: Walking. The best, easiest, and safest exercise is plain old walking, which almost anyone can do almost anywhere. This is the least expensive and least punishing exercise. You need no special facility or special equipment except good walking shoes. The risk of injury is negligible. So go to it. Work up to at least 30 minutes a day. You need not walk briskly, but you should not stop and start. In other words, walking in a mall and stopping to buy a sweater doesn't count. For the greatest aerobic effect, bend your elbows at a 90-degree angle and move them straight back and forth, slightly brushing your sides as you walk.

Treadmill. A treadmill in your family room or basement means you can easily walk at any time of day or in any season. Turn on the news or a video and walk as you watch. Treadmills range from expensive, hi-tech electric to relatively inexpensive manual. If you are into super-hi-tech, expensive equipment, treadmills are made that can be programmed to simulate different terrains, move at variable speeds, speak six different languages, and tell you your fortune (well, almost). You can also get an excellent workout using nonmotorized, self-propelled treadmills. Nancy works up a heavy sweat walking for 30 minutes on her Nancy-motorized treadmill.

Aerobic Dance. Many local YMCAs, hospitals, adult education programs, and spas offer aerobic dancing classes. Aerobic dancing is fun and, for some, paying for a class that meets on a schedule is a strong incentive to exercise. These classes are offered at both low-impact and high-impact levels. Even if you are not Jane Fonda, there is a class you can handle. Be sure, however, that you keep moving even when the music stops. Starting and stopping will cause your body to burn more glucose and less fat.

Video Exercise. If you prefer to do aerobic dancing in the privacy of your home, an extensive array of exercise and aerobic workout tapes is available at your local video store and through catalogues. They range from low-impact to strenuous and are adjusted to different ages and exercise goals. Try to schedule your video exercise at the same time each day. Make it part of your routine.

Bicycling. Riding a bicycle outdoors in fair weather is a visual and emotional treat as well as excellent exercise. If your neighborhood has accessible bike paths or streets that are safe for riding and you enjoy biking, budget half an hour of bicycling into your daily routine.

However, if the weather or terrain of your area does not encourage biking out of doors, a stationary bicycle is an excellent substitute. You can do it at any time and in any weather. For those with bad feet, bikes offer the advantage of supporting your weight as you exercise. The drawback to riding a stationary bike is that it is boring. Some exercisers overcome this problem by watching TV or reading while they bike.

Swimming. Swimming is a particularly effective exercise for those who find walking difficult. Swimmers use muscles in both the upper and lower body, but their weight is supported by the water. The disadvantage of swimming is that, as most of us do not own our own pools, we must fight kiddies for swimming lanes in the summer and endure overchlorinated indoor pools in the winter. As with any exercise, start slowly and work up to swimming 30 minutes of laps five times a week.

Jogging. We do not recommend jogging for most people. Jogging incurs a high risk of injury to backs, knees, legs, and ankles. Walking is much safer and more effective in terms of fat loss.

Trampolining. For about $20, you can buy a mini-trampoline (about 3 feet in diameter) that sits about 9 inches above the ground. Rain or shine, snow or sleet, you can jog in place in comfort with little or no risk of damaging yourself.

Exercise Equipment. We have just described the safest and easiest forms of aerobic exercise that require the least equipment and expense and can be easily incorporated into anyone's routine and budget. Of course, you can buy exercise equipment such as treadmills, stationary rowing machines, and cross-country skiing machines. All of these machines provide perfectly acceptable aerobic exercise if used properly and regularly.

TIP: Don't buy cheap exercise equipment. It won't last. It was constructed for people who like the idea of exercise but not the actual practice of it.

Cross Training. You may choose to swim in the summer and walk or bike in the winter or do aerobic dancing in the morning and walk in the evening. Whatever exercise or combination of exercises you choose, be sure you do some aerobic exercise every day. Soon exercise will become a part of your routine like taking a shower or bath and you will miss it sorely when circumstances force you to forgo it.

The BEST Exercise. As one exercise equipment salesman told us, the best exercise is the one that you will continue to do. And only you know which one that is. Experiment. Try a variety of exercises until you find the one or the combination that suits you.

THE TRUTH ABOUT EXCUSES

"Of course I know how important exercise is and I'd love to do it but . . . "

Choose one of the following excuses:

- ❑ I'm too busy to exercise.
- ❑ Exercise takes time away from my family.
- ❑ I get sweaty.
- ❑ I look funny in leotards.
- ❑ I get out of breath.
- ❑ It's too dark, too light, too cold, too wet to exercise.
- ❑ My Aunty Syl is coming to town.
- ❑ I (fill in the blank)______________________________

Have you found your favorite excuse(s) for not exercising? Here's why your excuses are no excuse:

❑ **I'm too busy to exercise.** No one is too busy to exercise. The president finds time to exercise, so you can, too. You can always fit it in. Get up 30 minutes early and ride a stationary bike or take a walk before you leave for work. Walk during your lunch hour. Take a walk before you go to bed. What do you do that is more important than exercising? Watch TV? Work overtime? Exercise is essential to your well-being, both physical and mental.

❑ **Exercise takes time away from my family.** Include your family when you exercise. Take a walk after dinner with your daughter, son, husband, wife. Leave the car at home and walk your spouse to and from the bus, subway, or train. Take a bike ride with your family on the weekend. Exercise won't hurt them a bit.

❑ **I get all sweaty.** Walking at a moderate pace should not make you sweaty. If your chosen exercise causes excessive perspiration, exercise at a place where showering is possible.

❑ **I look funny in leotards.** Almost everyone looks funny in leotards. They exaggerate what you want to flatten and flatten what you want to exaggerate. Don't wear leotards. Wear comfortable shorts. If you look funny in shorts, who cares? The exercise is for you. And, if you eat right and exercise, you will soon look great in shorts.

❑ **I get out of breath.** Take it easy. After all, if you haven't exercised in a long time you are probably out of shape. If you are overweight you are carrying a lot of extra pounds. Start slowly, a few minutes a day, and work up to 30 minutes of walking a day. However, you should check with your physician before you begin any exercise program.

❑ **It's too dark, too light, too cold, too wet to exercise.** Come on! You can find a solution to any problem if you try. If the weather is bad, take a walk in a covered mall, exercise at home on a stationary bike, mini-trampoline, NordicTrack, rowing machine, or treadmill or pop an exercise tape into your VCR and move with the beat.

❑ **My Aunty Syl is coming to town, I have a big job to finish, my son skinned his knee, it's my dentist's birthday.** You make time for what is important to you. If someone gave you a free ticket to a football game or a concert, you'd be able to fit it in. You must fit exercise in. Aerobic exercise is an essential component of good health and weight control.

TIPS TO KEEP YOU ON THE STRAIGHT AND NARROW

It is so easy to put off exercise. The following tips will help you incorporate exercise into your life and keep it there.

1. Make Exercise a Part of Your Routine

Schedule exercise at a specific time so that you will really do it. Reserve 30 minutes every weekday morning for a ride on your stationary bicycle. Walk for 30 minutes every day after dinner. Work out to a videotape every morning at 7:00. Make exercise a habit. Do it even if you're not in the mood. Force yourself. You'll be glad you did. Eventually you won't resent it; it will just be a normal part of your life (a very important part).

Walking Fits in Everywhere. Incorporating walking into your schedule is probably the easiest and most practical exercise program. Make it an integral part of your lifestyle, rather than an activity grafted onto an already busy schedule. For example, walk to work or part way by getting off the bus early or parking your car 30 to 40 minutes away.

Walk on Your Lunch Break. Walk for 30 minutes at lunch time. It's good to get out of the office and into the fresh air. You can combine the walk with errands. But, remember, the exercise *must be continuous.* Walking 5 minutes to the bank and standing in line for 15 minutes doesn't count as exercise.

2. Do Exercise You Enjoy

Choose an exercise you enjoy or make the exercise you do more enjoyable. If you find a stationary bike too strenuous, walk. If you find walking boring, plug in earphones, turn on your Walkman, and listen to your favorite music or talk show. Slide in language tapes and teach yourself Italian. Watch television as you stride along your treadmill, ride your bike, or jump on your mini-trampoline.

3. Exercise with a Buddy

Exercising with another person (spouse, child, neighbor, friend) can be more fun than exercising alone. Make it a family affair. Instead of spending an hour every evening gathered around the kitchen table nibbling and chatting, spend the time taking a walk and chatting. Let it become a ritual. Exercising with a friend can also give you the added nudge you need. How can you say, "Sorry, I'd rather sleep," when Barbara comes by to drive you to the 9 o'clock aerobics class?

However, if you exercise with a partner, do not use his or her absence (due to illness, other commitments) as an excuse to stop. Either find another partner or go it alone. If you know you don't have the discipline to exercise on your own, join a spa or exercise group. But do whatever it takes.

4. Exercise Alone

Exercise can be a special time for you to be by yourself. Many busy people call the time they exercise "my time." No phones, no children, no distractions.

5. Wear the Best Equipment: Good Shoes

Be sure to wear comfortable shoes with good support. Buying $125 sneakers with bubble-filled soles and foam-layered heels especially constructed for walking on pavement in temperate climates is not a requirement. In fact, you might find tie-up leather shoes more comfortable for walking because they don't get as hot as sneakers. Whatever type of shoe you prefer, make sure they fit and offer support.

The Bottom Line

You can lose weight if you just reduce fat in your diet. But if you want to lose fat and not muscle and you really want to keep it off, you MUST exercise aerobically every day. MUST!

STRENGTH OR RESISTANCE TRAINING

Aerobic exercise is essential to facilitate your fat loss as well as to keep your body in good working order throughout your life. Strength training will help you to remain independent, strong, and flexible as you age.

As its name implies, strength training is exercise which builds up your strength by building up your muscles. Strength training involves lifting weights such as dumbbells, stretching rubber bands, and/or using a Nautilus or multi-gym equipment.

Independence

You probably are amused to imagine yourself as a muscle-bound body builder and wonder why we think this form of exercise is so important. We are not urging you to become Mr. or Ms. Universe, but strength training is essential if you want to remain independent and strong throughout your life. Many adults, especially sedentary ones, lose muscle through disuse and, as a result, when they get old they lose the strength to lift objects, including themselves, to walk distances, sometimes even just to move across the room. They lose their sense of balance and frequently fall and break their hips (often a death sentence) or other bones. Many of the people confined to old age homes wouldn't be there if they had done strength training and aerobic exercise.

Do it NOW!

If you want to be independent through old age, you need to start building your muscles now. You need to make it a habit . . . *now.* Waiting until you are old is too late. Not because it is too late to build muscle — it's never too late to build muscle — but because when you are old, you won't listen to people who advise you to exercise so that you'll feel better. Old people get stubborn and work against their own self-interest. If you develop the habit of strength training and stretching now, you'll continue to do it.

You don't have to lift heavy weights or spend hours at the gym every day. In fact, you should give your muscles a rest every other day.

Elastic Bands

Although not really weights, elastic bands can have the same effect on your muscles as lifting weights. These 4-inch wide, 4-foot long strips of rubber are inexpensive (about $10 for three different elasticities), portable, and easy to use. We have had positive experience with Dyna-Bands (Fitness Wholesale, 895-A Hampshire Road, Stow, OH 44224, 800-537-5512), but other sources may also be good.

Get Expert Advice

To learn the correct way to lift weights to avoid hurting yourself and to maximize the benefit, you need to consult a personal trainer, get guidance at a gym, fitness center, or YMCA, or read an appropriate book.

STRETCHING

As you age, you lose your flexibility. You get stiff and may find moving your body difficult and sometimes painful. So to keep yourself flexible and limber for your entire life, you need to stretch your muscles. Do it *now.* It won't take much time and it will feel good.

You should stretch every day. It is best to warm up your muscles by using them a few minutes before stretching to prevent damage to the tendons, which may tear if stretched while cold. Hold the stretch for a count of 20 and do NOT bounce or jerk while stretching as this may increase the chance of injury.

Again, you need to get professional advice or read an appropriate book so that you know your stretching exercises will not hurt you and that you are getting the most for your effort.

THE MESSAGE

Daily aerobic exercise both helps you lose weight and keep it off. It gives you energy and helps you sleep. It protects you against many chronic diseases. It enhances your life.

Strength training builds muscle, improves circulation, and can strengthen your bones. It makes you strong and independent. It improves your balance. It enhances your life.

Stretching improves your flexibility, balance, and lessens your aches and pains. It feels good. It enhances your life.

Start now. Schedule them in. You'll be glad you did — and your body and psyche will be ecstatic.

Recommended Reading

To learn more about exercise, check the reference list for Chapter 10 on page 600. In particular, the books by Kenneth Cooper, M.D., and by Joyce L. Vedral, Ph.D., are excellent resources.

Remember:

1. Regular aerobic exercise is a necessary part of *Choose to Lose* because it
   a. protects your muscle from being burned for energy and builds new muscle which burns fat
   b. maximizes the burning of stored fat
   c. increases your metabolism
   d. increases your energy expenditure
   e. helps you reach and maintain your desirable and healthy body weight
   f. improves cardiovascular fitness, reduces risk of osteoporosis (if weight-bearing) and other diseases, and enhances your sense of well-being.
2. Walking is a safe, convenient, and inexpensive form of aerobic exercise. Alternatives include bicycling, swimming, rowing.
3. Make exercise a regular part of your daily schedule.
4. Add strength-training and stretching to ensure good health, independence, and a heightened quality of life.

12

It's Up to You

"With *Choose to Lose,* eating has become a normal life function and not an obsession."

— Joey Jacobson, Bigfork, Minnesota

COMMITMENT

Now it is up to you. You have a Fat Budget, so you know your fat calorie ceiling. You know how to find fat calories in foods so that you can make choices. You can even fit in an occasional high-fat splurge so that you will never feel deprived. You can (must!) eat a lot of low-fat, nutritious food so that you won't ever be hungry.

Choose to Lose is easy and not punishing. However, for *Choose to Lose* to work, you have to do it. It takes commitment. You need to throw yourself into your new lifestyle full force.

TO MAKE *CHOOSE TO LOSE* WORK

Don't Think DIET

You will need to get rid of your diet mentality. If you reduce fat and total calories to a bare-bone minimum, how long will you continue to choose to lose? How long will two broccoli spears and a baked potato for dinner satisfy you before you run screaming to Roy Rogers for a burger? This is a healthy way of eating for life, not starvation for a few weeks or months.

Fit in Your Favorites . . .

Don't forget: the reason you have a Fat Budget is so that you can occasionally splurge on a thick steak or cherry cheesecake. You really should splurge occasionally, because if you don't you will feel you are on a diet and your commitment will be temporary. The big surprise is that what you now consider the most delectable treat in the world may seem revolting in a couple of months. Your tastes are going to change and you won't crave the high-fat foods you once did.

. . . But Not Too Often

If you are fitting in too many high-fat foods, you will be denying your taste buds a chance to change. A Chooser to Lose told us that she once considered McDonald's French fries ambrosia fit for the gods. After a few months following *Choose to Lose,* she couldn't even look at those greasy yellow sticks. If she had included French fries in her daily meal plan, she never would have lost her taste and craving for them. Eating a low-fat diet would have been a struggle for her instead of simply a way of life.

Keep *Choose to Lose* Your #1 Priority

Eating à la *Choose to Lose* and daily aerobic exercise are top priorities. Nothing, save being captured by terrorists or being sent to jail, can interfere.

People have told us that they went "off" *Choose to Lose* because their husband changed his job, their mother had dental work, their Uncle Mo stubbed his toe. We find this hard to understand. *Choose to Lose* is an eating plan for life. It is not an exercise in starvation or balancing exchanges that you can follow for only a few weeks. When you follow *Choose to Lose,* you eat lots of healthy food. You reduce your fat, but you can still fit in an occasional high-fat favorite. If you go "off" *Choose to Lose,* how seriously were you ever "on" it? In crisis situations, like having a child in the hospital or moving to a new city, there is no reason to abandon *Choose to Lose.* You have to eat, so why not eat healthfully? You can still make wise choices. In fact, having control in this part of your life will help you cope with the rest of the chaos.

There is no good reason to fall off the wagon. But if you fall off, you can immediately climb back on. Forget the sins of the past. Start all over again. It's not hard to follow *Choose to Lose* and it is worth it.

Get Rid of Temptation

For many, entering their front door is the beginning of their weight-control problems. This doesn't have to be. Your home is where you can take control. You can start in your kitchen.

Food Cues

Are you driven to eat by subconscious forces? Think about it. What makes you feel like eating besides hunger? Make a list in your mind. Does sitting at your desk make you yearn for the vending machine? Does turning on TV make your fingers itch for a bowl of chips? When you talk on the phone, does the refrigerator beckon?

Solutions: Figure out your food cues and then figure out how to overcome them. If your subconscious thinks the best way to postpone working is to eat, bring a large bag of fruit to keep in your desk drawer. Take a snack of a banana, apple, or orange (*while* you work) instead of the high-fat garbage offered in the vending machine. Don't turn on TV; do something productive. Write your memoirs. Read. If reading makes you want to snack, make 10 cups of air-popped popcorn and munch away. If the kitchen offers too much temptation, talk on the phone in the living room.

Fat-Proof Your House. The amount of high-fat food that lingers in your home reveals your commitment to your new eating habits. Are there hidden fat stores? Do you keep ice cream in the freezer? Do you keep crackers in the cupboards? Do caches of high-fat foods offer irresistible temptation? Get rid of them. You can't devour a gallon of ice cream if it isn't in your freezer. You can't demolish a bowl of macadamia nuts if you never buy them. Keep high-temptation foods out of your house. Splurge elsewhere. If you have accumulated a stock of high-fat goodies, give it away to neighbors. Feed it to your dog. Take a deep breath and throw it into the garbage can. To be safe, throw it in your down-the-street neighbor's garbage can.

Down and Out for Dining Out. Eating out and carrying-in make losing weight a slow, uphill project. First, it is difficult to ensure that what you are eating is low-fat. Even if you ask the right questions and make the right requests, you *really* don't know what is happening in the kitchen.

Second, and maybe even more important, when you eat out you see all the high-fat food you *aren't* eating. You see that you *could* be eating fettuccine Alfredo or ribs. You could be eating salad drowning in 6 tablespoons of blue-cheese dressing. You become envious and bitter and feel sorry for yourself. This attitude leads to binge eating and, eventually, abandonment of your weight loss/healthy eating goals. Self-pity doesn't happen if you are at home enjoying a full, tasty, low-fat home-cooked meal.

Plenty. Instead of fixating on the high-fat food you are eliminating, pamper yourself with an abundance of delicious, nutritious low-fat food. How about starting dinner with quick, tasty, and completely fat-free *Tomato-Rice Soup* (page 267)? Take time (and it takes very little) to marinate skinless chicken breasts in a mixture of grainy mustard, honey, and soy sauce to make *Grainy Mustard Chicken* (page 274) and then broil them until cooked through. Boil sweet potatoes until tender and whip with orange juice to create *Sweet Potatoes à L'Orange* (page 284). Sprinkle vinegar, basil, salt, and pepper over sliced tomatoes. You can create a feast in almost no time. Eat well the low-fat way and you'll watch the pounds melt away.

FIND THE PROBLEMS AND FIX THEM

Before they become a problem, root out the underlying causes for uncontrolled eating.

Fatigue

When you come home from work do you use what little energy you have to drag yourself to the refrigerator? Is your self-control so exhausted you binge on everything in sight?

Fatigue need not be your undoing if you take the following advice.

Have only low-fat food around so if you binge the binge is benign. When you arrive home, don't go directly to the kitchen; go to your bedroom and take a nap. Some Choosers to Lose have told us they like to take a walk or run as soon as they return from work. This activity seems to energize them for the remainder of the evening.

Ask yourself why you are so tired. Is it because you are hungry? Did you eat lunch at 3 P.M.? Remember to take a low-fat snack — a piece of fruit or a nonfat yogurt during the afternoon or before you leave work. And be sure you never, ever skip breakfast or lunch.

Hunger

Does your hunger drive you to lose all control and devour everything in sight?

Skipping Meals: A Dieter's Undoing. Do you nobly skip breakfast and/or lunch to save calories? Don't. Treat yourself as well as you treat your car. Would you drive your car on empty? Your car (and you too) needs energy to get going. Skipping meals just slows your metabolism, makes you exhausted and irritable, and HUNGRY. What do hungry people do? They gobble up dinner like there is no tomorrow. They eat three times their Fat Budget without stopping for a gasp of air. Be sensible and satisfied. Eat a large low-fat breakfast and a large low-fat lunch and low-fat snacks in between, and you'll be cool and collected when you dig into your delicious, low-fat dinner.

I Never Eat Breakfast. Are you one of those people who claim you can't stomach food or even the thought of food in the morning? This is why. If you eat no breakfast (and sometimes no lunch), by the time dinner comes around, you are famished. As a result, you gorge yourself until bedtime. No wonder you can't think about food in the morning. You are stuffed from the night before.

Breaking the Cycle. You *can* break the cycle. Start tomorrow with a light breakfast, and then eat a normal lunch and dinner. Eat fruit or air-popped popcorn for night-time snacks and stop snacking at least an hour before bedtime. Continue this practice for several days until it becomes routine. Soon you will be so hungry at breakfast time that you will eat a hearty low-fat breakfast with delight.

Boredom

Many people eat because they are bored or lonely. They spend their evenings and weekends watching television. To satisfy their feeling of emptiness, they fill up on food. Of course, the food industry makes thinking about high-fat foods as easy as flicking on the TV. You can't help fantasizing about food you shouldn't eat when it pops out at you every ten minutes.

How about overcoming your boredom by taking up a hobby? It's never too late. How about photography, oil painting, or ceramics? Take an adult education course in Italian or woodworking at your local high school or Y. What about square dancing, folk dancing, or yoga? Join a barbershop quartet or a chess club. Work for your local politician or volunteer at your library or neighborhood hospital. Take up biking or birding and explore the great outdoors. Get out, meet people, get involved. You'll feel better about yourself and you won't have time to fixate on high-fat food.

Overbooking

At the other extreme are those whose eating problems stem from overscheduling. Do you get up at 5 A.M., drive 50 miles to work, work frantically all day, rush through lunch (if you eat at all), and then, when you arrive home, drive three carpools? No time even to think of preparing dinner. You stop at a fast-food restaurant, pick up dinner, and eat while driving the kids to soccer, baseball, football, hockey (you fill in the correct sport) practice. The evening is spent helping your children with their homework, attending choir practice, chairing PTA committees, bowling (you fill in the activity). You drag your exhausted body to bed after falling asleep watching TV. You probably sleep poorly because, although you feel you have been running around all day, your most vigorous exercise has been sitting in front of the steering wheel. The next day you are up before the sun rises and another harried day begins.

With a hectic life like this, you are certainly justified in feeling stressed out. But are *you* playing a part in creating your own hell?

Have you organized your life this way — overcommitting yourself and your family, cramming so many activities into your day that you have no time to think, much less take control of your life? It may seem easier to let your life control you, but do you feel very good about yourself? Are you worried about the effects on your health and the health of your family?

Sit down and make a list of your activities. Evaluate and eliminate. Do your kids really need to participate in two sports? Shouldn't they do their homework on their own? Do you have time for the PTA? Eliminate the activities that are killing you. Simplify your life so that you have time to take care of your own needs. Make time to prepare dinner and sit down with your family to eat it. Eating together makes for mental and emotional health.

A Poor Body Image

A poor body image can lead people to self-defeating eating behaviors. People generally develop a poor body image because they are overweight and out of shape. The ridicule often starts early. Overweight kids are teased, excluded from the "in" crowd, and chosen last for any team. Feelings of low self-esteem develop and last through adulthood. People often gorge themselves with every high-fat food in sight because they feel so unhappy about their appearance and the cycle continues. But it doesn't have to be.

A formerly obese Chooser to Lose told us that a complete stranger spat at her when she was at her heaviest. She was so hurt she felt like killing herself. But instead of sinking into a depression and turning to the refrigerator, she decided to take action. She followed a friend's recommendation and started *Choose to Lose* — eating lots of low-fat, nutritious food and exercising. She turned from a couch potato into a spa lady. It took many months before she achieved her dream shape but during that time she didn't dwell on the fat that she still had to lose; she congratulated herself on her new healthy lifestyles.

The best way to change your poor body image and lack of self-esteem is to change your lifestyle behavior. And that is what you are doing by following *Choose to Lose*. You need to be patient. You are not going to be a size 10 tomorrow (or maybe ever), but if you continue following *Choose to Lose,* you will eventually be lean and fit. Don't dwell on what your body looks like now. Instead, congratulate yourself for making a huge step toward becoming healthier and leaner.

Sabotage

When you decide to lose weight, does your boyfriend bring you a box of chocolates? Does your wife prepare your favorite whipped cream strawberry pie? Does your husband take you out to Arby's for dinner? Does your mother tell you you're getting too thin? Is this love? Whether your loved ones know it or not, this behavior is not without design — they want to thwart your new lifestyle changes. They may fear that a new, lean you will be attractive to others; a new, lean you may be independent and not need them. They may feel threatened because they should be doing something about their own weight problem. They may just not want to find a grilled chicken breast on their plate instead of a lamb chop.

This ostensibly generous behavior is triply hard to resist. First, the gift is an irresistible treat — of course you'd love to eat a box of Lady Godiva chocolates or dine at a fine French restaurant. Second, a guilt trip comes along free of charge: "If you don't eat what I give you, you don't love me." Third, refusing the gift means an uncomfortable confrontation with the saboteur. But if you want to be a successful Chooser to Lose, you must resist, reject, and confront. Be strong. You can do it.

Solution: Communicate. The best way to effect a change is to communicate. Talk to the saboteurs. They may not be aware of what they are doing. Tell them that you appreciate their thoughtfulness, but you are staying away from fat. You still love them, but at this moment losing weight and eating healthfully are the number one priorities in your life. Tell them that you don't feel good being fat; you are ready to make changes and you need their help. They should bear with you and keep high-fat foods out of the house and be open to eating new low-fat dishes. If you are thinner and healthier, they will benefit. Ask them for their support.

If they don't get the message and continue to shower you with unwanted fat packages, get tougher. Say, "No thanks." If they brought you arsenic, would you eat it to be polite? Tell them it's hard enough to change habits that you've developed and nurtured for years — you don't need to deal with their guilt trips, too. Ask them why they are trying to sabotage your efforts. If they still turn a deaf ear to your needs, don't give up your resolve. When you are a lean person glowing with self-confidence and good health, you will know why you were uncompromising in the pursuit of your new lifestyle.

Self-Sabotage

Does it turn out that you are your own best saboteur? Do you eat the mocha truffles your husband brought you to spare hurting his feelings, or do you just want an excuse to eat them? Do you prepare spare ribs for your relatives because it is their favorite dish, or because it is yours? Do you take your kids to a fast-food restaurant because it's convenient, or do *you* want a Big Mac? Don't kid yourself. If any high-fat wind that comes along blows away your resolve, you aren't really interested in changing your lifestyle. If you *really* want to eat healthfully, you can.

Be Realistic

Now that you have the tools to help you reach your fat-loss goal, you are probably feeling unconquerable. You can see the New You slipping into slinky black dresses or skin-tight blue jeans, or you can imagine looking down and seeing your feet rather than your belly. It is important to feel optimistic and to know that you will eventually reach your goal weight, but if you start out with grandiose expectations, you will only be disappointed. If your only goal is to lose a lot of weight quickly and you don't lose it all in six weeks, you will feel like a failure and give up *Choose to Lose.* You will give up just when your body is starting to make real changes.

One Step at a Time

Set goals that are under your control. Make them small, realistic, and achievable. Don't think, "50 pounds thinner"; think, "Today I will eat two vegetables with dinner." Don't think, "Size 12 pants"; think,

"This week I'll walk for 30 minutes every day at lunch time." By meeting small goals, you will more easily reach your large goals. You might want to sit down at the beginning of the day and list your goals for the day. Keep the Success Triangle in your mind's eye and attack each goal. For example, you might list the following:

1. Low-fat diet: Bring lunch. Bring a bowl of last night's *Chili Non Carne* and heat it in the office microwave. Add whole-wheat toast, carrot sticks, an apple, and a peach.
2. Adequate intake of nutritious food: Prepare *Cucumber Soup* and refrigerate before you go to work.
3. Aerobic exercise: Walk dog 30 minutes after dinner.

Choose to Lose is not a quick fix. It is a way of eating for life and a lifetime commitment. Allow yourself only one grandiose goal, and that is to reduce your risks of many diseases because you are eating a healthy diet. Relax, take a day at a time, and enjoy your new lifestyle of great eating, good health, and more energy.

Psychology

The emphasis of *Choose to Lose* is not on restructuring psyches. The emphasis is on knowledge. Knowing you don't have to starve, can fit in high-fat favorites, and can make smart choices will remove a lot of psychological stumbling blocks. You may have developed feelings of inferiority and self-hate because you are overweight, but deep-rooted psychological problems probably didn't cause you to become fat. Like millions of Americans, you were exposed to the American food system and didn't know the rules. Now you know the rules. You can take control and your self-esteem will grow as you shrink.

If you feel your weight problem has a psychological basis, you should probably seek professional counseling. You may find, however, that when the pounds start coming off easily and naturally without much thought, food will no longer be an obsession. You will feel so much better about yourself, you'll be able to reach your goal — and stay at that size.

Remember:

1. Throw yourself into *Choose to Lose.*
2. Make *Choose to Lose* work:
   a. Don't think *diet;* think eating plan for life.
   b. Fit in an occasional high-fat splurge.
   c. Make *Choose to Lose* your #1 priority.
   d. Fat-proof your house.
   e. Stock up on delicious low-fat, nutritious foods.
   f. Keep dining out to a minimum.
   g. Eat wonderful home-made meals.
   h. Eliminate causes for destructive eating such as fatigue, hunger, boredom, overscheduling, food cues, poor body image.
   i. Confront sabotage:
      1. Communicate with saboteur.
      2. Know thyself.
   j. Set small, achievable goals.

13

Ensuring a Balanced Diet

> **"This is absolutely the best program available for *anyone* wishing to eat healthy. Weight loss is a side benefit of how one feels while on your diet."**
>
> **— Thomas and Pauline Going, Chesapeake, Virginia**

CHOOSE TO LOSE is not just about reducing fat in your diet. It is also about eating an abundance of nutrient-dense, delicious foods. Eating a well-balanced diet should be easy to do if you replace your excess fat calories with fruits, vegetables, whole grains, and nonfat dairy.

THE FOOD GUIDE PYRAMID

The Food Guide Pyramid was proposed by the Department of Agriculture to replace the original four basic food groups wheel. Although both diagrams stress the importance of eating foods from all groups, the pyramid emphasizes the importance of eating more of some foods and less of others. The base gives you a healthy foundation. You should consume whole-grain breads and cereals, rice, and pasta, vegetables and fruits in the greatest quantities. Milk products and meat may be eaten in smaller amounts, and fats, oils, and sweets (not even part of the original basic food groups) should be eaten sparingly. In general, Americans have turned this pyramid on its head and have, as a result, been totally out of balance for years.

Meeting the Recommendations

To maintain your health, it is essential that you eat at least the minimum recommended number of servings of these foods as illustrated on

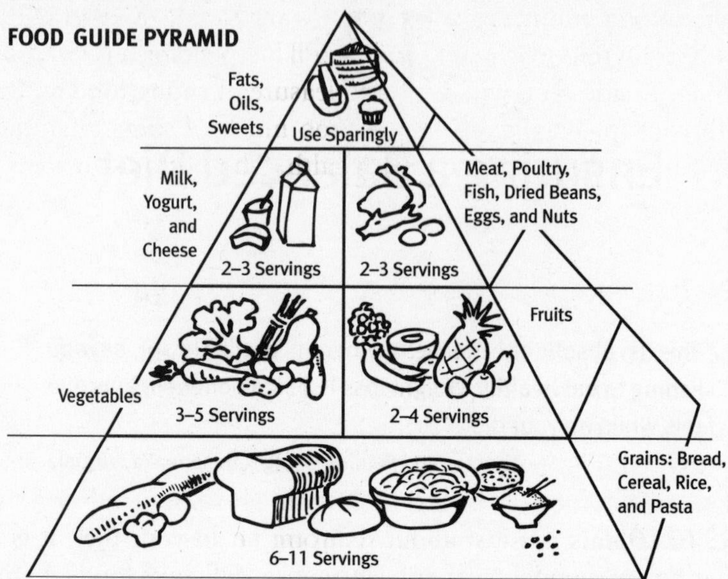

the Food Guide Pyramid (at the end of the chapter you will find a table that summarizes the recommended servings and serving sizes of foods from each group). Just meet these standards and you are set. Consuming at least 6 to 11 servings of grains, 2 to 4 servings of fruits, 3 to 5 servings of vegetables, and 2 to 3 servings of nonfat or low-fat dairy ensures that you will get enough vitamins, minerals, protein, and fiber for good health. It also means you won't be hungry. You will be too full from all these healthy, nutritious foods to have room for ersatz fat-free fluff or high-fat junk.

The recommended foods contain the vitamins, minerals, and fiber that contribute to your good health. Eating them will make your skin glow, your hair shine, and your teeth, gums, and bones strong. These nutrients will help your body maintain and build healthy tissue, increase resistance to infection, and help wounds heal. They are essential for proper function of the heart, nerves, and muscles.

The various micronutrients, enzymes, and phytochemicals in food work *together* to produce energy, assist in protein synthesis, and protect you from diseases — all those wonderful things that keep your body functioning and healthy. So if you want to take a short cut and just pop vitamin pills instead of food, you'll be spending a lot of money for nothing. Besides you will miss the pleasure of eating food and pills won't fill you up. Finally, consuming too much of some vitamins or minerals can be dangerous to your health. The bottom line: eat the food.

WHOLE-GRAIN BREADS, CEREALS, RICE, AND PASTA

For good health and weight control, you should consume 6 to 11 servings of whole-grain breads, cereals, rice, and pasta per day. (See the table on page 223 for sizes of servings.)

The following whole-grain products are rich in fiber, thiamine, riboflavin, niacin, folic acid, vitamin E, iron, phosphorus, magnesium, zinc, and other trace minerals. Choose from:

- Brown rice
- Buckwheat groats
- Bulgur
- Oatmeal
- Pumpernickel bread
- Whole-wheat bread and rolls
- Whole-wheat pasta

Warning: Before you count a food as meeting your grain requirement, be sure it is mostly complex carbohydrate. For example, don't count a croissant as a grain. A croissant is made of flour (complex carbohydrate) and lots of butter (fat — about 110 calories of fat). Croissants are not exactly in the complex carbohydrates category. Don't include cookies, cake, or crackers, either. Don't include foods that have all the nutrients processed out of them, such as pretzels. (There is nothing wrong with eating a pretzel or two, but it isn't really adding to your health.)

Don't count ice cream or cheese as dairy. Count it as fat. Don't count frozen yogurt, even nonfat frozen yogurt, as dairy. It is often more sugar than dairy. Remember: you want to meet the Food Guide Pyramid requirements to benefit your health, not to impress anyone. You want to eat an abundance of foods that are rich in fiber, vitamins, and minerals — foods that will enhance your long-term health.

The following grain products are enriched with thiamine, riboflavin, niacin, and iron. Eat these foods less often. Whole-grain products are far more nutritious.

Bagels	French bread	Pasta
Cereal (ready-to-eat)	Noodles	Rice

VEGETABLES

Vegetables contribute vital fiber, vitamins, and minerals to your diet. They should be varied and eaten in abundance, at least 3 to 5 servings a day.

The best way to prepare vegetables is to steam them for a short time so they are tender but still crunchy (overcooking causes loss of vitamins and taste). In restaurants, beware of vegetables cooked in butter or drowned in high-fat sauces. To keep them delicious, healthy, and non-fattening, ask that vegetables be steamed with no fat and served with sauces on the side.

Dark green vegetables should be included several times a week. They are an excellent source of vitamins A, C, and E, riboflavin, folic acid, iron, and magnesium. Choose from:

Beet greens	Chicory	Romaine lettuce
Bok choy	Endive	Spinach
Broccoli*	Escarole	Turnip greens
Chard	Kale	Watercress

Deep yellow vegetables are an excellent source of vitamin A. Choose from:

Carrots	Sweet potatoes
Pumpkin	Winter squash

Other vegetables also contribute varying amounts of vitamins and minerals. Choose from:

---

*This cruciferous vegetable may reduce your risk of some cancers.

Artichokes
Asparagus
Beets
Brussels sprouts*
Cabbage*
Cauliflower*
Celery
Chinese cabbage
Cucumbers
Eggplant
Green beans
Green peppers
Lettuce
Mushrooms
Okra
Onions
Radishes
Summer squash
Tomatoes
Turnips*
Zucchini

Starchy vegetables are rich in fiber, vitamin B_6, folic acid, iron, magnesium, potassium, and phosphorus. Choose from:

Corn
Green peas
Lima beans
Potatoes (white)
Rutabaga*
Sweet potatoes

Dried beans and peas are excellent sources of protein, soluble fiber (known to reduce blood cholesterol levels slightly), calcium, magnesium, phosphorus, potassium, iron, and zinc.

Choose from these dried beans and peas, which should be included as a starchy vegetable several times a week. Dried beans and peas are a healthy substitute for meat, poultry, fish, and eggs. They may be included in both the vegetable and meat food groups in the Food Guide Pyramid.

Black beans
Black-eyed peas
Chickpeas
Kidney beans
Lentils
Lima beans
Navy beans
Pinto beans
Split peas
Other types of dried beans and peas

FRUITS

Fruits contain vitamins, minerals, and fiber. Eat a variety — at least 2 to 4 servings of fruit daily. Have fruits available around the house instead of nuts, cookies, crackers, or candies. Bananas, oranges, pears, apples, strawberries, peaches, nectarines, cantaloupes . . . all fruits make sweet or crunchy treats. Take a bag of your favorite fruits to

*These cruciferous vegetables may reduce your risk of some cancers.

work: when hunger or boredom strikes, dip into the bag instead of heading for the vending machine and its high-fat temptations.

Citrus, melon, and berries are rich sources of vitamin C, other vitamins, folic acid, and minerals. Choose from:

Blueberries
Cantaloupe
Grapefruit
Grapefruit juice
Honeydew melon
Kiwi fruit
Lemon
Orange
Orange juice
Raspberries
Strawberries
Tangerine
Watermelon
Other citrus fruits, melons, and berries

Other fruits are also a rich source of vitamins and minerals. Choose from:

Apples
Apricots
Bananas
Cherries
Grapes
Nectarines
Peaches
Pears
Pineapples
Plums
Prunes
Raisins
Other fruit
Fruit juices

MILK, YOGURT, AND CHEESE

Milk products are rich in protein, calcium, riboflavin, vitamin B_{12}, magnesium, vitamin A, thiamine, and, if fortified, vitamin D. Milk products need not be high in fat to contain high amounts of calcium. In fact, skim milk contains slightly more calcium than its high-fat counterparts. Do not skimp on low- or nonfat milk products. Even as an adult, you need all the nutrients they supply. Choose from these foods daily:

Buttermilk
Nonfat or low-fat (1%) cottage cheese
Low-fat or nonfat yogurt
Skim milk

Calcium is also found in dark green vegetables.

Women, especially prepubescent girls, need more calcium than men in order to avoid developing the bone-thinning disease osteoporosis.

Minimum Basic Nutritional Requirements

FOOD GROUP	DAILY SERVINGS	EXAMPLES	SERVING SIZE
Fruits	2–4	*All* fruits*	½ to 1 cup 1 fruit
Vegetables	3–5	*All* vegetables**	½ to 1 cup
Dairy	2–3 (adults) 3–4 (children)		
		1% low-fat cottage cheese	½ cup
		Dry-curd cottage cheese	½ cup
		Skim milk	1 cup
		1% low-fat milk	1 cup
		Nonfat yogurt	1 cup
Grain	6–11		
		Whole-grain bread	1 slice
		Bagels	2
		Rice	½ cup
		Bulgur	½ cup
		Whole-grain cereals	1 oz.
		Pasta	1 cup
Meat	2		
Poultry		Chicken breast, skinless**	(2–3 oz)
		Turkey breast, skinless	
Fish		All fin fish**	(3–4 oz)
		Shellfish**	
		Tuna, water-packed	
		Sardines (in fish oil)	
Meat alternatives		Dried beans, peas, lentils	½ cup

*Fresh preferred over canned. Not avocado.
**Not fried.

MEAT, POULTRY, FISH, DRIED BEANS, EGGS, AND NUTS

Meats, poultry, fish, and eggs are good sources of protein, phosphorus, niacin, iron, zinc, and vitamins B_6 and B_{12}. However, Americans eat more meat than is necessary for good health. Limit your meat consumption to no more than 4 to 6 ounces a day. Choose often from the following:

White meat chicken with no skin	Fish
White meat turkey with no skin	Shellfish

Choose less often (or not at all) from the following:

Beef	Lamb	Veal
Ham	Pork	Eggs

Eat organ meats such as liver, pancreas, and brain rarely, if at all. They are loaded with cholesterol. Pancreas and brain are also high in fat.

Cooked dried beans or peas, as well as nuts and seeds, may be substituted for meat, poultry, fish, and eggs. However, as these plant foods lack vitamin B_{12}, this vitamin must be supplied by other foods. Nuts and seeds are high in fat and should be eaten in moderation, if at all.

Remember:

1. To ensure a well-balanced and healthy diet, you should consume daily at least:
   a. 6–11 servings of grains, preferably whole grains
   b. 3–5 servings of vegetables
   c. 2–4 servings of fruits
   d. 2–3 servings of nonfat or low-fat dairy
2. You should consume 25–30 grams of fiber daily.

14

Choose to Lose for Children

"I can't believe how easy good nutrition and good eating habits are to obtain. My children are now interested in what they eat. I am a true advocate of *Choose to Lose.* So is my husband."

— Mary Rose Robbins, Derry, New Hampshire

IF YOU ARE OVERWEIGHT, there is an excellent chance that your children and your spouse are too. You probably share the same taste in foods. Start your children on the road to thinness early; good eating habits begin in childhood. Don't let your plump toddlers become your chubby children and your obese adolescents. It's not fun to be a fat kid. Fat children are often ostracized. Fat children often suffer discrimination. Fat children almost always become fat adults. If you live *Choose to Lose* and bring your family into your new lifestyle, your children will grow up lean and healthy. It sounds easy and it is. But it will take an effort on your part. There's a big, fat world out there waiting to corrupt your children, and you will have to give them the tools to keep it at bay.

Fitting into a Fat Culture

Your children are not unique in being overweight. The United States is filled with fat kids. Over one-fourth of all children are obese. Your children became fat because they were born into a fat, sedentary culture. They entered a world where fast foods, snack foods, convenience foods, and carry-out are the basic fare. Kids no longer run up to the rec center every afternoon for a pickup game of softball, football, or basketball. They are driven to organized soccer games once a week.

Facts about Obesity in Children and Teenagers

- Between 25 percent and 30 percent of children are obese, about a 54 percent increase in the last 15 years.
- Between 20 percent and 25 percent of teenagers are obese, about a 40 percent increase in the last 15 years.
- Between 50 percent and 95 percent of obese adolescents become obese adults. The chances of an obese teen attaining normal weight in adulthood is 1 in 28.
- Obese children and teenagers have an increased prevalence of irregular periods, hypertension, respiratory problems, and hyperinsulinemia. They often suffer from peer and adult discrimination, social isolation, low self-esteem, depression, and a negative body image.

Walking or riding a bike is unheard of. Children are chauffeured from activity to activity. Television programs and video games offer hours of fascinating inactivity. No wonder so many children are overweight and physically unfit.

Lifestyle Puts Your Children at Risk

The two main reasons your children became fat and flabby — eating a high-fat diet and lack of exercise — are lifestyle behaviors that, if carried through adulthood, will increase their risks for all sorts of horrendous adult diseases. Even normal-weight adults are more likely to suffer heart disease if they were fat children. Does the future look bleak? Before you gather up your children and jump off the nearest cliff, rest assured that turning the situation around is not difficult and will even be fun and delicious. What's more, it will benefit not only your overweight children but you and the rest of your family as well.

A Family Affair

The most effective way to change your overweight children's eating habits is to ease the whole family into a new low-fat regime. Do this subtly and gradually. Don't single out and focus attention on the child with the problem. Don't serve special low-fat food to the overweight child while everyone else gorges on high-fat fare. Feed everyone the same food. Make low-fat eating a family affair. Prepare the *Choose to Lose* and *Eater's Choice Low-Fat Cookbook* recipes, but make no grand announcement that you are cooking healthfully. Just present the finished product and, within a few minutes, you'll hear requests for seconds.

Real Men Eat Low-Fat. If you have a recalcitrant spouse ("Diet food is for the birds!") don't tell him that you are cooking low-fat — just do it. Serve him *Chickenburgers (Eater's Choice Low-Fat Cookbook)* or *Grainy Mustard Chicken* (page 274). He'll lick the plate clean. Make him *Potato Skins* (page 284). He'll ask for seconds and thirds. When he realizes that the large watermelon he called his belly has become flat as a board, that his cholesterol has dropped 50 points, and that he has never eaten so well, he won't complain.

If you are a man reading this and you cook, the above advice applies to dealing with your significant other.

Involve the Troops. If you have a family that will take on a new healthy lifestyle as an adventure, disregard the last paragraphs. Involve them. Have them choose low-fat recipes. Better yet, have them prepare the recipes, too.

Reward the Whole Family. Don't worry about resistance. The food will sell itself. You'll love *Tomato-Rice Soup* (page 267) because it's so quick and easy to make. They'll love it because it tastes so good. They'll eye *Indonesian Chicken with Green Beans* (page 272) with suspicion, but watch them vie for the last biteful.

Add new vegetables to your repertoire. Steam them until they are crisp, but still tender. Don't even think butter and cheese. Reduce the fat until you are using none. A tasty topping is a mixture of Dijon mustard and nonfat yogurt. Sprinkle steamed vegetables with balsamic vinegar.

Try low-fat soups. They're filling and satisfying and will keep your kids from leaving the table hungry for high-fat snacks.

For your daughters in particular, be sure to include lots of low- or nonfat dairy and green, leafy vegetables. The years of childhood and adolescence are the time females need to consume a lot of calcium to build strong bones to reduce their risks of osteoporosis later in life.

When you're old, calcium is much less effective in preventing osteoporosis.

If the most advanced cooking you have ever attempted is opening a frozen food carton, get ready for a pleasurable experience. Cooking is fun. Have a positive attitude and you will enjoy it. It's one job where you always get a reward — your culinary creation.

Rewards of Eating Together. A major cause for poor nutrition is drive-by eating. People grab a burger and fries at a fast-food restaurant. They gobble up Chinese food out of paper cartons. Their porcelain is styrofoam. Their silver is plastic. They never really enjoy a meal because they don't focus on the food. They drive and eat, watch TV and eat, shop and eat. Sad, but not hopeless. You don't have to live that way. One of the joys of eating real food is that you sit down as a family and eat together. You actually talk to one another. Eating together at the same time may seem impossible because no one ever seems to be at home at the same time. Again, all you have to do is plan. Instead of scheduling yourself (and your children) to the hilt so that you have no time to breathe, make dinner with the family a priority. You'll find time.

You're the Boss. Even though you have introduced a divine cuisine and your children are hungry, you may meet resistance. BE FIRM. If they balk at a new type of soup or chicken dish, ask them to try a bite. They'll change their tune after a taste. If they refuse to eat, don't force them, but don't make them another dish or feed them later. You will only be rewarding their defiant behavior and spoil their chance to discover how wonderful low-fat food can be. Don't get into a power

The Snack Law

VERY IMPORTANT!! Make sure that your kids don't eat snacks — even healthy snacks — too close to dinner. A full child will reject all but junk food. A hungry child will be willing to try your luscious, low-fat meals. If you have a fussy eater, snacking may be the reason.

struggle over food. Look at it this way. Not eating dinner one night won't kill them, but eating a diet of high-fat snacks may.

Food Is for Eating, Not Controlling

Food should never be used as a reward, bribe, or punishment, for this both encourages children to have an exaggerated desire for high-fat foods and diminishes the activity that inspired the bribe. For example, the bribe "If you do your homework, I'll take you to McDonald's" gives two messages: McDonald's is a desirable place to eat and homework is not worth doing. A better bribe would be, "If you do your homework, you'll learn the subject matter." A better message is that doing homework is worthwhile because it teaches you something.

Here's another example: the threat "Eat your chicken or you won't get dessert." Messages: chicken is bad, dessert is good. Better strategy: "Eat your chicken, it's delicious!"

High-Fat Not Allowed

Don't sabotage your efforts to introduce healthy eating by allowing high-fat competition to interfere. Don't bring high-fat chips, crackers, cookies, ice cream, cake, cold cuts, into the house. Don't let your kids bring them in either. They don't need it. Tell them your house is a junk-food-free zone. If you hear complaints, remember that you're the boss. Do keep plenty of fruit, nonfat flavored yogurt, carrot sticks, nonfat pretzels, and air-popped popcorn for them to snack on (but not too close to dinner).

TV: A High-Fat Habit

Television is a convenience just like fast-food and frozen dinners. You plop your children in front of the TV and can forget about them for hours. But watching TV is a fattening habit. Scientific studies have shown that the prevalence of obesity in 12- to 17-year-olds increases by 2 percent for each hour of TV watched daily. This is no surprise. Children who watch TV are getting no exercise except for the constant movement of their hand from the bowl of chips to their mouth. Limit the amount of time your children sit in front of the television if you want them to be lean and healthy.

Just Say No

The ads on Saturday morning cartoons are not for carrots or whole-wheat bread. Limiting your children's TV watching time will probably also limit their high-fat demands. Remember, your goals for your children are not the same as the goals of the advertising industry. Just because your kid has been brainwashed to ask for a certain candy bar, why should you buy it? When your children whine that they want that cute Lunchables Fun Pack (100–300 fat calories) or the fried chicken frozen dinner with cartoons on the front (230 fat calories), just say no. Would you buy them bars of strychnine? You don't have to give them everything they want. Think of their health first. They'll still love you even if you don't buy them everything they ask for.

But Not for Me

You may be reading this section and instead of fantasizing about all the wonderful food you will be eating, you are thinking, "I can't do this. I don't have time." If you are truly interested in helping your overweight children become thin and in ensuring your family's long-term good health, you have to make time. You can do it. It just takes planning. Plan ahead so that you have lots of good, healthy food to make satisfying meals. They don't have to be elaborate. Enlist your family's help in preparing meals and cleaning up after. You will love your new life.

A MORE DIRECT APPROACH

In addition to quietly changing your family's meals and snacks into nutritious treats, you may want to confront your children's weight problem more directly. This may be tricky. Never make children feel that they are bad or inferior because they are overweight. Be matter-of-fact. Remember at all times that this is *their* problem. Eating is one area over which they have complete control. You cannot force them to eat right, but you can help them develop skills to make better choices. If, at any time, you feel a resistance to your help, stop. Weight loss is an area where pressuring and pushing have disastrous consequences.

It's Their Problem

For some of you, a nonintervention policy will take a strong resolve. Either you have been fat all your life and don't want your children to suffer the emotional pain you have, or you are thin and don't want your children to be unhappy. In either case, it doesn't matter what you want. The child has to want to lose weight. Let go and your child will be much more successful.

Be Positive

Instead of rebuking your children for their eating habits or making negative comments about them in front of other people, build up their confidence. Be positive about their strengths. Make them feel good about themselves. Positive self-esteem will go a long way to helping them overcome their eating problems.

Teens

If you have overweight teenagers, encourage them to read this book. (Even thin teenagers can learn a lot from reading *Choose to Lose.*) Knowing that fat makes fat and how much fat lurks in the food they eat will have a profound effect on their food choices. By giving your teens control over their diet, *Choose to Lose* will free you from being a food cop and improve your relationship. You may even find your roles reversed with your teens keeping you true to *your* Fat Budget.

If your teens have stopped growing, they can use the adult tables in this book to determine their own Fat Budget. Otherwise, they can read the instructions below on determining a Fat Budget.

Children over Nine

If your children are capable and interested, try to explain why fat makes them fat and carbohydrates don't. Use the instructions on page 232 and work with them to determine their Fat Budgets. Then introduce them to the fast-food and snack sections of the Food Tables for a real eye-opener. Go over other relevant sections of the Food Tables to give them a better understanding of why their food choices are so important. Use pages 140–160 to teach them how to read food labels.

Children under Nine

You may want to determine a Fat Budget and keep food records of your young child's intake so that you know how much fat he or she is really consuming. This could be helpful in identifying foods or food patterns to change. However, if keeping track means counting the peas on his spoon as they enter his mouth, a more casual approach will probably be more effective.

DETERMINING A FAT BUDGET

Because there are no ideal height-weight tables for children as there are for adults, you will have to figure out your child's Fat Budgets or, if old enough, your child can figure out his or her own. You will start by discovering approximately how many total calories and fat calories your child eats a day. Then you will take 20 percent of the difference between the total calorie intake and the fat calorie intake to determine your child's Fat Budget.

1. Keep a record (or have your child keep a record) of your child's food intake for 3 days (2 weekdays and 1 weekend day). Read the Appendix, "The Nitty-Gritty of How to Keep a Food Record" (page 588) to find out exactly how to do it. (You may want to order an inexpensive *Choose to Lose* Passbook for keeping the food record. See the order form at the back of the book.) Be exact. Make sure your child's diet is typical of his or her normal daily food intake. Your child should not try to look "good."

2. Figure out the average daily total calorie intake by adding together the final total calorie subtotals* for the 3 days and dividing by 3.

   Your child's **average daily total calorie intake is** ________.

---

*If you have not kept a total calorie subtotal, add up the total calories that your child ate each day to get a total caloric intake for the whole day. Then add these 3 total caloric intakes together and divide by 3 to get an average daily total caloric intake.

3. Figure out the average daily fat calorie intake by adding together the final fat calorie subtotals for the 3 days and dividing by 3.

   Your child's **average daily fat calorie intake** is ________.

4. Subtract the average daily fat calorie intake from the average daily total calorie intake.

   average daily total calorie intake
   − average daily fat calorie intake
   adjusted total calorie intake

   Put the adjusted daily total calorie intake here: ________

5. To determine your child's Fat Budget, multiply the adjusted total calorie intake by 20 percent.

   20 percent × ________ (adjusted daily total caloric intake) = ________ fat calories = Fat Budget

   Your child's **Fat Budget:**________ fat calories a day.

   Here is an example to help clarify the procedure.

Kate's Fat Budget Worksheet

a. Kate looked at her final total calorie subtotal entry for Thursday and saw that she had consumed 1700 total calories for that day. She found she had consumed 1800 total calories on Friday and 1900 total calories on Saturday.

b. She added together the final total calorie subtotals for the 3 days: 1700 + 1800 + 1900 = 5400 total calories.

c. To find an average total calorie intake for the 3 days, she divided the 5400 total calories by 3. 5400 ÷ 3 = 1800 average daily total calories.

   Kate's **average daily total calorie intake** is **1800** total calories.

d. To find an average fat calorie intake for the 3 days, she added the fat subtotals for three days and divided by 3:

600 fat calories + 500 fat calories + 700 fat calories = 1800 fat calories.
1800 ÷ 3 = 600 average daily fat calories

Kate's **average daily fat calorie intake** is **600** total calories.

e. Kate subtracted her average daily fat calorie intake from her average total calorie intake.

1800 total calories
− 600 fat calories
1200 adjusted daily total calorie intake

f. To find her Fat Budget, she multiplied 20 percent times her average adjusted daily total calorie intake of 1200.

20 percent × 1200 (average total calorie intake) = **240** fat calories = Kate's Fat Budget

g. Kate must reduce her fat intake to no more than 240 fat calories a day (less would be fine), but she must eat at least 2000 total calories a day. She must replace the fat she is eliminating with calories of nutritious complex carbohydrates.

Note to Zealous Parents

When you reduce the fat in your child's diet, be sure to replace the fat removed with nutrient-dense/fiber-rich calories (not empty calories) so that your child gets enough calories to grow and develop normally and so he or she is *never* hungry. Because complex carbohydrates have twice as much bulk as fat for a given number of calories, your child will be eating much more food. Don't be alarmed. Low-fat, nutrient-dense/fiber-rich foods are not fattening.

Fat Budget Applied

When your child knows his or her Fat Budget, leafing through the Food Tables will be fun. Having a Fat Budget makes children judge foods more critically and make better choices. Your child may also enjoy learning how to read a food label (see page 140). Knowing their Fat

Budgets, children quickly learn how to read a nutrition label and evaluate if a food is a healthy choice. Your child probably will stop at this level of expertise, but you may be surprised how ably she or he applies the knowledge gained to new situations.

Keeping Track. If you have children (more likely teenagers) who want to keep track of their fat intake, they should be encouraged to do so. They will probably like the idea of having a passbook (it looks like an adult's checkbook) for their own. Have them read the Appendix, "The Nitty-Gritty of How to Keep a Food Record" (page 588) or read it yourself and help them get started. They should follow the instructions above to determine their Fat Budgets.

Let your children be in control of adhering to their Fat Budgets. To be blunt, don't nose into their food records. What they write down is their business alone. They must own the problem and accept responsibility. Only if they are in control will they be successful.

You do not want to make eating into a BIG THING. You want eating healthfully to be a natural part of your children's lives. You can do your part by being supportive and by providing an ample variety of delicious, nutritious low-fat foods to keep your children full and happy.

Note to Overzealous Parents

In your zeal to keep your children thin, don't eliminate all fat from their diets. For proper growth and development, young children require a diet that is about 20 percent total fat and filled with vitamins, minerals, and fiber found in fruits, vegetables, and whole grains. (Frankly, drastically reducing children's fat intake is a rare problem. The more widespread and serious problem is allowing children to eat 40 to 50 percent of their calories as fat.)

Remember:

1. Feed the entire family nutritious foods.
2. Don't single out overweight children for special attention or foods.
3. Don't make food an issue.

4. Don't feed children even healthy snacks too close to meal times.
5. Use *Choose to Lose* adult tables to determine Fat Budgets for children who have stopped growing.
6. For children who are still growing, base Fat Budget on total caloric intake minus fat intake as determined from three-day food records; be sure to replace fat with complex carbohydrates and not let total calories drop.

15
Frequently Asked Questions

Insulin Misinformation

Q. *Choose to Lose* is based on eating a lot of carbohydrates, but many diet books say eating carbohydrates raises your insulin and that is bad.

A. Baloney! You need to produce insulin to live! When you eat carbohydrates, your body produces insulin to move the carbohydrates into the cells where it is burned (not converted into fat). Insulin production in response to eating carbohydrates is as normal as an increased pulse rate is to exercise. People hear insulin and think diabetes, but the production of insulin doesn't cause diabetes. In fact, the inability to produce insulin is a sign of diabetes.

Other Carbohydrate Misinformation

Q. Are dietary carbohydrates converted to fat and stored?

A. The answer to this is a resounding NO under normal eating conditions and yes under only very special circumstances that do not apply to most people. Here are the facts.

Dietary Carbohydrate Is NOT Converted to Fat. Research on animals and humans shows that carbohydrate stimulates its own oxidation (burning). Carbohydrate storage (in the form of glycogen in the liver and the muscles) is minimal — about 800 to 1000 calories. The glycogen in the muscles serves as a source of quick energy (that is, for spurts of activity as in competitive sports). The glycogen stored in the liver is used primarily to provide an even, uninterrupted source of glucose to the brain so that the brain is not flooded with glucose during and immediately after meals and then starved for it hours later. This flow is carefully regulated because the brain can burn only glucose and needs a constant supply to maintain consciousness.

When you eat, a small amount of the carbohydrate is diverted into the liver and muscles to top up the glycogen stores that have been partially emptied between meals. The rest (the great bulk) is moved via the action of insulin into cells where it is burned. The energy released is captured and used to fuel the body's functions and activity or converted to heat. NONE is converted into fat in the human under normal conditions.

Humans Are Not Rats. The persistent belief that carbohydrates are converted into fat in humans has several sources. One source is metabolic research done on lab rats and mice. The mistake here is applying the results to humans. Rats and mice metabolize food very differently than we do. Rodents *do* convert dietary carbohydrate into fat under normal conditions. This is probably essential for their survival due to the combination of their very high metabolisms, little room for glycogen storage, and the uncertainty of steady food supplies.

Second, many of the metabolic studies that show that carbohydrates turn to fat have been done on liver smashes (not on living humans), so all the control mechanisms of the intact human are lost.

Under normal circumstances, a few hours after eating carbohydrates the net energy balance is zero. That is, hours after a meal there is no net storage of carbohydrate either as carbohydrate or as fat. On the other hand, a few hours after eating fat the net storage of fat is almost exactly equivalent to the amount eaten.

One can force dietary carbohydrates to be stored as fat by a technique called glycogen loading. Marathon runners do this before a race. It involves eating several thousand calories of carbohydrate above their normal intake for more than seven days at a time. If you think this is easy or fun, try it.

Diet Drugs

Q. If you can take a pill that keeps you from being hungry so that you don't eat, what's wrong with that?

A. Everything. First, food is wonderful. Not only do fruits, vegetables, whole grains, nonfat dairy, poultry without skin, and seafood satisfy your palate, they contain vitamins, minerals, and fiber that you need for good health. Food is the fuel your body needs to operate. You need to eat to keep up your metabolism. Eating is POSITIVE. It is healthy (and human) to be hungry and to eat. Starvation is always a short-term and unhealthy diet solution.

Second, diet drugs have been shown to be extremely dangerous. Phen-fen and dexfenfluramine were pulled off the market because they cause heart problems. They also cause a fatal disease called primary pulmonary hypertension in some people, and in probably everyone they cause some neurological damage. The newest drug on the market, Meridia, was approved by the FDA even though an advisory panel of experts found it too dangerous to approve. It causes high blood pressure in some people and may increase the risk of stroke and heart disease. The new FDA fast-track approach to drug approval means the diet pill you are taking may not have been tested long enough or carefully enough to see if it is really safe. Why endanger yourself when the best way to lose weight is to eat healthfully and exercise?

Even if the pills were safe, you would never get the benefits — a nutritious diet, feeling full and satisfied, reduction of risks for chronic diseases, more energy, sense of well-being, strength, independence, and balance — that you get from eating nutritious, low-fat foods and exercising.

"We Are Eating Less Fat. Why Are We Getting Fatter?"

Q. I keep reading that as a nation we are getting fatter and fatter, and yet we are eating less fat. How can this be true if fat makes fat?

A. It is frustrating to hear this statement made over and over when it is so wrong. The problem is that people, even the so-called experts, don't understand percentages.

The fallacious argument that we are eating less fat yet are getting fatter is based on a misunderstanding of the information released by the National Center for Health Statistics in 1994. According to the NHANES studies, 14 to 18 years ago Americans were eating on average 36 percent of their total calories as fat. Ten years later, they were eating 34 percentage of their total calories as fat. The conclusion was that since the percentage is lower, we are eating less fat, but that is not true. We are eating a smaller percentage of fat, but the actual amount of fat is larger.

Let's look at the actual amount of fat: In the first NHANES analysis people were eating 709 fat calories, which was 36 percent of 1969 total calories; in the second NHANES study they ate 748 fat calories, which was 34 percent of 2200 total calories. The percentage of fat may be less, but the absolute amount is greater. Americans are fatter because they are eating more fat (and exercising less). See Chapter 4, "Where's the Fat?" for an explanation of why this is happening.

Reaching a Plateau

Q. I have been following *Choose to Lose* for several weeks and seem to have reached a standstill in my weight loss. Should I give up?

A. Never! In the first few weeks of following *Choose to Lose* you will be replacing the fat you are losing with the new muscle you are building from exercising. In the long run, added muscle will help you burn fat. In the short run, because the muscle you are building may weigh as much or more than the fat you are losing, you may not see the loss registered on the scale. Throw away your scale. Your fat loss will be evident in the way your clothes fit.

Do not give up now. First of all, no matter what your weight loss, *Choose to Lose* is the healthiest way to eat. You are reducing your risks for all sorts of diseases. Second, you *will* lose weight. This is only a temporary plateau. Hang on. See the next answer.

Not Losing

Q. I have been doing *everything* you recommend. Why am I not losing weight?

A. The explanation for your weight-loss standstill may be found in your answers to the following questions.

1. Are you keeping track of everything you eat? If you don't write down everything you eat honestly and accurately, you may *think* you are eating a perfect diet, but you may be consuming more fat than you realize. Say you go to Au Bon Pain and order a turkey sandwich. You choose a four-grain roll, mustard instead of mayo, and tomato and hot peppers. Perfect! You assume the sandwich is low in fat because all of the ingredients are. When you look in the Au Bon Pain entries in the **RESTAURANT FOODS** section of the Food Tables, you find that the four-grain roll has 99 fat calories — 99 fat calories! A plain baguette, which would have also tasted great, has 18 fat calories. What a waste of fat calories. Fat is pervasive (and hidden) throughout our food supply. It accumulates fast and furiously.

2. Are you eating above your minimum total caloric intake? *You must eat to lose weight.* This is the hardest concept for chronic dieters to accept. They are so afraid that eating a lot of food — even low-fat, nutritious food — will make them fat they still (no matter how many times we beg and plead and jump up and down and yell) limit the amount of calories they eat. As a result, their metabolism slows down and they are hungry and dissatisfied. They lose much more slowly and usually give up before they see any results. Take a chance. Eat! (Of course, not empty calories or high-fat calories.) What have you got to lose but fat?

3. Are you meeting your minimum total caloric intake with nutrient-dense foods or with a lot of nonfat ersatz food? If you fulfill the minimum nutritional requirements as suggested by the Food Guide Pyramid, you will be a success at losing fat. Besides being chock full of vitamins, minerals, and fiber, these foods will keep you full. If you fulfill your minimum total caloric intake with essentially empty calories (nonfat cakes, cookies, frozen yogurt, and so on), you may find you lose very slowly or not at all.

There are several reasons for this. First, the calories in these empty foods add up quickly to very large amounts. Your body is busy burning all those calories instead of raiding your fat stores, so your weight is stalled.

Another problem with eating too many fat-free cookies, cakes, crackers, and ice cream is that they take the place that should be filled by more nutritious foods. You are denying yourself the vitamins, minerals, and fiber that you need for good health. Fiber is almost totally missing in these products. Fiber fills you up so that you don't crave high-fat foods.

Third, some nonfat food may not be as nonfat as we are led to assume. One of our readers called because although she was following *Choose to Lose* to the letter and maintaining her weight, she wasn't losing. The mystery was solved when she told us she ate three large "fat-free" muffins that she bought at a local shop every day. Eating three large muffins is pushing the system. First of all, she really had no way of knowing how much fat they contain. Second, even if they are truly fat-free, the three muffins add up to a lot of total calories for her body to burn instead of her fat stores.

4. Are you eating out too often? Even though we can advise you what questions to ask and what requests to make to optimize low-fat eating when you dine out, it is still a dangerous activity. You never *really* know whether your vegetables were steamed with no fat as you asked. You can't be sure the rice is not riddled with butter. For the scary truth about eating out, see Chapter 4, "Where's the Fat?" and **RESTAURANT FOODS** and **FAST FOODS** in the Food Tables. If you are eating out once a week or more (eating out isn't limited to going to a special restaurant; carry-out, breakfast in the office cafeteria, and lunch at a sub shop all count as eating out), you may find your weight-loss progress going nowhere.

5. Are you exercising aerobically for at least 30 minutes every day? Exercise is an essential part of fat loss. The more muscle you build by exercising aerobically, the more fat you will burn. You can lose without exercising, but it is much slower and much less likely to be permanent.

Weighing In

Q. How often should I weigh in?

A. We recommend weighing yourself once every six months. Better yet, once a year. Best yet, not at all. The point is that weighing yourself doesn't show you anything about your fat loss. Short-term changes in your weight don't discriminate between loss of water, muscle, or fat. In fact, because the muscle you are building by exercising may weigh as much or more than the fat you are losing, you may reach a plateau and not lose weight for a while. If you just looked at the scale, you might become discouraged and go back to your old high-fat ways when, in fact, you should feel encouraged because you have succeeded in building fat-burning muscle. The best advice is to throw away your scale.

But if throwing away your scale would be like tearing out a rib, at least try to overcome the urge to weigh yourself daily (or more likely, hourly). Weigh yourself no more than once a week, or once every other week for a more realistic reflection of your fat loss. Try to weigh yourself on the same day of the week, at the same time of the day, wearing the same amount of clothing, and using the same scale.

Your best bet is to take *Choose to Lose* Leader Jan Bishop's advice: measure your fat-loss progress by trying on a pair of slacks that are too tight to wear in public. In a few weeks, try them on again. When this pair fits perfectly, find another pair that is snug and watch it become looser and looser. This way you will be seeing your real progress, not a number on a scale.

Who Can Eat That Much?

Q. I know you say I must eat more than my minimum daily total caloric intake, but at the end of the day I find I hardly ever top 1200 total calories. Do I have to force myself to eat enough food?

A. Surprisingly, eating enough food is *the* most difficult part of following *Choose to Lose.* Start slowly. After all, you probably have been eating almost no food in terms of bulk (but not in terms of fat!) for many years. Set the minimum nutritional requirements of the Food Guide Pyramid as your goal. Add a fruit, a grain, a vegetable, one at a time, until you are fulfilling the recommendation. Nancy Ochsner of Woodland Park, Colorado, thought of this great idea: Each morning, place your daily requirements of fruit, vegetables, grain, and dairy on a plastic tray in the refrigerator and make sure you have emptied it by the end of the day.

Are you eating real dinners? We have soup, a chicken, turkey, fish, or seafood entrée on rice, a steamed vegetable or two, sometimes a sweet potato or a white potato. Are you eating only a salad for dinner? Do you see why we have no trouble accumulating enough calories? Are you including healthy snacks? We enjoy 10 cups of air-popped popcorn every day. Try harder and you will reach and surpass your minimum daily total caloric intake in no time.

Caveat: empty calories don't count. See "Not Losing," earlier in this chapter.

Too Little Fat?

Q. I have cut down my fat intake to almost nothing. Is it dangerous to eat too little fat?

A. From a health standpoint it is almost impossible to eat too little fat. Rural Chinese eat a 10 percent fat diet and they are much healthier than we. If you are eating according to the recommendations of the Food Guide Pyramid, you are probably getting adequate fat. The problem with eating almost no fat is that such martyr-like behavior cannot be endured for long and soon you're off *Choose to Lose* and back to Choose to Gain.

Cholesterol

Q. I have high blood cholesterol. Will it hurt me to follow *Choose to Lose*?

A. Following *Choose to Lose* is a great way to lower your blood cholesterol level. By reducing total fat, you will automatically be reducing saturated fat, which is the main culprit in raising blood cholesterol. We recommend reading our book, *Eater's Choice: A Food*

Lover's Guide to Lower Cholesterol, to learn everything you need to know about heart disease, diet, and cholesterol, but after you read *Choose to Lose.* Cook the low-fat, low-sat-fat recipes from *Choose to Lose* and *Eater's Choice Low-Fat Cookbook* and you'll be set for life — a long and delicious life.

Fading Away to Nothing

Q. If I keep following *Choose to Lose,* will I eventually become emaciated?

A. You only wish! The Fat Budget you determined is the maximum amount of fat you should consume to reach your goal weight and stay there.

Men — It Just Ain't Fair!

Q. How come my husband loses 10 pounds for every 5 pounds I lose?

A. It really isn't fair. Men naturally have more muscle mass than women, so they burn fat more readily. Men's fat often congregates around their middles (apple shape), which may make it easier for them to lose than it is for women, whose weight more often accumulates around the hips and thighs (pear shape). In addition, many men don't have the hangups many women have about eating and losing weight. They just do it.

Diet Drinks Versus Regular

Q. Fat makes fat, not sugar. So can I *really* drink regular cola without gaining weight?

A. Absolutely. Neither regular nor diet drinks are fattening, but neither are good drink choices either. Diet drinks have the added disadvantage of containing unhealthy additives. If you have to have a soda, limit yourself to no more than one a day. Try the best drink choices instead — skim milk, water, and 100 percent fruit juices.

Keeping Records

Q. I hate keeping any type of records. Do I *really* have to keep track of the foods I eat forever?

A. It is a pain in the neck to write down everything you eat. However, only if you record what you eat and determine the cost in fat calories, will you know what foods are making you fat and if you are making enough changes to lose weight. You will *not* have to keep records forever. Pretty soon you will know the combinations of foods you like to eat that fit into your Fat Budget. You can stop keeping records when you are consistently eating comfortably within your budget and your records no longer teach you anything.

Boiled Chicken

Q. I love to eat. How can I stay on this diet if I have to eat boiled chicken and jello?

A. We hate to be rude, BUT DIDN'T YOU READ THIS BOOK?!!! We never eat boiled chicken and jello. We eat *Salmon with Grainy Mustard Sauce* and *Cajun Chicken* and *Calzone*. Low-fat cooking does not have to be bland and dull. With herbs and spices, and garlic and onions and ginger and vegetables and fruits and vinegars, you can make sublime low-fat food.

Fruit and/or Vegetables

Q. I don't like fruit or vegetables. What can I eat?

A. When is the last time you tried fruit or vegetables? Are you remembering the overcooked green beans or peas that your mom used to make? Try steaming fresh vegetables until they are crisp but tender. Bake sweet potatoes or butternut squash. Do you really dislike baked potatoes or watermelon? Maybe you like more produce than you realize. Be open. Try fruit and vegetables you haven't eaten before. Try bok choy and bosc pears. If you want to live a long, healthy life, you need to eat fruit and vegetables. Eat them even if you think you don't love them. Soon you really will love them.

Chocolate

Q. Once a month I get a mad craving for chocolate. How can I handle it?

A. The beauty of *Choose to Lose* is that you never have to feel deprived or guilty. Unless you are a chocoholic who loses control after one whiff, limit yourself to a small piece or two during that time of the month. You can balance the cost with lower-fat choices the rest of the day. See "Splurging" (page 181).

High-Fat Cravings

Q. Will I ever get over the need for high-fat foods?

A. You won't believe it now, but if you are like many, many Choosers to Lose, you will lose your high-fat taste. The French fries you thought were divine will seem greasy; the creamy New England chowder you adored will turn your stomach. It takes about 12 weeks to change your tastes. If, however, you are constantly subjecting yourself to high-fat-tasting foods, even if they are low-fat or nonfat, you may never give up the high-fat taste.

Bingeing

Q. On every diet I've tried (and there is not one I haven't) I always end up bingeing and giving up the diet. Will this happen to me on *Choose to Lose*?

A. Although we can't guarantee you won't become completely unhinged and eat an entire bag of potato chips, including the bag, we strongly believe that this will never happen when you follow *Choose to Lose. Choose to Lose* staves off this type of behavior in two ways.

1. You are encouraged to eat a tremendous amount of food so that you are not hungry. On the other diets you followed, you limited calories and thus were starving. You were justified in going crazy and eating everything in sight because you were ravenous.

2. You are allowed to fit in high-fat splurges. You can fit in a few chocolate chip cookies, so you don't need to fantasize about the chocolate chip cookies you aren't eating and then madly devour the whole jar of peanuts to compensate. You know you can have that piece of chocolate candy tomorrow, so you don't have to eat the whole box today.

Water

Q. Every other diet I've been on required drinking at least eight glasses of water a day. Why don't you stress drinking water?

A. Drinking water is an essential part of many diets for two reasons. The first is that when you eat a low-calorie, high-protein diet, you need to flush out the excess toxic nitrogen waste from your kidneys.

The second reason is to fill you up, because on low-calorie diets you are always famished. Although we see nothing wrong with drinking water, *Choose to Lose* does not require drinking any special amount or any at all. When you are hungry, we want you to eat real food, which provides you with vitamins, minerals, and fiber, and not take a few unsatisfying sips from your water jug. You also get a lot of water by consuming fruits and vegetables because they are mainly water.

Portion Control

Q. I was taught that limiting portion size is the key to weight loss. Why are you against portion control?

A. We are not against portion control if the foods you are eating are high-fat or full of empty calories. These should be strictly limited. We are against limiting low- or nonfat nutrient-dense/fiber-rich foods because eating too little food will defeat your fat loss goals. It will slow your metabolism, reduce your intake of vitamins, minerals, and fiber, and make you hungry. If you are eating low- or nonfat nutrient-dense/fiber-rich foods, these foods will limit themselves. Could you eat five potatoes? How many carrots can you consume? If you are eating highly processed foods like pasta, bagels, white bread, you do have to set limits. Use sense. Two cups of pasta, two bagels, two cups of low-fiber cereal are not too much, but more would be.

Slowing Down to Eat Less

Q. Many other diet programs teach behavioral modification techniques to help you eat more slowly. Why don't you discuss these?

A. Putting your fork down between mouthfuls, chewing your food 50 times between bites — these devices are tedious, boring, quickly abandoned, and, most important, unnecessary. They have been developed to help people eat less food. We believe in eating a lot of food (see the answer to the previous question) as quickly or slowly as you wish.

Of course, that is not to say that we don't think you should pay attention to what you are eating. We do. Try to taste every bite and you'll enjoy it more.

Young Children

Q. If I follow *Choose to Lose* and feed my young children a low-fat diet, will it stunt their growth?

A. The American Heart Association recommends that all children above the age of two eat a low-fat diet. This is because the conditions that precipitate diseases such as heart disease begin in childhood. The plaque that clogs your arteries and gives you a heart attack at age 55 started accumulating when you were a young child. Children develop food tastes and food habits at a young age. Teach them to eat healthfully at a young age, and you will be giving them a precious gift that will last a very long lifetime.

Personal experience: our two sons were raised on the *Eater's Choice* recipes — a low-fat diet — and grew up to be 6 foot 2 and 6 foot 3. So much for low-fat diets stunting growth!

Fat Calories Are Part of Total Caloric Intake

Q. When I eat a food, I write down the number of total calories and the number of fat calories it contains. Are those fat calories a *part* of my total caloric intake, not in addition to it?

A. Yes. The number of fat calories you eat is a part of your total caloric intake. Say you ate 2000 total calories today. These total calories are made up of calories of fat, calories of protein, and calories of carbohydrates. You are concerned about reducing your fat calorie intake, so you are keeping track of your fat calories separately, but those fat calories are still a part of your total caloric intake. If, for example, you ate 300 fat calories today, those 300 fat calories would be part of the 2000 total caloric intake. The rest of your total calorie intake for today (2000 total calories − 300 fat calories = 1700 total calories) would be divided between carbohydrate and protein calories.

Total Daily Caloric Intake: 2000

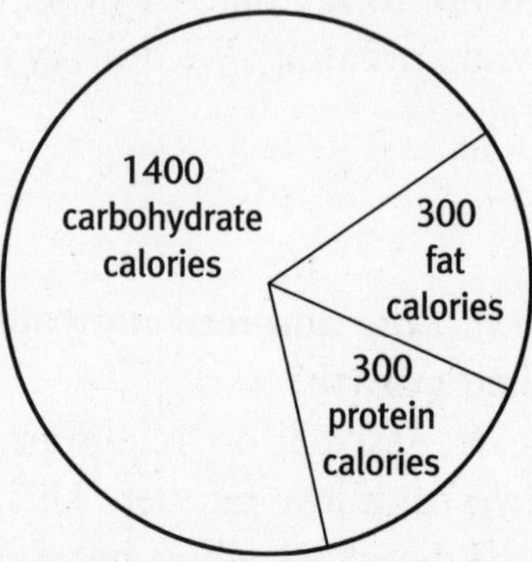

Individual foods are also made up of carbohydrate, protein, and fat calories. For example, a food that has 200 total calories might have them distributed as follows: 55 fat calories + 45 protein calories + 100 carbohydrate calories = 200 total calories.

Total Calories of a Food: 200

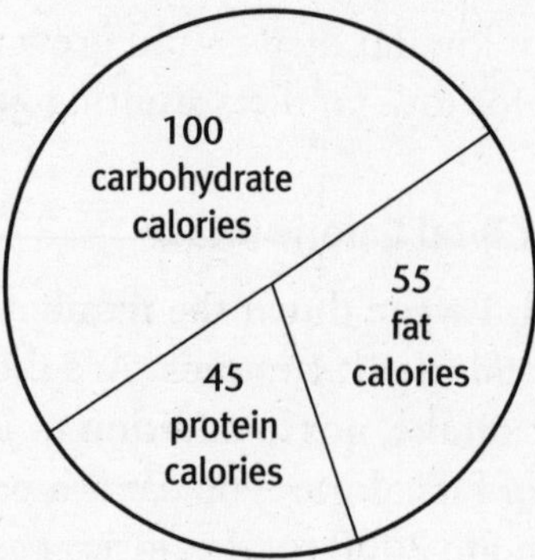

Epilogue
Choose to Lose for Life

> "I have followed this diet for the last 8 months and have lost 75 pounds and feel great. I have tried so many diets, but yours made so much sense and seemed to be so healthy. It was something I could follow and, better yet, it is a way of eating that I could continue for the rest of my life so I can keep the weight off. Thanks again. I feel terrific. I have more energy and, best of all, I look better than I ever looked before."
>
> — Barbara La Sante, New York

> "Guess what? I still have all the weight off and that is after four and a half years! Everyone I know keeps asking me how I have kept the weight off and I just say, read the *Choose to Lose* book and you will find out."
>
> — Barbara La Sante, New York

CHOOSE TO LOSE is more than a diet. It is an education. It is a way of life. Not only will you learn how to make the right food choices to keep you lean forever, but a whole new world of eating will open up to you as you replace fat with complex carbohydrates. Your life will become richer and your palate will expand as you explore new foods and recipes that fit into your Fat Budget. Eating lots of delicious, low-fat, nutrient-dense/fiber-rich food and little fat as well as exercising will increase your energy and make you feel great.

You will be amazed at how you will change, not only in size and shape, but in attitude. At first you may find yourself hungrily eyeing a croissant. But instead of gobbling it down, you will see a number pop out at you: *croissant = 110 calories of fat,* and then you'll ask yourself, "Is that soggy croissant worth almost half my Fat Budget? Wouldn't

French bread taste just as good and leave not a trace of fat in my fat stores?" You'll begin to prefer low-fat choices. You may find it hard to believe, but you may eventually lose your craving for high-fat food.

At first, you may find it almost impossible to eat enough food. At the end of the day you will even be happy to reach your minimum daily caloric intake. Very soon you will wolf down loads of nutritious, low-fat foods naturally, without effort — only delight.

You may start out thinking exercise is a drag, but eventually you will look forward to those walks in the crisp, morning air or in the evening with a friend or loved one. It may take you a while to stop thinking starvation, DIET, quick fix, size 12 dress, and start thinking full meals, delicious way of eating for life, and good health. But soon your whole attitude will change. Soon you'll be a bona fide Chooser to Lose for life.

KEEP THOSE MUSCLES MOVING

As part of your new start on life you have included daily aerobic exercise. Be sure to keep exercising even after you achieve your goal size. Not only will exercising help you maintain the new you, it is beneficial for the many reasons discussed in Chapter 11, "Exercise: Is It Necessary?"

PREDICTORS OF SUCCESS

Make the following list a part of your life and you will greatly increase your success in achieving your weight loss goals and creating the New You.

✔ Focus on health, not weight loss.
✔ Focus on body size changes, not scale changes.
✔ Replace high-fat foods with low-fat foods.
✔ Eat enough total calories by eating more fruits, vegetables, whole grains, and low- or nonfat dairy.
✔ Eat real, full meals.
✔ Eat within your Fat Budget.

✔ Budget in an occasional high-fat splurge.
✔ Eat out rarely.
✔ Cook! Prepare *Choose to Lose, Eater's Choice Low-Fat Cookbook* or other low-fat recipes.
✔ Exercise aerobically 30 minutes every day.
✔ Have patience!

IT'S UP TO YOU

Soon you will be seeing the New You in the mirror. Not only will the New You be lean, the New You will be full of energy, vim, and vigor. The New You will be full of confidence and high self-esteem because taking control of your diet is the first step to taking control of your life.

You've got the tools. It's your move. Remember, this is for life. Go to it and have fun!

What Next?

After you achieve your desired weight, you should continue eating the *Choose to Lose* way for the rest of your life. Following the three *Choose to Lose* strategies — reducing fat in your diet by eating below your Fat Budget (the Fat Budget you use to reach your desirable and healthy weight is also your lifetime maintenance budget); eating large amounts of nutritious, low-fat foods such as fruit, vegetables, low- or nonfat dairy, and lean meats; and exercising aerobically every day for 30 minutes — you will be able to maintain your new weight forever.

If, however, you are beginning to gain back the weight you lost, have no fear. It is easy for fat to sneak back into your diet if you have not been paying close attention. Even though you *know* you have defatted your food choices, you really must begin keeping track again. Reread *Choose to Lose* and start keeping food records — just for a few days or weeks. Look for the sources of fat in your diet so that you can make the right choices and changes. Be sure you are eating below your Fat Budget and exercising aerobically every day. Be sure you are eating enough nutritious, high-fiber foods. Within a short time you will be lean once again.

One Week of Meal Plans

WE CREATED the following meal plans to show you how easy it is to fit delicious food into your Fat Budget and how to save fat calories for a splurge. You will notice that these menus are chock full of wonderful food, but are extremely low in fat, ranging from 96 to 147 fat calories.

These are only SAMPLE meal plans, not menus designed to be repeated 52 times each year until you are 99 years old. It is your body and your taste buds, and you will have to devise meal plans that suit you and your Fat Budget. Don't worry. It will be fun.

The meal plans illustrate how to save up calories for a splurge following a sample budget of 315 fat calories (a Fat Budget based on a goal weight of 135 pounds). Remember, it is not always necessary to go on a high-fat bender at the end of the week. Don't feel compelled to spend all you have saved. Including an occasional high-fat favorite is important for your psyche but the less fat you consume without being a martyr, the sooner you will reach your fat-loss goal.

At the bottom of each meal plan you can see how it (and you, if you eat it) meets the minimum nutritional requirements of 2 to 4 servings of fruit, 3 to 5 servings of vegetables, 6 to 11 servings of grain, and 2 to 3 servings of dairy.

The recipes for dishes printed in italics can be found in *Choose to Lose*.

Fat Budget: 315 Minimum Total Calorie Intake: 1579

Sunday

	AMOUNT	CALORIES TOTAL	CALORIES FAT
Breakfast			
Buttermilk waffle	1	220	21
Maple syrup	1 Tbsp	50	0
1% cottage cheese	½ cup	90	10
Honeydew melon	⅕	92	0
Orange juice	1 cup	111	0
Coffee	1 cup	0	0
Lunch			
Tomato-Rice Soup	1½ cups	140	0
Broiled Ginger Fish	1 fillet	194	35
Whole-wheat bread	2 slices	120	20
Steamed broccoli	1 cup	40	0
Dijon/Yogurt Sauce	1 tbsp	8	0
Sliced fresh peaches	1 cup	74	0
Coffee	1 cup	0	0
Snack			
Nonfat fruit yogurt	8 fl oz	200	0
Dinner			
Grapefruit	½	38	0
Chili Non Carne	1½ cups	176	8
Rice	½ cup	100	0
Baked potato	1 small	75	0
Nonfat yogurt	2 Tbsp	14	0
Zucchini Matchsticks	1 cup	34	0
Onion Flat Bread	2 slices	166	8
Coffee	1 cup	0	0
Snack			
Skim milk	1 cup	80	0
Cheerios	1 cup	110	15
Banana	1	105	5
F: **5,** V: **4,** G: **7,** D: **3**	**TOTALS:**	2237	**122**
		Fat Budget:	315
		Saved:	193

Fat Budget: 315 Minimum Total Calorie Intake: 1579

Monday

	AMOUNT	CALORIES TOTAL	CALORIES FAT
Breakfast			
Oatmeal	⅔ cup dry	100	33
Whole-wheat toast	1 slice	60	10
Jelly	1 tsp	18	0
1% cottage cheese	½ cup	90	10
Strawberries, fresh	1 cup	45	0
Skim milk	1 cup	80	0
Coffee	1 cup	0	0
Lunch			
Chickpea Sandwich	1	195	35
Whole-wheat pita bread	1	180	10
Tangerine	1	35	0
Carrot sticks	2	62	0
Orange juice	1 cup	111	0
Coffee	1 cup	0	0
Snack			
Nonfat fruit yogurt	8 fl oz	200	0
Dinner			
Tortilla Soup	1½ cups	216	15
Indonesian Chicken with Green Beans	1 serving	135	14
Rice, white	1 cup	200	4
Acorn squash	1 sm	140	0
Cauliflower, steamed	1 cup	92	0
Dijon/Yogurt Sauce	1 tbsp	7	0
Nonfat yogurt	1 cup	110	0
Blueberries, fresh	1 cup	80	0
Coffee	1 cup	0	0
Snack			
Popcorn, air-popped	6 cups	180	0
F: **3,** V: **3,** G: **7,** D: **4**	**TOTALS:**	2336	**131**
		Fat Budget:	315
		Saved:	184
		Carry-over:	193
		New carry-over:	377

Fat Budget: 315 Minimum Total Calorie Intake: 1579

Tuesday

	AMOUNT	CALORIES TOTAL	CALORIES FAT
Breakfast			
Whole-wheat toast	1 slice	60	10
Jelly	1 tsp	17	0
Shredded wheat cereal	2 biscuits	160	5
Skim milk	1 cup	80	0
Banana, sliced	1	105	5
Orange juice	1 cup	111	0
Cantaloupe	½	94	0
1% cottage cheese	½ cup	90	10
Coffee	1 cup	0	0
Lunch			
Tuna sandwich:			
Tuna, water-packed	3 oz	111	18
Nonfat mayonnaise	2 tsp	30	0
Rye bread	2 slices	130	20
Carrot	1	31	0
Orange	1	60	0
Coffee	1 cup	0	0
Snack			
Nonfat fruit yogurt	8 fl oz	200	0
Dinner			
Asparagus Soup	1½ cup	116	0
Turkey Mexique	1 serving	146	16
Rice, white	1 cup	200	4
Sweet potato	1	118	0
Spinach, cooked	1 cup	41	0
Pear	1	98	0
Skim milk	1 cup	80	0
Coffee	1 cup	0	0
Snack			
Popcorn, air-popped	6 cups	180	16
Skim milk	1 cup	80	0
F: **4,** V: **4,** G: **8,** D: **4**	**TOTALS:**	2338	**104**

Fat Budget:	315
Saved:	211
Carry-over:	377
New carry-over:	588

Fat Budget: 315 Minimum Total Calorie Intake: 1579

Wednesday		CALORIES	
	AMOUNT	**TOTAL**	**FAT**
Breakfast			
Whole-wheat toast	2 slices	120	20
Jelly	2 tsp	34	0
Wheat Chex cereal	1 cup	180	10
Skim milk	1 cup	80	0
Banana	1	105	5
Orange juice	1 cup	111	0
Lunch			
Turkey sandwich:			
Turkey breast	3 oz	114	6
Lettuce, tomato		10	0
Mustard	2 tsp	10	0
Whole-wheat bread	2 slices	120	20
Apple	1	81	0
Carrot sticks	1 cup	31	0
Coffee	1 cup	0	0
Snack			
Nonfat fruit yogurt	8 fl oz	200	0
Dinner			
Ginger-Carrot Soup	1½ cups	120	9
Shrimp Curry	1 serving	181	33
Rice, white	1 cup	200	4
Zucchini, steamed	1 cup	18	0
Vinegar	1 tsp	17	0
Butternut squash	½	40	0
Fresh sliced peaches	½ cup	37	0
Coffee	1 cup	0	0
Pudding Brownies	2	166	4
Snack			
Nonfat vanilla yogurt	1 cup	170	0
Fresh strawberries	½ cup	22	0
F: **5,** V: **3,** G: **6,** D: **3**	**TOTALS:**	2157	**111**

Fat Budget:	315
Saved:	204
Carry-over:	588
New carry-over:	792

Fat Budget: 315 Minimum Total Calorie Intake: 1579

Thursday

	AMOUNT	CALORIES TOTAL	CALORIES FAT
Breakfast			
Apricot Oat Muffin	1	114	10
Grapefruit	½	38	0
Wheatena	⅔ cup dry	300	10
Skim milk	1 cup	80	0
Coffee	1 cup	0	0
Lunch			
Ted Mummery's Turkey Barbecue	1 cup	262	10
Whole-wheat pita	1 large	165	10
Carrot	2	62	0
Peach	1	37	0
Coffee	1 cup	0	0
Snack			
Nonfat fruit yogurt	8 fl oz	200	0
Dinner			
Cucumber Soup	1½ cups	117	0
with walnuts		129	10
Bar-B-Que Chicken	1 breast	173	13
Mashed potatoes	1 cup	160	0
Indian Vegetables	1 cup	72	7
Cocoa Meringue Kisses	4	80	0
Coffee	1 cup	0	0
Snack			
Popcorn, air-popped	6 cups	180	16
Banana	1	105	5
F: **3,** V: **4,** G: **7,** D: **2**	**TOTALS:**	2157	**91**
		Fat Budget:	315
		Saved:	224
		Carry-over:	792
		New carry-over:	1016

Fat Budget: 315 Minimum Total Calorie Intake: 1579

Friday

	AMOUNT	CALORIES TOTAL	CALORIES FAT
Breakfast			
Buttermilk Waffles	1	220	21
Maple syrup	2 tbsp	105	0
Skim milk	1 cup	80	0
Pineapple slices (¾″ thick)	2	84	0
Coffee	1 cup	0	0
Lunch			
Asparagus Pasta	1 serving	253	26
Tomato	1	24	0
Cumin Wheat Bread	2 slices	214	30
Blueberries	1 cup	41	0
Coffee	1 cup	0	0
Snack			
Nonfat fruit yogurt	8 fl oz	200	0
Dinner			
Zucchini Soup	1½ cups	120	0
Cajun Chicken	1 serving	196	28
Rice, white	1 cup	200	4
Steamed carrots	1 cup	35	0
Dijon/Yogurt sauce	2 tbsp	14	0
Whole-wheat bread	1 slice	60	10
Cucumber Salad	1 cup	20	0
Nonfat frozen yogurt, vanilla	1 cup	160	0
Strawberries, fresh	½ cup	22	0
Coffee	1 cup	0	0
Snack			
Popcorn, air-popped	6 cups	180	16
Skim milk	1 cup	80	0
F: **4,** V: **3,** G: **9,** D: **3**	**TOTALS:**	2308	**135**

Fat Budget:	315
Saved:	180
Carry-over:	1016
New carry-over:	1196

Fat Budget: 315 Minimum Total Calorie Intake: 1579

Saturday		CALORIES	
Splurge Day	**AMOUNT**	**TOTAL**	**FAT**
Breakfast			
Oatmeal	⅔ cup dry	100	33
Pumpernickel toast	2 slices	160	20
Jelly	1 Tbsp	55	0
Orange	1	60	0
Nonfat yogurt	1 cup	200	0
Skim milk	1 cup	80	0
Coffee	1 cup	0	0
Lunch			
Pasta with Spinach-Tomato Sauce	1 serving	162	17
Italian bread	2 slices	150	20
Pear	1	98	0
Carrot sticks	1 cup	15	0
Skim milk	1 cup	80	0
Coffee	1 cup	0	0
Snack			
Nonfat fruit yogurt	8 fl oz	200	0
Dinner			
Sirloin steak	6 oz	354	132
Baked potato	1	145	0
Sour cream	2 Tbsp	60	50
French bread rolls	2	200	0
Tossed salad	2 cups	15	0
Blue cheese dressing	4 Tbsp	340	300
Steamed broccoli	1 cup	40	0
Cheesecake	1 slice	280	162
Coffee	1 cup	0	0
Snack			
Pepto Bismol	2 Tbsp	0	0
F: **2,** V: **3,** G: **8,** D: **3**	**TOTALS:**	2994	**734**
		Fat Budget:	315
		Overspent:	−419
		Carry-over:	1196
		New carry-over:	777

Recipes

WE LOVE TO EAT, our sons love to eat, our families love to eat, our guests love to eat — that is why we developed the delicious, low-fat recipes in *Eater's Choice* and *Choose to Lose.* You don't have to eat Jell-O, boiled chicken, or broth to lose weight. You can eat *Cajun Chicken* and *Ginger-Carrot Soup* and *Potato Skins* and *Turkey with Capers.* The following recipes are generally quick and easy to make. Try them. They will make reaching and maintaining your desirable and healthy weight a pleasure. Be sure to check out the recipes in *Eater's Choice Low-Fat Cookbook* too. You will find an additional 270 quick and easy, tasty, low-fat treats.

Be advised: Because the Goor oven is not necessarily calibrated to your oven, and your chicken breasts or swordfish steaks may be thicker or thinner than ours, etc., consider the baking time in the recipes an estimate. If the recipes tell you to bake a dish for 45 minutes, check your oven at 30. Your food may be ready.

A Note to Sodium and Sugar Watchers

If you are monitoring your sodium or sugar intake, you may want to cut down on the amount of salt or sugar in some of these recipes.

Soups

ASPARAGUS SOUP CHEZ GOOR

Asparagus Soup is simply delicious. We say "simply" because it couldn't be simpler or more delicious. You can use more or less of any ingredient and this soup will still be a success.

3 pounds fresh asparagus
2 medium potatoes, cut into ½-inch cubes (about 1½ cups)
3 green onions sliced
3 cans (10¾ oz each) chicken broth, defatted
Freshly ground black pepper to taste
½–¾ cup nonfat yogurt

Snap off the stalk ends of the asparagus and discard. Wash asparagus well.

Cut off asparagus tips and place in small saucepan. Cover with water. Set aside.

Cut stalks into 1-inch pieces.

Place asparagus, potatoes, and green onions in a large soup pot. Add broth to cover. Add pepper to taste.

Bring soup to a boil. Lower heat and simmer for about 10 minutes or until asparagus and potatoes are tender.

Meanwhile, bring to a boil, then simmer asparagus tips for 1–3 minutes, until tender but crisp. Set aside.

In a blender, purée soup mixture until smooth. Add yogurt and blend. Add asparagus tips and serve.

9 one-cup servings

Calories per serving: 77 total; 0 fat; 0 sat-fat

GINGER-CARROT SOUP

This soup can be eaten either cold or warm. The lime and ginger give it an interesting flavor.

- 1 tablespoon minced ginger root
- 2 cloves garlic, minced
- ½ cup chopped onion
- 1 teaspoon olive oil
- 5 cups sliced carrots
- 2 cans (10¾ oz each) chicken broth, defatted, + 1 can water
- ¼ cup fresh lime juice (approximately)
- Nonfat yogurt, for garnish

In a soup pot or large casserole, sauté ginger, garlic, and onion in olive oil. Add a splash of water and cook vegetables until tender.

Stir in carrots.

Add chicken broth and water and simmer until carrots are tender (about 20 minutes).

Add lime juice and purée soup in blender until smooth.

Chill or serve warm. Top each soup bowl with a large dollop of yogurt.

Do not mix the yogurt into the soup. For a wonderful combination of tastes and textures, take a little bit of yogurt with each spoonful of soup.

7 one-cup servings

Calories per serving: 80 total; 0 fat; 0 sat-fat

CAULIFLOWER SOUP

Cauliflower Soup is a treasure. It is simple and quick to make and has a wonderful, delicate taste.

- 1 large head cauliflower
- 1 can chicken broth (10¾ oz), defatted
- 2 stalks celery, diced
- 3 green onions, sliced
- 1 teaspoon olive oil
- 2 tablespoons unbleached white flour
- 1–2 cups water

Remove and discard cauliflower stem. Steam flowerets until tender. Reserve about ½ to 1½ cups and set aside. (You need about 4–5 cups cauliflower for the soup itself. Make sure you don't reserve too much!)

Purée cauliflower and chicken broth in blender. Set aside.

Sauté celery and green onions in oil until tender.

Reduce heat to medium and stir in flour.

Stir in cauliflower purée. Slowly mix in 1 cup of water, stirring constantly until soup thickens. If soup is too thick, stir in remaining water, ¼ cup at a time. Continue cooking until warmed through.

Cut reserved flowerets into bite-size pieces, add to soup, and serve.

6 one-cup servings (approximately)

Calories per serving: 47 total; 7 fat; 1 sat-fat

CUCUMBER SOUP

Cucumber Soup gets a gold star for excellence. It is as refreshing as a dip in a cool lake on a hot summer day. It is simple, quick, and impressive. Be sure to chill the soup thoroughly.

2 cucumbers
2 cups nonfat yogurt
1 can (10¾ oz) chicken broth, defatted
1 clove garlic, crushed
Walnuts, for garnish

Peel cucumbers and cut into bite-size cubes. Salt heavily and set aside.

Spoon yogurt into a medium casserole and stir until smooth.

Stir in chicken broth.

Mix in garlic.

Rinse salt off cucumbers and add them to yogurt mixture. Add salt to taste.

Chill in refrigerator for several hours. Garnish with chopped walnuts.

5 one-cup servings

Calories per serving: 78 total; 0 fat; 0 sat-fat

Calories per walnut half: 12 total; 10 fat; 0 sat-fat

TOMATO-RICE SOUP

If you hate wasting the juice from canned tomatoes, collect it in a storage container in the freezer and save it for Tomato-Rice Soup. Then, when you have forgotten to plan for soup and have no fresh vegetables in the house, you can raid the freezer for peas, corn, and the juice you have collected. Add some onion, cloves, and rice. *Voilà*!

6 cups juice from canned tomatoes
1 tablespoon grated onion
1 pinch ground cloves
½ cup long-grain rice
1 cup frozen peas
1 cup frozen corn

Combine tomato juice, onion, cloves, and rice and bring to a boil. Reduce heat and simmer for about 25 minutes.

Add peas and corn and cook about 5 minutes more.

8 one-cup servings

Calories per serving: 93 total; 0 fat; 0 sat-fat

TORTILLA SOUP

This instantaneous soup always gets rave reviews.

6 corn tortillas*
1 medium onion, diced
3 cloves garlic, minced
1 teaspoon olive oil
2 tablespoons chili powder
1 teaspoon oregano
1 large can (28 oz) concentrated crushed tomatoes
1 can (10¾ oz) chicken broth, defatted, + 1 can water
1 green pepper, diced
1 cup frozen corn
Salt and pepper to taste

About 15 minutes before you serve soup, heat tortillas in a slow oven (325°F), until crisp.

*Corn tortillas can be found in the refrigerator section of most supermarkets. Be sure they contain no lard or other saturated fat.

In a soup pot, sauté onion and garlic in oil. Add a splash of water and cook vegetables until soft.

Stir in chili powder and oregano.

Stir in tomatoes, chicken broth, and water.

Bring to a boil and simmer for a few minutes.

Add green pepper and corn.

Add salt and pepper to taste.

For each serving, break a tortilla into small pieces and place at the bottom of a soup bowl. Ladle soup over the tortilla and serve.

7 one-cup servings

Calories per serving: 144 total; 10 fat; 1 sat-fat

ZUCCHINI SOUP

One of the greatest recipes known to mankind. Hot or cold, summer or winter, for family or company, zucchini soup is delicious — and easy. You can even prepare it 20 minutes before you eat. If you want to make more and freeze it, just add a few more zucchini and more broth. This recipe is very flexible. Add more or less of any ingredient, and it will still taste superb.

3 large or 4 medium zucchini (or more or less), sliced
½ cup chopped onion
¼ cup long-grain rice
Chicken stock to cover zucchini, or 2 cans (10¾ oz each) chicken broth, defatted, + water to cover zucchini
1 teaspoon curry powder (approximately)
1 teaspoon Dijon mustard (approximately)
½–1 cup nonfat yogurt

In a large soup pot, combine zucchini, onion, rice, chicken stock, and water (add more water, if necessary, to cover zucchini).

Simmer for 15 minutes or until zucchini are tender.

Purée in blender, adding curry powder, mustard, and yogurt to taste.

Eat warm or cool.

This soup freezes well. Reheat frozen soup for best results. Eat immediately or cool for later.

8 (approximately) one-cup servings

Calories per serving: 80 total; 0 fat; 0 sat-fat

Chicken and Turkey

APRICOT CHICKEN DIVINE

One-quarter cup of nonfat yogurt (0 fat calories) replaces ¼ cup of sour cream (108 fat calories) to create this divine chicken.

4 boneless, skinless chicken breasts
¼ cup unbleached white flour
½ teaspoon salt (optional)
¼ cup apricot preserves
½ tablespoon Dijon mustard
¼ cup nonfat yogurt
1 tablespoon slivered almonds

Preheat oven to 375°F.

Shake chicken in a plastic bag filled with flour and salt until chicken is coated.

Place chicken in a single layer in a shallow baking pan and bake for 25 minutes.

Combine apricot preserves, mustard, and yogurt.

Spread apricot mixture on chicken and bake for 10–15 minutes more or until done.

Just before serving, brown almonds lightly in toaster oven.

Sprinkle almonds over chicken and serve over rice.

4 servings

Calories per serving: 215 total; 23 fat; 5 sat-fat

BAR-B-QUE CHICKEN

Bar-B-Que Chicken is a popular main dish with food lovers of all ages. It also makes a great sandwich. Place the barbecued chicken on a slice of bread, top with sauce, a slice of tomato, a few slices of onion, and another slice of bread.

2 tablespoons Dijon mustard
¼ cup vinegar
¼ cup molasses
½ cup ketchup
½ teaspoon Worcestershire sauce
2 cloves garlic, minced
Dash of Tabasco sauce
8 skinless chicken breasts or thighs*

In a large bowl, mix together Dijon mustard, vinegar, molasses, ketchup, Worcestershire sauce, garlic, and Tabasco for marinade.

Pour about half of the marinade over the chicken and refrigerate the rest to use later.

Marinate chicken for an hour (or less, if you haven't planned ahead).

Remove chicken and reserve marinade.

Barbecue, broil, or use a cast-iron pancake griddle to grill chicken for 10 minutes or until fully cooked.

Turn and coat chicken with marinade and continue cooking until done.

Combine reserved marinade with the marinade you refrigerated. Heat to boiling and simmer for 2 minutes. Spoon over chicken before serving.

8 servings

Calories per serving:	*Total*	*Fat*	*Sat-fat*
1 breast	173	13	4
1 thigh	126	24	6

---

*For a barbecue, you may want greater quantities of chicken. Just double, triple, quadruple, etc., the recipe.

CAJUN CHICKEN

8 boneless, skinless chicken breasts
½ cup unbleached white flour
½ teaspoon salt, optional

Seasoning Mix

(Use more or less of the peppers for a hotter or milder taste.)

¾ teaspoon oregano
½ teaspoon thyme
½ teaspoon basil
½ teaspoon salt
½ teaspoon paprika
¼ teaspoon freshly ground black pepper
½ teaspoon cayenne pepper
¼ teaspoon white pepper

3 cloves garlic, minced
¾ cup chopped onion
2 stalks celery, diced
1 green pepper, chopped
1 tomato, coarsely chopped
1 tablespoon olive oil
1 can (10¾ oz) chicken broth, defatted
1 small can (8 oz) tomato sauce
1 potato, peeled and diced
2 bay leaves

Preheat oven to 350°F.

Shake chicken in a plastic bag filled with flour and salt until chicken is coated.

Place chicken in a single layer in a shallow baking pan and bake for 25–30 minutes until cooked through.

While chicken is baking, combine oregano, thyme, basil, salt, paprika, and peppers, and set aside.

In a large frying pan, sauté garlic, onion, celery, green pepper, and tomato in oil. Add a splash of water and cook vegetables until soft.

Mix in seasoning mixture and let simmer for about a minute.

Mix in broth and tomato sauce. Bring to a boil, then simmer for 5 minutes.

Lower heat and add chicken breasts, potato, and bay leaves. Cook until potatoes are tender.

NOTE: If you find the finished dish too "hot" for your taste, dilute with an additional can of tomato sauce.

8 servings

Calories per serving: 196 total; 28 fat; 2 sat-fat

INDONESIAN CHICKEN WITH GREEN BEANS

A beautiful dish — green beans set against ocher-colored sauce — and very tasty. This is one of our all-time favorites.

TIP: use about a cup of Chinese snow peas and ½ pound of mushrooms, sliced, instead of green beans. Add raw snow peas and mushrooms with the chicken as the last step.

6 boneless, skinless chicken breasts
½ cup unbleached white flour
¼ teaspoon freshly ground black pepper
½ teaspoon salt (optional)
1 pound green beans, washed and cut into bite-size pieces
10 cloves garlic, minced
1 tablespoon minced ginger root
1 small onion, chopped
1 teaspoon olive oil
Juice of 1 lime (preferred) or lemon
1 tablespoon double black soy sauce*
2 teaspoons brown sugar
2 teaspoons turmeric
1 teaspoon salt (optional)
½ cup water

Preheat oven to 375°F.

Shake chicken in a plastic bag with flour, pepper, and ½ teaspoon salt until chicken is coated.

Place chicken in a single layer in a baking pan and bake until fully cooked, about 20–30 minutes. Set aside. You may bake the chicken the night before and refrigerate it until ready to add to sauce.

Cook green beans in a pot of boiling water for 5–10 minutes until tender but still crisp. Set aside.

---

*Double black soy sauce is available at Asian food stores and some supermarkets. You can make your own by mixing 2 teaspoons soy sauce with 1 teaspoon dark molasses.

Cut chicken into bite-size pieces. Set aside.

Sauté garlic, ginger, and onion in olive oil until soft.

Add lime juice, soy sauce, brown sugar, turmeric, 1 teaspoon salt, and ¼ cup of the water.

Slowly add remaining water, if necessary. Sauce should not be watery.

Add chicken and green beans and stir until completely covered with sauce.

Serve over rice.

8 servings

Calories per serving: 135 total; 14 fat; 4 sat-fat

LEMON CHICKEN

A delicate blending of tart and sweet, Lemon Chicken can't help but become one of your most popular family or company dishes.

4 boneless, skinless chicken breasts
Juice of 1 lemon
¼ cup unbleached white flour
¼ teaspoon freshly ground black pepper
½ teaspoon salt (optional)
¼ teaspoon paprika
1½ tablespoons grated lemon peel
3 tablespoons brown sugar
3 tablespoons lemon juice
1 tablespoon water
1 lemon, sliced thin

Place chicken in a bowl or casserole. Pour lemon juice over breasts and marinate in refrigerator for several hours or overnight, turning chicken periodically.

Preheat oven to 350°F.

Combine flour, pepper, salt, and paprika in plastic bag.

Remove chicken breasts from marinade and coat each with flour by shaking it in the plastic bag.

Place chicken in a baking pan in a single layer.

Either peel the yellow (zest) from a lemon and chop it fine in your food processor (a mini food chopper makes a perfect grater), or grate the zest with a hand grater. Mix 1 tablespoon of the grated peel with the brown sugar.

Sprinkle the lemon zest–sugar mixture evenly over the chicken breasts. Combine 3 tablespoons lemon juice and water and sprinkle evenly over chicken.

Put 1 lemon slice on each chicken breast and bake chicken for 35–40 minutes or until cooked through.

4 servings

Calories per serving: 176 total; 13 fat; 4 sat-fat

GRAINY MUSTARD CHICKEN

This simple, elegant dish could not be simpler to make. The recipe is for 4 breasts, but you could just as easily make 20. Just increase the mustard, honey, and soy sauce in the same proportions to accommodate larger amounts.

4 boneless, skinless chicken breasts
3 tablespoons grainy Dijon mustard*
3 tablespoons honey
1½ teaspoons soy sauce

Place chicken breasts in a glass or ceramic container.

In a small bowl, mix mustard, honey, and soy sauce.

Spoon 3 tablespoons of the mustard mixture over chicken, making sure all surfaces are covered, and refrigerate the rest to use later. Marinate for at least 20 minutes but preferably for several hours in the refrigerator.

Preheat the broiler for about 10 minutes.

Place breasts in one layer on a broiler pan. Pour remaining marinade over them. Spread the reserved sauce over them.

Broil about 3–4 inches from the heating coil for 10–15 minutes, checking to see if done after 10 minutes.

4 servings

Calories per serving: 193 total; 13 fat; 4 sat-fat

---

*We use Maille L'Ancienne Dijon Mustard. Choose a Dijon mustard that is full of seeds.

CHICKEN WITH RICE, TOMATOES, AND ARTICHOKES

3 boneless, skinless chicken breasts
2 cloves garlic, minced
½ cup chopped onion
1 teaspoon olive oil
1 large can (28 oz) tomatoes, chopped
2 cups water
½ teaspoon thyme
½ teaspoon oregano
1 teaspoon salt (optional)
¼ teaspoon freshly ground black pepper
1 bay leaf
1½ cups uncooked long-grain rice
1 jar (11½ oz) artichoke hearts, packed in water

Cut chicken into bite-size pieces and set aside.

In a large casserole, sauté garlic and onion in olive oil. Add a splash of water and cook vegetables until soft.

Stir in tomatoes and their liquid, water, thyme, oregano, salt, pepper, and bay leaf and bring to a boil.

Add chicken and rice, cover casserole, and reduce heat to low.

Cook for 25 minutes or until rice is tender, most liquid is absorbed, and chicken is cooked through.

Stir in artichokes and serve.

8 one-cup servings

Calories per serving: 220 total; 15 fat; 3 sat-fat

SESAME CHICKEN BROCHETTES

6 boneless, skinless chicken breasts
¼ cup soy sauce
½ cup dry white wine or vermouth
1 clove garlic, minced
1 tablespoon sesame seeds, lightly toasted

Cut chicken breasts into 1-inch cubes and place in a bowl or casserole.

Combine soy sauce, wine, and garlic and pour over chicken. Marinate in refrigerator for at least 30 minutes.

Preheat oven to broil or prepare grill.

Remove chicken from marinade and reserve marinade.

Skewer chicken and broil or grill until cooked through, basting occasionally with marinade.

Sprinkle sesame seeds over chicken and serve over rice.

Heat marinade and spoon over chicken and rice.

6 servings

Calories per serving: 160 total; 20 fat; 5 sat-fat

TURKEY WITH CAPERS

You may sauté turkey cutlets, but this takes a lot of fat. Flouring turkey cutlets, then grilling them on a cast-iron griddle with no added fat produces cutlets that are also tender but have almost no fat calories.

1 pound turkey breast cutlets
½ cup unbleached white flour
1 clove garlic, minced
¼ cup chopped onion
1 teaspoon olive oil
3 tablespoons capers, drained
3 tablespoons red wine vinegar
2 tablespoons Dijon mustard
1 cup chicken broth, defatted
2 tablespoons tomato paste
⅓ cup chopped parsley

Place cutlets between two pieces of wax paper and pound with meat mallet or rolling pin until thin.

Place flour on a plate. Dip each cutlet in flour, coating it on both sides, and set it on a large plate.

Heat cast-iron griddle. Cover it with cutlets. Turn cutlets when their edges start turning white. Remove cutlets as soon as they are no longer pink. Don't let them overcook! Turkey can become tough if cooked too long. Set them aside.

In a large skillet, sauté garlic and onion in olive oil until soft.

Stir in capers, vinegar, Dijon mustard, chicken broth, and tomato paste. Add parsley.

Add turkey cutlets and mix until they are covered with sauce.

4 servings

Calories per serving: 174 total; 18 fat; 5 sat-fat

TURKEY MEXIQUE

A great dish with a chili taste.

1 cup chopped onion
2 teaspoons minced garlic
1 teaspoon olive oil
1–2 tablespoons chili powder
1 tablespoon cumin seed
½ teaspoon salt (optional)
1 tablespoon unbleached white flour
1½ cups chicken broth, defatted
3 tablespoons tomato paste
3 cups diced turkey breast, raw or cooked
1 green pepper, diced
2–3 cups sliced mushrooms (optional)
¼ cup stuffed green olives, sliced
½ cup water

In a large skillet, sauté onion and garlic in olive oil. Add a splash of water and cook until vegetables are soft.

Stir in chili powder, cumin seed, salt, and flour.

Add chicken broth and tomato paste and blend well.

Cook for 5 minutes over low heat.

Stir in turkey, green pepper, mushroons, and olives and heat through. (If turkey is raw, cook until turkey turns white and is fully cooked.)

If sauce is too thick, add water 2 tablespoons at a time until desired consistency.

Serve over rice.

6 servings

Calories per serving: 146 total; 16 fat; 4 sat-fat

Fish and Shellfish

BROILED GINGER FISH

- 1 cup flour
- 1 teaspoon salt (optional)
- ½ teaspoon freshly ground black pepper
- 4 six-ounce fish fillets (monkfish, haddock, etc.)
- 2 teaspoons margarine
- 4 teaspoons diced ginger root
- Lemon slices to cover fillets

Set oven to broil and grease broiler pan with oil.

Combine flour, salt, and pepper on a large plate.

Dredge fillets in flour, covering both sides.

Dot fillets with margarine, sprinkle them with ginger, and cover with lemon slices.

Broil for 5–15 minutes or until fish flakes easily.

4 servings

Calories per serving: 194 total; 35 fat; 4 sat-fat

SHRIMP OR SCALLOP CURRY

Delight your guests or family with this unusual curry. The apples and lime create a unique combination of sweet and sour tastes.

- 1 cup chopped onion
- 1 apple, peeled, cored, and diced
- 2 cloves garlic, minced
- 1–3 teaspoons curry powder
- 1 tablespoon olive oil
- ¼ cup unbleached white flour
- ½ teaspoon salt (optional)
- ¼ teaspoon cardamom
- ¼ teaspoon freshly ground black pepper
- 1 can chicken broth (10¾ oz), defatted
- 1 tablespoon fresh lime juice
- 1¼ pounds bay scallops or shrimp, shelled and deveined
- 1 cup sliced mushrooms
- 10–15 snow peas (optional)
- ½–1 cup water

In a large skillet, sauté onion, apple, garlic, and curry powder in olive oil until tender.

Remove skillet from heat and blend in flour, salt, cardamom, and pepper.

Stir in chicken broth and lime juice until curry sauce is well blended.

Bring curry sauce to a boil, reduce heat, and simmer, uncovered, for about 5 minutes. Stir occasionally.

Meanwhile, place scallops or shrimp in a pot of boiling water and cook until just tender (5–10 minutes). Drain and set aside.

When curry sauce is finished cooking, add shellfish, mushrooms, and snow peas. If sauce is too thick, add water, ¼ cup at a time until desired consistency is achieved. Serve curry over rice.

6 servings

Calories per serving:	*Total*	*Fat*	*Sat-fat*
Shrimp	181	33	6
Scallops	164	27	3

SPICY SHRIMP LOUISIANA

This dish also tastes great with bite-size pieces of chicken instead of shrimp.

1–1¼ pounds shrimp
¼ teaspoon ground white pepper
⅛ teaspoon freshly ground black pepper
¼ teaspoon cayenne pepper
½ teaspoon basil
¼ teaspoon thyme
¼ teaspoon salt (optional)
¼ cup water
¼ cup unbleached white flour
2 teaspoons olive oil
½ cup chopped onion
1 green pepper, chopped
2 stalks celery, chopped
2 cloves garlic, minced
1 can (10¾ oz) chicken broth, defatted
1 tablespoon tomato paste
¼ cup chopped green onions

Shell, devein, and clean shrimp. Cook in boiling water for about 3 minutes. Drain and set aside.

Combine the white, black, and cayenne pepper, basil, thyme, and salt. Set aside.

Slowly add water to flour and mix into a paste. Set aside.

Heat oil in wok or large frying pan until hot.

Stir in onion, green pepper, celery, and garlic and cook until soft.

Mix in spice mixture.

Stir in broth and tomato paste.

Stir in flour mixture and cook until sauce thickens.

Add shrimp and green onions and serve over rice.

6 servings

Calories per serving: 88 total; 28 fat; 2 sat-fat

Vegetables

INDIAN VEGETABLES

Indian Vegetables is one of our all-time favorite recipes. This large pot of colorful, tasty vegetables may be eaten warm or cold. Try it as a main dish served over rice.

2 teaspoons olive oil
1 teaspoon black mustard seeds*
3 cloves garlic, chopped
1 medium onion, chopped
1 green pepper, chopped
2–3 potatoes, peeled and cubed
1 small eggplant, peeled and cubed
1½ teaspoons turmeric
1 teaspoon salt (optional)
¼ cup water
1 teaspoon cumin
1 teaspoon coriander
1 teaspoon garam masala*
Any vegetables, for example:
1 head broccoli, cut into flowerets (about 3 cups)
1 cup or more cauliflower flowerets
6 carrots, sliced
1 cup or more green beans, cut in half
1 cup sliced celery
1 zucchini, sliced
1 cup water

*Available at Indian or Mideastern food stores.

In a large pot, heat olive oil and add black mustard seeds.

When the mustard seeds begin to pop, add garlic, onion, and green pepper and cook until soft.

Stir in potatoes and eggplant.

Add turmeric and salt and mix until vegetables are covered with turmeric sauce.

Add ¼ cup water, reduce heat to low, cover the pot, and cook for 10 minutes.

Stir in cumin, coriander, garam masala, and vegetables.

Add 1 cup water and increase heat to medium.

After 10 minutes, lower heat and cook until vegetables are tender.

12 servings

Calories per serving: 72 total; 7 fat; 1 sat-fat

STEAMED VEGETABLES WITH DIJON-YOGURT SAUCE

Vegetable, about ¾–1 cup per person

1 cup plain yogurt

2–4 teaspoons Dijon mustard

The Dijon-Yogurt Sauce is such a super easy and delicious sauce, you can eat it every day with any steamed vegetable.

Vegetable

Prepare vegetable for steaming by peeling and slicing if necessary or cutting off the stalk and cutting into bite-size flowerets.

Place about 1 inch of water in a saucepan and turn heat to high. Put in metal steamer, spreading its "petals." Place vegetable on steamer.

Cover saucepan and heat water to boiling. Lower heat and steam vegetable until tender.

Sauce

When making the sauce, mix 2 teaspoons mustard into the yogurt and taste. If you want a stronger taste, add more mustard. Place a tablespoon or two of the sauce over cooked vegetable.

1 cup of sauce

Calories per tablespoon: 8 total; 0 fat; 0 sat-fat

Cooking with Eggplant

Choose the blackest eggplant you can find. Peel it, cut it into bite-size pieces, and salt them heavily. Place a heavy plate on top of the pieces to help squeeze out the bitter juices. In 30–60 minutes, wash off the salt and gently squeeze eggplant pieces to rid them of bitter juices.

Steam eggplant in a vegetable steamer until tender. There are two advantages to steaming eggplant: steamed eggplant needs no oil (eggplant absorbs an enormous amount of oil) and thus saves hundreds of fat calories; and you can test the precooked eggplant for bitterness before using it in a recipe.

HOT AND GARLICKY EGGPLANT

1 medium eggplant (about 1 pound)
5 small, dried black Chinese mushrooms*
1 tablespoon chili paste with garlic*
1 tablespoon vinegar
½ tablespoon soy sauce
½ tablespoon double black soy sauce*
2 tablespoons dry sherry
½ teaspoon sugar
1 large green pepper, chopped
1 teaspoon olive oil
½ cup water

Prepare eggplant as explained in the box.

Place mushrooms in a small bowl and cover with boiling water.

After about 15 minutes remove mushrooms. Squeeze out the excess water and discard stems. Slice mushrooms. Set aside.

Combine chili paste with garlic, vinegar, soy sauces, sherry, and sugar and set aside.

In a large skillet, sauté green pepper and mushrooms in oil. Add a splash of water and cook vegetables until tender.

Stir in eggplant.

---

*Available at Asian food stores.

Mix in soy sauce mixture until vegetables are covered, and then stir in water.

Simmer for about 5 minutes.

8 half-cup servings

Calories per serving: 25 total; 5 fat; 1 sat-fat

CURRIED WHIPPED POTATOES

You need not add butter or margarine to make delectable whipped potatoes. In this recipe, sautéed onions, mustard seed, and cumin make them special and exotic. Leave the onions out of this recipe and you have plain but scrumptious whipped potatoes at 0 fat calories per serving.

4 potatoes (about 2 pounds)
¾ cup chopped onions
1 teaspoon olive oil
½ teaspoon mustard seed
½ teaspoon cumin
1–4 tablespoons skim milk
Salt and freshly ground black pepper to taste

Peel potatoes, slice thinly, and place in a medium saucepan with water to cover. Bring to a boil and cook until tender.

Meanwhile, sauté onions in olive oil. Lower heat to medium and stir in mustard seed and cumin. Cook a few moments until onions are soft.

Drain potatoes. Beat with an electric mixer. Add milk, one tablespoon at a time, until potatoes are whipped.

Mix in onions and salt and pepper to taste.

4 servings

Calories per serving: 168 total; 10 fat; 1 sat-fat

POTATO SKINS

Leslie Goodman-Malamuth of the Center for Science in the Public Interest devised this recipe as a substitute for the fat-filled potato skins you often find in restaurants. Not only is this recipe quick and easy, the resulting potatoes are scrumptious.

4 large potatoes
1 teaspoon olive oil (optional)
Paprika to taste

Preheat oven to 450°F.

Scrub potatoes well, cut them lengthwise into six wedges the size and shape of dill pickle spears, and dry them on a paper towel.

In a large bowl, toss potato spears with olive oil until they are well covered.

Spread potatoes on a baking sheet, dust them with paprika, and bake for 20–30 minutes or until fork-tender.

6 servings (so good, two people can easily finish them off!)
Calories per serving: 66 total; 7 fat; 1 sat-fat

SWEET POTATOES À L'ORANGE

Keep sweet potatoes on hand so you can whip up this wonderful, EASY recipe at the last moment. In the time it takes to boil the potatoes, you have created a healthy, delicious side dish.

2 pounds sweet potatoes
½ cup orange juice
1 orange (optional)

Peel and cut sweet potatoes or yams into ½-inch slices. Place in saucepan with water to cover and boil until tender (about 20–30 minutes). Drain.

With an electric beater, begin beating sweet potatoes.

Add orange juice (slowly) and beat until creamy. Add more orange juice than you think you will need so the potatoes will be fluffy, but do it slowly so that you don't drown them.

Cut orange into pieces and mix by hand into whipped sweet potatoes.

8 half-cup servings

Calories per serving: 125 total; 0 fat; 0 sat-fat

ZUCCHINI MATCHSTICKS

So light, so tasty — you don't even need to add salt, spices, or fat. You may want to crush garlic into 1 teaspoon and combine it with the vegetables.

2 small zucchini (or ½ small zucchini per person)
1 thick carrot, peeled
1 teaspoon margarine (optional)
1 clove garlic (optional)

Cut zucchini and carrot into 2-inch lengths.

Place a zucchini section on a cutting surface, skin-side down.

Holding the sides of the section, slice lengthwise at ⅛-inch intervals. Hold slices together.

Roll the section one-quarter turn, making sure the slices stay together.

Again, make parallel slices, ⅛ inch apart lengthwise.

Result: zucchini matchsticks.

Repeat for remaining sections of zucchini and carrot.

Place zucchini sticks on top of carrot sticks in a vegetable steamer and steam until just tender (about 1 or 2 minutes).

4 servings

Calories per serving:	*Total*	*Fat*	*Sat-fat*
	17	0	0
With margarine	24	7	2

Focaccia, Chili, and Pasta

FOCACCIA

The whole-wheat flour in this pizza not only makes it healthier but also gives it a toasty taste and crunchy texture. Try it! It is easy and makes great snacks. Freeze it in snack-size slices.

Dough

1½ teaspoons dry active yeast
½ teaspoon honey
1 cup warm water
2½ cups whole-wheat flour
¾ teaspoon salt (optional)
1 tablespoon olive oil

Tomato Sauce

1 large can (28 oz) tomatoes, or 3 large tomatoes
2 cloves garlic, minced
1 small onion, chopped
1 teaspoon olive oil
½ teaspoon oregano
¼ teaspoon basil
Ground hot cherry peppers (optional)

The Dough

Place yeast, honey, and warm water in food processor or large bowl. Let proof.

Add flour, salt, and olive oil and process or knead until smooth and elastic, adding flour if needed.

Place dough in an oiled bowl, cover with a towel, and let rise in a warm place for about 1 hour.

Punch down dough and let it rest on floured counter for 10 minutes.

Grease a 10×15-inch cookie sheet with oil. Roll out dough (or press with your hands) onto the cookie sheet.

Pinch a rim around the edge. Cover with a towel and let rise for 30 minutes.

Meanwhile, make the tomato sauce.

The Tomato Sauce

If using canned tomatoes, drain liquid and chop tomatoes. If using whole tomatoes, chop fine.

In a medium skillet, sauté garlic and onion in olive oil until tender.

Stir in tomatoes, oregano, and basil and let simmer until thick (about 10–15 minutes).

Let cool.

Baking the Focaccia

Preheat oven to 400°F.

Spread the sauce over the dough.

Add hot cherry peppers, if desired.

Bake for 20–25 minutes.

12 pieces

Calories per piece: 113 total; 18 fat; 3 sat-fat

CHILI NON CARNE

This is a great chili recipe. It is filled with nutritious vegetables that provide texture but do not interfere with the delicious chili taste. Eat it hot in a bowl mixed with chopped onion, tomatoes, and lettuce over rice or spoon it into pita bread with chopped onion, lettuce, and tomatoes.

¾ cup chopped onion
2 cloves garlic, minced
1 teaspoon olive oil
2 tablespoons chili powder
¼ teaspoon basil
¼ teaspoon oregano
¼ teaspoon cumin
2 cups finely chopped zucchini
1 cup finely chopped carrot
1 large can (28 oz) tomatoes + 1 small can (14½ oz) tomatoes, drained and chopped
1 can (15 oz) kidney beans, *undrained*
2 cans (15 oz each) kidney beans, *drained* and thoroughly rinsed
Chopped onions, tomatoes, lettuce, green peppers, for garnish

In a large pot, sauté onion and garlic in olive oil. Add a splash of water and cook vegetables until soft.

Mix in chili powder, basil, oregano, and cumin.

Stir in zucchini and carrots until well blended. Cook for about 1 minute over low heat, stirring occasionally.

Stir in chopped tomatoes, undrained and drained kidney beans.

Bring to a boil. Reduce heat and simmer for 30–45 minutes or until thick.

Top with chopped onions, tomatoes, lettuce, or green peppers.

8 one-cup servings

Calories per serving: 117 total; 5 fat; 1 sat-fat

ASPARAGUS PASTA

Asparagus Pasta makes a delightful luncheon dish, first course for an elegant meal, or light dinner. No one consuming this dish will believe how incredibly simple it is to make.

1 pound asparagus, sliced into 1–2-inch pieces
3 tablespoons Dijon mustard
1 tablespoon olive oil
¼ cup thinly sliced shallots
1 clove garlic, minced
2 anchovy fillets
¼ teaspoon thyme
2 tablespoons chopped parsley
¾ pound very thin spaghetti or pasta of your choice
Salt and freshly ground black pepper to taste
1 cup sliced mushrooms

In a large pot of boiling water, cook asparagus until tender and still bright green (about 3 minutes).

Combine mustard, olive oil, shallots, garlic, anchovies, thyme, and parsley. Set aside.

Cook pasta. Drain, but reserve 1 cup of the cooking water.

Combine pasta with dressing. Add asparagus and mushrooms. Mix well. Add some of the cooking water if pasta is too dry. Add salt and pepper to taste.

Variation: Follow the original recipe but add two boneless, skinless chicken breasts that have been steamed and cut into bite-size pieces.

4 servings

Calories per serving:	*Total*	*Fat*	*Sat-fat*
	253	26	4
With chicken	296	30	5

PASTA WITH SPINACH-TOMATO SAUCE

One, two, three, and you have a delicious, healthy meal.

4 cloves garlic, pressed
1½ cups chopped onions
1 teaspoon olive oil
1 large can (28 oz) crushed tomatoes
1 tablespoon balsamic vinegar
½ teaspoon rosemary
¼ teaspoon cayenne pepper
¼ teaspoon basil
¼ teaspoon oregano
Freshly ground black pepper to taste
5 cups spinach, torn
1 tomato cut into bite-size chunks
Pasta of your choice

In a large skillet, sauté garlic and onions in olive oil until soft. Add a splash of water to help cook the vegetables.

Stir in crushed tomatoes. Lower heat to medium.

Mix in balsamic vinegar, rosemary, cayenne, basil, oregano, and pepper.

Add spinach and stir until spinach is covered with sauce.

Add tomato.

Prepare pasta as directed on package. Drain.

Spoon sauce over pasta.

6 cups

Calories per cup of sauce: 52 total; 7 fat; 1 sat-fat

Salads and Sandwiches

BLACK BEAN AND CORN SALAD

4 cups frozen corn
2 cans (15 oz) black or other beans
1 cup chopped red and/or green pepper
½ cup white vinegar
½ teaspoon salt (optional)
¼ cup pickled jalapeño pepper (optional)
½–1 teaspoon freshly ground black pepper
½–1 teaspoon chili powder

Cook corn until tender according to package directions.
Drain and rinse beans and place in bowl or casserole.
Mix in peppers.
Drain cooked corn and add to bean mixture.
Combine vinegar and spices and pour over corn and beans.
Mix well. Cover and refrigerate several hours.
12 half-cup servings
Calories per cup: 102 total; 10 fat; 2 sat-fat

CUCUMBER SALAD

2 cucumbers, peeled and sliced thin
1 medium onion, sliced thin
½ teaspoon salt (optional)
1 teaspoon sugar
1 teaspoon dillweed
1 cup white vinegar

Mix cucumbers and onion together in a ceramic or glass bowl.

Add salt, sugar, and dillweed to vinegar and pour over cucumbers and onion.

Chill 1 hour.

6 servings
Calories per serving: 20 total; 0 fat; 0 sat-fat

RED POTATO SALAD

4–5 red potatoes
½ cup nonfat yogurt
3 tablespoons low-fat mayonnaise
1 teaspoon tarragon
1 tablespoon white vinegar
1 teaspoon Dijon mustard
½ teaspoon salt (optional)

Scrub potatoes thoroughly. Steam until tender.
Meanwhile, combine remaining ingredients. Set aside.
Cut potatoes into chunks (do not remove skin).
Pour sauce over potatoes so they are thoroughly coated.
Serve warm or cold.
6 servings
Calories per serving: 95 total; 5 fat; 0 sat-fat

CHICKPEA SANDWICH OR DIP

1 can chickpeas
2 cloves garlic
⅓ cup parsley
1 tablespoon tahini (sesame seed paste)*
Juice of 1 lemon (about ¼ cup)
4 six-inch whole-wheat pita bread pockets
Chopped tomatoes, green onions, and lettuce, for garnish

Drain chickpeas. Reserve liquid. In blender or food processor, chop garlic and parsley.

Add chickpeas, tahini, and lemon juice.

Blend until smooth, adding more chickpea liquid if spread is too stiff.

Make sandwiches by spooning chickpea spread into pita bread pockets.

Garnish with chopped tomatoes, green onions, and lettuce.

4 sandwiches

*Available at Mideastern food shops and many supermarkets.

Use as a dip for vegetables, crackers, squares of pita, etc.

About 1½ cups

Calories	*Total*	*Fat*	*Sat-fat*
Per sandwich	195	35	4
Per tablespoon	20	4	1

TED MUMMERY'S TURKEY BARBECUE

We discovered a most wonderful turkey barbecue while at a health conference in Steven's Point, Wisconsin. Ted Mummery, the owner of La Claire's Frozen Yogurt, Inc., has kindly allowed us to share his recipe with you. It's delicious in a whole-wheat pita pocket, on an onion bagel, or any other bread.

5–6 lbs turkey breast
4 cans (14½ oz each) stewed tomatoes
2 large cans (12 oz each) tomato paste
1 cup water
½ cup brown sugar
⅓ cup dark molasses
½ cup apple cider vinegar
1 teaspoon hickory salt* (optional) or 1 teaspoon salt
1 heaping teaspoon garlic powder
1½ heaping teaspoons onion powder
1 heaping teaspoon basil
1½ heaping teaspoons oregano
1 heaping teaspoon cayenne
1 teaspoon all-purpose seasoning
½ teaspoon white pepper

Preheat oven to 350°F.

Remove skin and fat from turkey breast.

Bake turkey with breast down until fully cooked, about 1½–2½ hours. (You may do this the night before.) Refrigerate.

Mix tomatoes, tomato paste, and water in blender or food proces-

---

*Ted recommends 3 teaspoons of hickory salt. If you are eating this barbecue often, I suggest eliminating the hickory salt as it contains an unhealthy ingredient — smoke (charcoal).

sor for a few moments. You want to chop the tomatoes, not purée them.

Place in 5-quart Crockpot or slow cooker. (You may also cook barbecue in a large pot on the stove.) Stir in remaining ingredients. Set Crockpot on automatic or cook at low heat for about 2 hours.

Pull turkey from bones and cut into bite-size pieces. Add turkey to sauce and cook for about another hour.

30 half-cup servings
Calories per serving: 131 total; 5 fat; 1 sat-fat

Breads, Muffins, and Waffles

CUMIN WHEAT BREAD

Although this is a totally whole-wheat bread, it is not the least bit heavy. It is a favorite with young children as well as adults.

1 tablespoon dry yeast
¼ cup warm water
2 tablespoons honey
3½ cups whole-wheat flour
1 teaspoon salt
½ teaspoon whole cumin seeds
1½ tablespoons olive oil
1 cup skim milk

In a large mixing bowl or the bowl of a food processor, combine yeast, water, and honey. Let proof.

Mix in 2 cups of the flour, the salt, and cumin seeds.

Add oil and milk and mix or process well, adding more flour until dough is fairly stiff.

Knead dough 10 minutes or process for 15 seconds or until dough is smooth and elastic, adding more flour if necessary.

Form dough into a ball. Place in an oiled bowl. Cover and let rise for about 1 hour.

Roll dough into a loaf and place in loaf pan greased with margarine. Cover with towel and let rise for 1 hour.

Preheat oven to 375°F and bake for 50–60 minutes or until bread is golden and sounds hollow when tapped.

1 loaf

Calories	*Total*	*Fat*	*Sat-fat*
Per ½-inch slice	107	15	2
Per loaf	1814	252	35

ONION FLAT BREAD

1 tablespoon active dry yeast
1 pinch sugar
1 cup warm water
2½–3 cups unbleached white flour*
1 teaspoon + a few shakes salt
1 teaspoon olive oil
1 cup chopped onions
1 teaspoon paprika

Place yeast, sugar, and water in large bowl or a food processor. Let proof.

Mix in 2 cups flour and salt. Knead for 10 minutes or process for 15 seconds, until smooth and elastic, adding flour if necessary.

Place dough in an oiled bowl, cover with a towel, and let rise in a warm place for an hour or until doubled in bulk.

Punch dough down and split in half. Let rest for 5 minutes.

Meanwhile, grease two 9-inch cake pans with margarine.

Press dough into cake pans.

Spread olive oil over the tops and press onions into the surface.

Let rise about 45 minutes or until doubled in bulk.

Preheat oven to 450°F.

Sprinkle tops with paprika and a few shakes of salt.

Bake 20–25 minutes, until lightly browned.

2 loaves, 8 slices per loaf

Calories	*Total*	*Fat*	*Sat-fat*
Per slice	83	4	0
Per loaf	667	32	3

*For a slightly different taste, use 1 cup of whole-wheat flour, along with 1½–2 cups unbleached white flour.

APRICOT OAT MUFFINS

Healthy and delicious — a good snack or breakfast on the run.

½ cup orange juice
1 cup dried apricots, chopped (easily done in food processor)
¼ cup brown sugar
1 cup oat bran
¼ cup wheat germ
¾ cup whole-wheat flour
1 teaspoon baking powder
1 teaspoon baking soda
2 tablespoons applesauce
½ cup skim milk
1 egg

Preheat oven to 400°F. Grease a 12-cup muffin tin with margarine.

In a small saucepan, heat orange juice until boiling. Mix in apricots and brown sugar.

Remove saucepan from heat and cool apricot mixture slightly.

In a medium bowl, combine oat bran, wheat germ, whole-wheat flour, baking soda, and baking powder. Set aside.

In a mixing bowl, beat together applesauce, skim milk, and egg.

Add dry ingredients and apricot–orange juice mixture to milk mixture and mix until just moistened.

Fill muffin tin and bake for 15 minutes or until muffins are golden brown and a cake tester comes out clean.

12 muffins

Calories per muffin: 114 total; 10 fat; 1 sat-fat

CINNAMON FRENCH TOAST

2 egg whites
3 tablespoons skim milk
½ teaspoon vanilla
½ teaspoon ground cinnamon
Pinch of grated nutmeg
3 slices whole-wheat bread or French bread
2 teaspoons margarine (optional)

In a shallow dish, mix egg whites, skim milk, vanilla, cinnamon, and nutmeg.

Soak both sides of bread in mixture.

Heat a large frying pan or cast-iron griddle.* Spread margarine over it if necessary.

Add bread. Reduce heat to medium.

Turn bread after 2 minutes. Cook until golden brown and crispy.

3 servings

Calories per serving	*Total*	*Fat*	*Sat-fat*
Without margarine	75	6	0
With margarine	95	26	4

BUTTERMILK WAFFLES

1 cup whole-wheat flour
1 cup unbleached white flour
2 teaspoons baking powder
½ teaspoon salt
1 cup buttermilk
1 cup skim milk
1 egg white
2 teaspoons olive oil

Spray waffle iron with cooking spray and preheat.

In a medium bowl, combine whole-wheat flour, white flour, baking powder, and salt.

Add buttermilk and skim milk. *Do not overmix.*

Fold in egg white and oil.

Place batter on waffle iron in amounts specified by your waffle-iron instructions. Cook accordingly.

5 Belgian waffles

Calories	*Total*	*Fat*	*Sat-fat*
Total batter	1102	107	16
Per square**	220	21	3

---

*Use a cast-iron griddle, which requires little or no margarine to keep the French toast from sticking, to reduce the sat-fat and total fat even more.

**Because waffle irons come in many sizes, this recipe may make more or less than 5 waffle squares. Divide 1102 by the number of waffle squares you make to figure out the total calories per square (for example, 1102 ÷ 5 = 220). Divide 107 by the number of waffle squares to figure out the fat calories per square (for example, 107 ÷ 5 = 21).

Desserts

THE YORK BLUEBERRY CAKE

½ cup brown sugar
1 teaspoon cinnamon
2 cups unbleached white flour
1 teaspoon baking soda
½ teaspoon salt
2 cups blueberries
½ cup applesauce
¾ cup sugar
1 egg
2 egg whites
1 cup nonfat yogurt
1 teaspoon vanilla

Preheat oven to 350°F and grease a 13×9-inch baking pan with margarine.

Combine brown sugar and cinnamon and set aside.

In a large bowl, combine flour, baking soda, and salt.

Mix ¼ cup of flour mixture with blueberries and set aside.

In a large mixing bowl, cream applesauce and sugar until fluffy.

Beat in egg and egg whites.

Stir in yogurt and vanilla.

Add remaining flour to batter and beat until smooth.

Stir in floured blueberries.

Spread half of batter into baking pan and sprinkle half of brown sugar mixture over it.

Cover with remaining batter (it will barely cover the entire surface) and top with remaining brown sugar mixture.

Bake for 30–45 minutes until cake tester comes out clean.

20 squares

Calories per square: 113 total; 3 fat; 1 sat-fat

LEMON CHEESECAKE

Cheesecake doesn't have to be packed with fat to be good. This delectable, light cheesecake is simple to make, low-low-low in fat, and delicious.

Lemon zest* of one lemon, chopped
3 egg whites
⅔ cup sugar
¼ cup cornstarch
¼ teaspoon vanilla
2⅔ cup low-fat (1%) cottage cheese
8 low-fat graham crackers (about 2 inches × 2 inches) to make ⅔ cup graham cracker crumbs**

Preheat the oven to 325°F and grease an 8-inch springform pan with margarine. You are going to set the cake pan in a pan of water to bake so you will also need an ovenproof pan large enough to hold the springform pan.

Fill a pot or teakettle with water and start heating it to a boil.

Place lemon zest, egg whites, sugar, cornstarch, vanilla, and cottage cheese in a blender and blend until very, very smooth (no lumps). Pour into springform pan.

Place springform pan in the middle of the larger pan.

Carefully pour boiling water into large pan so that it reaches about halfway up the sides of the springform pan.

Bake for 60 minutes or until cheesecake is set and cake tester comes out clean.

Remove pan from water and refrigerate cake in pan for at least 4 hours or overnight. You want it to be really cold.

In a food processor, turn graham crackers into ⅔ cup fine crumbs. After cake has been refrigerated 4 or more hours, spread crumbs evenly over the cake surface and gently pat them down. You may do this just before serving.

---

*The yellow of the lemon peel is called the zest. You can peel it off and chop it in a mini-chopper or your food processor, use a zester and chop by hand, or grate the lemon with a hand grater.

**Use commercial low-fat graham crackers or make *Honey Graham Crackers*, page 303.

Before serving, carefully run knife around edge of springform pan and remove sides.

8 slices

Calories per slice: 160 total; 10 fat; 3 sat-fat

PEACH SOUR CREAM CAKE

3 peaches
1 teaspoon sugar
2 cups cake flour
1 teaspoon baking powder
1 teaspoon baking soda
¼ teaspoon salt
½ cup applesauce
⅔ cup sugar
1 egg + 1 egg white
½ teaspoon vanilla extract
¼ teaspoon almond extract
1 cup nonfat sour cream

Preheat oven to 350°F. Grease an 8-inch springform pan with margarine or cooking spray.

Prepare peaches. Drop the peaches into a pot of boiling water. They should be completely covered. After about 1–2 minutes remove a peach and run it under cold water. Peel the peach. Follow the same procedure for the other peaches.

Halve each peach, take out pit, and thinly slice. Mix with teaspoon of sugar and set aside.

In a bowl, combine flour, baking powder, baking soda, and salt. Set aside.

Using an electric mixer, cream the applesauce and sugar until fluffy.

Add the egg and egg white and mix until smooth.

Add vanilla and almond extract and mix until smooth.

Beating on low speed, add sour cream and flour mixture alternately and mix until just combined. Don't over mix.

Pour into springform pan.

Place peach slices evenly over the batter so that they cover the entire surface.

Bake for 45–50 minutes or until an inserted cake tester comes out clean.

Remove rim and let cake cool.
10 slices
Calories per square: 174 total; 5 fat; 2 sat-fat

GRAHAM CRACKER CRUST

Not only do honey graham crackers make delicious cookies, the crumbs make a delicious pie crust, particularly for lemon meringue pie, *Key Lime Pie,* or *Lemon Cheesecake* (page 298). Commercial graham crackers, graham cracker crumbs, and ready-to-bake graham cracker crusts may have more fat than you want. Make your own Honey Graham Crackers, using the recipe on page 303. To reduce the fat in this recipe, use no margarine.

8 homemade graham crackers, crumbled (1¼ cups)
1 teaspoon sugar
1 tablespoon margarine

Preheat oven to 375°F.
In food processor or blender, finely crush 8 graham crackers.
Combine crumbs and sugar and pour into pie plate.
Melt margarine and mix into crumbs.
Press crumbs into pie plate to make a crust.
Bake 8 minutes, or until lightly browned.
Makes one 9-inch pie crust
Calories per crust: 600 total; 186 fat; 34 sat-fat
Without margarine: 510 total; 96 fat; 16 sat-fat

KEY LIME PIE

This pie is as pretty to look at as it is delightful to eat. It never fails to wow guests and please the most discriminating palate. Many Key lime pies call for 3 egg yolks and butter. You will use only 1 egg yolk and margarine to create this divine dessert. You can use lemon juice instead of lime juice and call this Lemon Meringue Pie.

9-inch Graham Cracker Crust (page 300)

Filling

¾ cup sugar
¼ cup unbleached white flour
3 tablespoons cornstarch
¼ teaspoon salt
½–⅔ cup fresh lime juice + water to equal 2¼ cups
1 egg yolk
1 teaspoon margarine

Meringue

5 egg whites, at room temperature
¼ teaspoon cream of tartar
½ cup + 2 tablespoons sugar

The Filling

In a medium saucepan, thoroughly mix sugar, flour, cornstarch, and salt.

Add ¼ cup lime water and blend into a smooth paste.

Add remaining lime water and mix until smooth.

Stir filling over medium heat until it begins to boil and thicken. Remove saucepan from heat.

In a small bowl, combine egg yolk with a small amount of filling and blend until smooth. Mix back into filling in saucepan. (This step is important. If you add the egg yolk directly into the hot filling, the egg yolk will curdle.) Stir margarine into the filling and pour into crust. Cool until filling gels.

The Meringue

Preheat oven to 375°F.

Whip egg whites with cream of tartar until stiff but not dry.

Add sugar and continue beating until whites form stiff peaks.

Gently cover lime filling with egg whites, making sure whites cover pie completely.

Bake for 8–10 minutes or until meringue is golden brown.

Cool at room temperature.

10 slices

Calories per slice: 201 total; 27 fat; 6 sat-fat

PUDDING BROWNIES

Batter

1 cup unbleached white flour
½ cup sugar
¼ cup unsweetened cocoa
2 teaspoons baking powder
½ teaspoon salt
½ cup skim milk
¼ cup applesauce
1 teaspoon vanilla extract

Topping

3 tablespoons unsweetened cocoa
⅓ cup brown sugar
1 cup hot water

Batter

Preheat oven to 350°F. Grease an 8×8-inch baking pan with margarine or a cooking spray.

In a large mixing bowl, mix together flour, sugar, cocoa, baking powder, and salt.

Stir in milk, applesauce, and vanilla extract.

Spoon batter into pan.

Topping

In a small bowl, combine cocoa and brown sugar. Mix in hot water. Pour mixture over batter in pan.

Bake for 30 minutes or until cake tester comes out clean.

Cool brownies in pan.

Remove to serving plate. Take excess pudding from pan bottom and "ice" brownies with it.

Serve warm or refrigerate.

16 servings

Calories per square: 83 total; 2 fat; 0 sat-fat

COCOA MERINGUE KISSES

These delectable morsels should be eaten right from the oven. Pop a whole kiss in your mouth and savor the taste of warm, melting chocolate.

2 egg whites
⅛ teaspoon cream of tartar
¾ cup sugar
1 teaspoon vanilla
2 tablespoons cocoa

Preheat oven to 325°F. Line a cookie sheet with foil, dull side up.

With an electric beater, whip egg whites and cream of tartar until foamy.

Add sugar a tablespoon at a time. Add vanilla. Beat until beater leaves stiff peaks.

Fold in cocoa.

Drop by teaspoonfuls in the shape of a candy kiss onto foil.

Bake 20 minutes. In a few minutes, slide a knife or spatula under the cookies and they will lift off foil.

Pop into mouth. (Use some restraint. Don't eat them all, as they add up to a lot of empty sugar calories.)

About 32 kisses

Calories per kiss: 20 total; 0 fat; 0 sat-fat

HONEY GRAHAM CRACKERS

1 cup whole-wheat flour
½ cup unbleached white flour
½ teaspoon baking powder
¼ teaspoon baking soda
Pinch of salt
2 tablespoons margarine
2 tablespoons light brown sugar
2 tablespoons honey
½ teaspoon vanilla
2 tablespoons skim milk

Preheat oven to 350°F and grease a cookie sheet with margarine.

Combine flours, baking powder, baking soda, and salt, and set aside.

Cream margarine, sugar, and honey in an electric mixer.

Mix in vanilla.

Add flour mixture and milk.

Gather dough together (add a drop or two of milk if too dry) and knead into a ball.

Roll dough onto cookie sheet in a rectangle, ⅛ inch thick.

If dough is too sticky, sprinkle it with flour.

Without moving dough, cut into 3-inch squares.

Lightly score a line through the center of each square and pierce each side several times with a fork.

Bake 10–15 minutes, until edges brown. Remove crackers and cool on a wire rack.

Crackers will become crisp as they cool.

18 crackers

Calories per cracker: 58 total; 12 fat; 2 sat-fat

Food Tables

CONTENTS

INTRODUCTION TO THE FOOD TABLES

The Food Tables in *Choose to Lose* contain both saturated-fat calories and total fat calories. Total fat is the sum of saturated fat, monounsaturated fat, and polyunsaturated fat. Each has a different effect on heart health, but all are equally fattening. Although in most cases reducing total fat will automatically reduce saturated fat, there are certain instances where also knowing saturated fat calories will help you make healthier choices. For instance, both olive oil, the most heart-healthy of all the fats, and butter, a highly heart-risky food, are 100 percent fat and fattening. A tablespoon of olive oil, which has 119 total fat calories, has only 16 sat-fat calories. A tablespoon of butter has 100 fat calories, but has 65 sat-fat calories. By using the Food Tables, you can quickly see that butter is a poorer choice because it is very saturated.

Because products are constantly changing, always look at nutrition labels for the most up-to-date nutrition information.

The abbreviation "NA" means the sat-fat calories are "not available."

BEVERAGES

		CALORIES		
FOOD	**AMOUNT**	**TOTAL**	**FAT**	**SAT-FAT**
ALCOHOLIC				
Amaretto di Saronno	1 fl oz	82	**0**	0
Anisette	1 fl oz	92	**0**	0
Beer & Ale	12 fl oz	145–165	**0**	0
Lite	12 fl oz	70–110	**0**	0
Bourbon	1 fl oz	70	**0**	0
Brandy	1 fl oz	65	**0**	0
Bloody Mary	5 fl oz	120	**0**	0
Brandy Alexander	3 fl oz	254	**52**	32
Brandy, liqueur, all fruit flavors	1 fl oz	88–100	**0**	0

BEVERAGES

FOOD	AMOUNT	CALORIES TOTAL	CALORIES FAT	CALORIES SAT-FAT
Champagne	5 fl oz	125	**0**	0
Cherry Heering	1 fl oz	80	**0**	0
Crème de banane	1 fl oz	95	**0**	0
Crème de cacao	1 fl oz	100	**0**	0
Crème de cassis	1 fl oz	90	**0**	0
Crème de menthe	1 fl oz	95	**0**	0
Daiquiri	2 fl oz	120	**0**	0
banana	3 fl oz	155	**2**	0
Drambuie	1 fl oz	110	**0**	0
Eggnog	8 fl oz	340	**170**	100
Gimlet	2 fl oz	110	**0**	0
Gin	1 fl oz	75	**0**	0
Gin and tonic	8 fl oz	195	**0**	0
Grasshopper	3 fl oz	295	**52**	32
Irish coffee	8 fl oz	210	**99**	60
Kahlúa	1 fl oz	105	**0**	0
Manhattan	4 fl oz	290	**0**	0
Martini	3 fl oz	225	**0**	0
Piña colada	6 fl oz	392	**105**	85
Port	4 fl oz	185	**0**	0
Rob Roy	3 fl oz	195	**0**	0
Rum	1 fl oz	65	**0**	0
Rye	1 fl oz	70	**0**	0
Scotch	1 fl oz	70	**0**	0
Screwdriver	6 fl oz	160	**0**	0
Sherry				
cream	4 fl oz	185	**0**	0
dry	4 fl oz	140	**0**	0
Sloe gin	1 fl oz	180	**0**	0
Sloe gin fizz	4 fl oz	70	**0**	0
Stinger	2 fl oz	145	**0**	0
Tequila	1 fl oz	80	**0**	0
Tequila sunrise	7 fl oz	220	**0**	0
Vermouth				
dry	1 fl oz	65	**0**	0
sweet	1 fl oz	75	**0**	0
Vodka	1 fl oz	65	**0**	0
Whiskey	1 fl oz	65–80	**0**	0

BEVERAGES

FOOD	AMOUNT	CALORIES TOTAL	CALORIES FAT	CALORIES SAT-FAT
Whiskey sour	1 fl oz	130	**0**	0
White Russian	3.5 oz	290	**26**	16
Wine				
red	4 fl oz	85–100	**0**	0
rosé	4 fl oz	90–130	**0**	0
white	4 fl oz	85–100	**0**	0
Mixers				
grenadine	1 fl oz	65	**0**	0
Snap-E-Tom	6 fl oz	38	**0**	0
CARBONATED DRINKS				
Club soda	12 fl oz	0	**0**	0
Cola type	12 fl oz	160	**0**	0
Ginger ale	12 fl oz	125	**0**	0
Lemon-lime	12 fl oz	150	**0**	0
Orange, grape	12 fl oz	180	**0**	0
COCOA				
Hershey's cocoa	1 tbsp	45	**5**	0
Hershey's chocolate				
milk mix	3 tbsp	90	**0**	0
made with whole milk	6 fl oz	135	**58**	34
made with 2% milk	6 fl oz	113	**35**	20
made with skim milk	6 fl oz	87	**5**	0
Hershey Hot Cocoa Collection				
Chocolate Raspberry	1 env. (35 g)	150	**25**	5
Dutch Chocolate	1 env. (35 g)	150	**25**	0
Fat-Free	1 env. (25 g)	90	**0**	0
Good Night Kisses	1 env. (35 g)	150	**30**	9
Nestlé Carnation				
Double Chocolate Meltdown	1 env. (35 g)	150	**30**	23
Fat-Free	1 env. (8 g)	25	**0**	0
Rich Chocolate	1 env. (28 g)	120	**30**	18
Nestlé Quick				
Chocolate	2 tbsp (22 g)	90	**5**	5
Strawberry	2 tbsp (22 g)	90	**0**	0

BEVERAGES

FOOD	AMOUNT	CALORIES TOTAL	CALORIES FAT	CALORIES SAT-FAT
Safeway Hot Cocoa				
Mix	1 packet (28 g)	110	**15**	0
Swiss Miss				
Chocolate Sensations	1 env. (34 g)	150	**35**	14
Cocoa & Cream	1 env. (34 g)	150	**45**	27
Marshmallow Lovers	1 env. cocoa + 1 env. marshmallow (34 g)	140	**25**	9
COFFEE (for coffee bar coffees, *see* **RESTAURANT FOODS)**				
Nescafé Cappuccino	1 env. (27 g)	110	**20**	5
International Coffees (General Foods)				
Cafe Français	8 fl oz prep.	60	**30**	9
Cafe Vienna	8 fl oz prep.	70	**20**	5
French Vanilla Cafe	8 fl oz prep.	60	**25**	5
Italian Cappuccino	8 fl oz prep.	60	**15**	5
Swiss Mocha	8 fl oz prep.	60	**20**	5
FRUIT DRINKS				
Noncarbonated				
canned	6 fl oz	85–100	**0**	0
frozen	6 fl oz	80	**0**	0
HOT CHOCOLATE				
Without whipped cream	8 fl oz	232	**122**	72
With whipped cream	¼ cup	334	**221**	133
INSTANT BREAKFAST SHAKES				
Carnation				
Creamy Milk Chocolate	1 env. (37 g)	130	**10**	5
Creamy Milk Chocolate (sugar-free)	1 env. (21 g)	70	**10**	5
Ovaltine (all types)	4 tbsp (21 g)	80	**0**	0
Postum	1 tsp (3 g)	10	**0**	0

BEVERAGES

FOOD	AMOUNT	CALORIES TOTAL	CALORIES FAT	CALORIES SAT-FAT
SOYBEAN DRINKS				
Soy Protein Drink	8 fl oz	90	**50**	9
Soy Drink Light	8 fl oz	90	**20**	5
TEA				
Tea	8 fl oz	0	**0**	0

Dairy drinks (milk, milk shakes, etc.): *see* **DAIRY AND EGGS; FAST FOODS.**
Fruit juices: *see* **FRUITS AND FRUIT JUICES.**

DAIRY AND EGGS

FOOD	AMOUNT	CALORIES TOTAL	CALORIES FAT	CALORIES SAT-FAT
BUTTER				
Regular	1 pat	36	**36**	23
	1 tbsp	100	**100**	65
	1 stick (½ cup)	813	**813**	515
Whipped	1 tbsp	67	**67**	38
	1 stick (½ cup)	542	**542**	344
light	1 tbsp	35	**30**	25

Margarine and other butter substitutes: *see* **FATS AND OILS.**

FOOD	AMOUNT	CALORIES TOTAL	CALORIES FAT	CALORIES SAT-FAT
CHEESE				
American	1 oz	106	**80**	50
Blue	1 oz	100	**73**	48
Bonbel (Laughing Cow)	1 oz	70	**50**	36
Brie	1 oz	100	**70**	54
Camembert	1 oz	90	**70**	27
Cheddar	1 oz	114	**85**	54
shredded	¼ cup	110	**80**	54
Colby	1 oz	112	**82**	54
Cottage cheese				
4% fat	½ cup	110	**43**	23
2% fat	½ cup	102	**20**	14
1% fat	½ cup	90	**10**	7
dry curd	½ cup	80	**9**	4

DAIRY AND EGGS

FOOD	AMOUNT	CALORIES TOTAL	CALORIES FAT	CALORIES SAT-FAT
Cream cheese				
regular	1 tbsp	52	**48**	26
with salmon or strawberries	2 tbsp	100	**80**	54
soft	2 tbsp	100	**90**	63
whipped	1 tbsp	37	**34**	19
Edam	1 oz	101	**71**	45
Feta	1 oz	75	**54**	38
with basil & tomato	1 oz	80	**60**	36
Farmer				
Friendship	1 oz	40	**27**	18
May-Bud	1 oz	90	**63**	41
Goat cheese (chèvre)	1 oz (1" cube)	80	**50**	41
Gouda	1 oz	101	**70**	45
Gruyère	1 oz	117	**83**	48
Limburger	1 oz	93	**69**	43
Mascarpone	2 tbsp	120	**120**	90
Monterey	1 oz	106	**77**	45
Mozzarella				
whole milk	1 oz	80	**50**	36
fresh, handmade	1 oz	90	**60**	41
shredded	1 oz	90	**65**	42
part skim	1 oz	72	**45**	27
Muenster	1 oz	104	**77**	49
Neufchâtel	1 oz	74	**60**	38
Parmesan	1 tbsp	23	**14**	9
	1 oz	129	**77**	49
Port du Salut	1 oz	100	**72**	43
Provolone	1 oz	100	**68**	44
Ricotta				
whole milk	½ cup	216	**145**	93
part skim	½ cup	171	**88**	55
Romano	1 oz	110	**70**	49
Roquefort	1 oz	105	**78**	49
String	1 oz	80	**50**	36
handmade	1 oz	90	**60**	54
Swiss	1 oz	107	**70**	45
Tilsit	1 oz	96	**66**	43

DAIRY AND EGGS

FOOD	AMOUNT	CALORIES TOTAL	CALORIES FAT	CALORIES SAT-FAT
CHEESE, FAT-FREE				
All brands	1 slice (21 g)	25–30	**0**	0
Cream cheese	2 tbsp	30	**0**	0
Mozzarella-type, shredded	¼ cup (28 g)	45	**0**	0
CHEESE, REDUCED-CALORIE OR LITE				
Alouette Lite Herbs & Garlic	2 tbsp	60	**35**	27
Bonbel Light Wedge (Laughing Cow)	1 piece (28 g)	50	**30**	18
Cheddar, shredded				
Kraft ⅓ Less Fat	¼ cup (31 g)	90	**50**	36
Sargento	¼ cup (28 g)	70	**40**	18
Cream cheese				
Philadelphia ⅓ Less Fat	2 tbsp (30 g)	70	**60**	36
Monterey jack				
Kraft	28 g	80	**45**	27
Dorman's	1 slice (43 g)	120	**60**	41
Mozzarella	¼ cup (28 g)	60	**30**	18
Ricotta	¼ cup (62 g)	75	**35**	18
Rondelé Soft Spreadable Lite	2 tbsp	60	**35**	23
String cheese (Poly-O)	1 piece (28 g)	80	**50**	36
Weight Watchers, all	1 slice (21 g)	50	**20**	0
CHEESE SPREADS				
Alouette				
Garlic & Spices	2 tbsp	70	**60**	41
Spinach	2 tbsp	60	**50**	32
Boursin	2 tbsp	120	**110**	45
Cheez Whiz (Kraft)	2 tbsp (33 g)	100	**70**	45
Rondelé Soft Spreadable				
Black pepper & garden vegetable	2 tbsp	90	**80**	54
Garlic & herbs	2 tbsp	100	**80**	54
CREAM				
Half-and-half	1 tbsp	20	**15**	10

DAIRY AND EGGS

FOOD	AMOUNT	CALORIES TOTAL	FAT	SAT-FAT
Light, coffee or table	1 tbsp	29	**26**	16
	1 cup	469	**417**	260
Nondairy				
frozen				
Rich's Coffee Rich	1 tbsp	25	**15**	0
powdered				
Coffee-Mate	1 tsp	10	**5**	5
	1 env. (3 g)	15	**10**	10
flavored	1⅓ tbsp	60	**25**	23
refrigerated				
Coffee-Mate				
fat-free	1 tbsp	10	**0**	0
Lite	1 tbsp	20	**10**	0
flavored	1 tbsp	40	**20**	0
fat-free	1 tbsp	25	**0**	0
Farm Rich				
fat-free	1 tbsp	10	**0**	0
light	1 tbsp	10	**5**	0
original	1 tbsp	20	**15**	0
International Delight				
all flavors	1 tbsp	35	**10**	0
fat-free	1 tbsp	30	**0**	0
Sour cream	2 tbsp	60	**50**	36
	1 cup	493	**434**	270
light	2 tbsp (30 g)	35	**25**	14
no fat	2 tbsp (30 g)	20	**0**	0
Whipping cream				
heavy, fluid	1 cup	821	**792**	493
whipped	½ cup	205	**198**	123
light, fluid	1 cup	699	**665**	416
nondairy (Cool Whip)	2 tbsp (8 g)	25	**15**	14
	½ cup	200	**120**	108
Pressurized topping				
Whipped Light	2 tbsp (10 g)	30	**20**	14
Reddi Whip	2 tbsp (8 g)	20	**15**	9

DAIRY AND EGGS

FOOD	AMOUNT	CALORIES TOTAL	FAT	SAT-FAT
MILK				
1% fat	1 cup	100	**20**	14
2% fat	1 cup	120	**45**	27
Buttermilk	1 cup	80	**0**	0
Chocolate				
2% milk	1 cup	190	**45**	27
whole milk (Nestlé Quick)	1 cup	230	**70**	45
Condensed, sweetened	1 tbsp	62	**15**	9
Evaporated				
skim	1 cup	200	**0**	0
	2 tbsp	25	**0**	0
whole	1 cup	300	**160**	100
	2 tbsp	40	**20**	14
Nonfat				
dry	¼ cup	109	**2**	1
instant	to make 1 qt	326	**6**	4
Skim	1 cup	80	**0**	0
Whole	1 cup	150	**75**	45
dry	¼ cup	159	**77**	48
YOGURT				
Custard-style				
Yoplait custard style all flavors	6 oz	190	**30**	18
Yoplait Adventure Pack all flavors	4 oz	120	**20**	14
Low-fat yogurt				
Breyers 99% Fat Free				
Black Cherry	8 oz	230	**15**	9
Blueberry	8 oz	220	**15**	9
Peach	8 oz	230	**15**	9
Peaches 'n' Cream	8 oz	240	**15**	9
Raspberry à la Mode	8 oz	220	**15**	9
Strawberry	8 oz	220	**15**	9

DAIRY AND EGGS

FOOD	AMOUNT	CALORIES TOTAL	CALORIES FAT	CALORIES SAT-FAT
Low-fat yogurt (cont.)				
Strawberry à la Mode	8 oz	280	**25**	14
Strawberry Cheese Cake	8 oz	230	**20**	9
Breyers 99% fat-free Smooth & Creamy				
Black Cherry Parfait	8 oz	230	**20**	9
Classic Strawberry	8 oz	230	**15**	9
Raspberries 'n' Cream	8 oz	230	**15**	9
Strawberry Cheesecake	8 oz	240	**15**	9
Colombo Low Fat all flavors	8 oz	120	**40**	23
Dannon Low Fat				
Danimals (all flavors)	4.4 oz	140	**10**	0
Double Delights Chocolate Cheesecake	6 oz	220	**10**	5
all other flavors	6 oz	170	**10**	5
French Vanilla with fruit	8 oz	270	**30**	14
Fruit on the bottom, all flavors	8 oz	240	**25**	14
Premium Low-Fat	8 oz	150	**35**	23
Dannon Sprinkl'ins				
Magic Crystals (vanilla)	4.1 oz	110	**10**	5
Rainbow Sprinkles	4.1 oz	130	**10**	5
La Yogurt 99% Fat Free				
all flavors except White Chocolate Almond	6 oz	170	**15**	9

DAIRY AND EGGS

FOOD	AMOUNT	CALORIES TOTAL	CALORIES FAT	CALORIES SAT-FAT
Low-fat yogurt (cont.)				
White Chocolate Almond	6 oz	170	**25**	18
La Yogurt Sabor Latino all flavors	6 oz	180	**15**	9
Safeway Lucerne	8 oz	250	**25**	14
plain	8 oz	160	**35**	23
all other flavors	8 oz	250	**25**	14
Yoplait	6 oz	190	**30**	18
Trix	4 oz	120	**20**	14
Yoplait Original 99% Fat-Free	6 oz	180	**15**	9
Nonfat yogurt				
Breyers Light	8 oz	130	**0**	0
fruit flavors	8 oz	220	**0**	0
other flavors	8 oz	170	**0**	0
Colombo Classic vanilla	8 oz	170	**0**	0
All other flavors	8 oz	200	**0**	0
Colombo Light Key Lime Pie	8 oz	100	**0**	0
Dannon Light	8 oz	100	**0**	0
with toppings	8 oz	140	**0**	0
chunky fruit	6 oz	160	**0**	0
blended (all flavors)	4.4 oz	110	**0**	0
Horizon Organic	6 oz	110	**0**	0
Lucerne Light all flavors	6 oz	90	**0**	0
Lucerne Nonfat Pre-Stirred all flavors	8 oz	180	**0**	0
SnackWell's Double Chocolate	6 oz	190	**0**	0
Milk Chocolate Almond	6 oz	160	**0**	0

DAIRY AND EGGS

FOOD	AMOUNT	CALORIES TOTAL	CALORIES FAT	CALORIES SAT-FAT
Nonfat yogurt (cont.)				
Milk Chocolate Cheesecake	6 oz	160	**0**	0
Milk Chocolate Peanut Butter	6 oz	180	**0**	0
Stonyfield Farm all flavors	8 oz	160	**0**	0
Weight Watchers Ultimate 90, all flavors	8 oz	90	**0**	0
Yoplait Light, all flavors	6 oz	90	**0**	0
EGG, CHICKEN (*see also* **FROZEN FOODS**)				
Whole, large	1 egg	79	**50**	15
white	1 white	16	**0**	0
yolk	1 yolk	63	**50**	15
EGG, DUCK				
Preserved	1 egg (55 g)	100	**60**	23
EGG SUBSTITUTE				
Egg Beaters (Fleischmann's)	¼ cup (60 g)	30	**0**	0
Healthy Choice Egg Product	¼ cup (56 g)	25	**5**	0
Scramblers (Morningstar Farms)	¼ cup (57 g)	35	**0**	0
Simply Eggs	½ cup (117 g)	80	**20**	9
	3 tbsp (50 g)	35	**10**	0

FAST FOODS

FOOD	AMOUNT	CALORIES TOTAL	CALORIES FAT	CALORIES SAT-FAT
ARBY'S				
Bacon platter	1	593	**297**	83
Baked potato, plain	1	355	**0**	0

FAST FOODS

FOOD	AMOUNT	CALORIES TOTAL	FAT	SAT-FAT
Baked potato (cont.)				
Broccoli & Cheddar	1	571	**180**	45
Deluxe	1	736	**324**	144
Margarine & Sour Cream	1	578	**216**	81
Biscuit				
Bacon	1	318	**162**	39
Ham	1	323	**150**	36
Plain	1	280	**135**	30
Sausage	1	460	**288**	85
Blueberry muffin	1	230	**81**	18
Cheesecake	1 serving (87 g)	320	**207**	126
Chicken sandwiches				
Breaded Chicken Fillet	1 (205 g)	536	**252**	45
Chicken Breast	1	445	**207**	27
Chicken Cordon Bleu	1 (240 g)	623	**297**	72
Chicken Fingers	2 pieces (102 g)	290	**144**	18
Grilled Chicken Barbecue	1	388	**117**	27
Grilled Chicken Deluxe	1	430	**180**	36
Roast Chicken Club	1 (241 g)	546	**279**	81
Roast Chicken Deluxe	1	433	**198**	45
Roast Chicken Santa Fe	1 (182 g)	436	**198**	54
Chocolate Chip Cookie	1 (27 g)	125	**54**	18
Cinnamon Nut Danish	1	360	**99**	9
Croissant				
Bacon & egg	1	430	**270**	139
Butter	1	260	**140**	94
Ham & cheese	1	345	**186**	109
Mushroom & cheese	1	495	**340**	137
Plain	1	220	**108**	63
Sausage & egg	1	520	**353**	167
Egg platter	1	460	**216**	65
Fish Fillet Sandwich	1 (220 g)	529	**243**	63
French Toastix	1 serving	430	**189**	45
Fries				
Cheddar Cheddar Curly fries	1 order	333	**162**	36
Curly fries	1 order (100 g)	300	**135**	27
Home-style French fries	1 order (71 g)	246	**119**	27
Ham 'n Cheese Sandwich	1 (169 g)	359	**126**	45

FAST FOODS

FOOD	AMOUNT	CALORIES TOTAL	FAT	SAT-FAT
Ham platter	1	518	**234**	72
Light sandwiches				
Roast Beef Deluxe	1 (182 g)	296	**90**	27
Roast Chicken Deluxe	1 (195 g)	276	**54**	18
Roast Turkey Deluxe	1 (195 g)	260	**63**	18
Panini sandwiches				
Roast Beef & Havarti	1 (423 g)	847	**306**	180
Roast Chicken & Pesto	1 (388 g)	855	**342**	108
Sicilian Meat & Cheese	1 (361 g)	825	**324**	126
Polar Swirl				
Butterfinger	1 (11.6 oz)	457	**162**	72
Heath	1 (11.6 oz)	543	**198**	45
Oreo	1 (11.6 oz)	482	**198**	90
Peanut Butter Cup	1 (11.6 oz)	517	**216**	72
Snickers	1 (11.6 oz)	511	**170**	63
Potato cakes	2 cakes (85 g)	204	**108**	20
Roast beef sandwiches				
Arby's Melt with Cheddar	1	368	**162**	36
Arby Q	1 (182 g)	431	**162**	54
Bac'n Cheddar Deluxe	1	539	**306**	90
Beef 'n Cheddar	1 (189 g)	487	**252**	81
Giant	1 (228 g)	555	**252**	99
Junior	1	324	**126**	45
Regular	1 (154 g)	388	**171**	63
Special	1 (126 g)	324	**126**	45
Super	1 (247 g)	523	**243**	81
Salads				
Chef salad (no dressing)	1	205	**86**	35
Roast chicken salad (no dressing)	1 (408 g)	149	**18**	5
Garden salad (no dressing)	1	61	**5**	0
Side salad (no dressing)	1 (142 g)	25	**3**	0
Sauces & Dressings				
Arby's Sauce	1 packet (14 g)	15	**2**	0
Bleu Cheese Dressing	1 packet (56 g)	290	**281**	54
Buttermilk Ranch	1 packet	349	**347**	50
Cheddar Cheese Sauce	1 packet (21 g)	35	**27**	9
Honey French Dressing	1 packet (56 g)	280	**207**	27

FAST FOODS

FOOD	AMOUNT	CALORIES TOTAL	FAT	SAT-FAT
Sauces & Dressings (cont.)				
Horsey Sauce	1 packet (14 g)	60	**45**	9
Italian, light	1 packet	23	**10**	1
Mayonnaise	1 packet (14 g)	110	**108**	63
Red Ranch Dressing	1 packet (14 g)	75	**54**	9
Tartar Sauce	1 tbsp	70	**70**	9
Thousand Island Dressing	1 packet (56 g)	260	**234**	36
Sausage platter	1	640	**370**	120
Shakes				
Chocolate	12 oz	451	**108**	27
Jamocha	12 oz	384	**90**	27
Vanilla	12 oz	360	**108**	36
Vanilla				
nonfat	455 g	470	**18**	14
Soups				
Boston Clam Chowder	1 cup	190	**81**	27
Cream of Broccoli	1 cup	160	**72**	36
Lumberjack Mixed Vegetable	1 cup	90	**36**	18
Old-Fashioned Chicken Noodle	1 cup	80	**18**	0
Potato with Bacon	1 cup	170	**63**	27
Wisconsin Cheese	1 cup	281	**162**	63
Sub Roll Sandwiches				
French Dip	1	475	**198**	72
Hot Ham 'n Swiss	1	500	**207**	63
Italian	1	675	**324**	117
Philly Beef 'n Swiss	1	755	**423**	135
Roast Beef	1	700	**378**	126
Triple Cheese Melt	1	720	**405**	144
Turkey	1	550	**243**	63
Turnover				
Apple	1	330	**126**	63
Blueberry	1	320	**180**	57
Cherry	1	320	**117**	45
ARTHUR TREACHER'S				
Chicken, fried	1 serving	369	**198**	36
Chicken patties	2 patties (136 g)	369	**194**	32

FAST FOODS

FOOD	AMOUNT	CALORIES TOTAL	CALORIES FAT	CALORIES SAT-FAT
Chicken sandwich	1 (156 g)	413	**173**	27
Chips (French fries)	1 serving	276	**117**	18
Cod fillet, "Bake 'n Broil"	1 serving (142 g)	245	**128**	NA
Chowder	1 serving	112	**45**	18
Cole slaw	1 serving (85 g)	123	**74**	10
Fish, broiled	5 oz	245	**126**	NA
Fish, fried	2 pieces (148 g)	355	**180**	27
Fish sandwich	1 (156 g)	440	**216**	38
French fries "Chips"	1 serving (114 g)	276	**119**	21
Hushpuppy "Krunch Pup"	1 piece (57 g)	203	**133**	33
Krunch Pup (batter-fried hot dog)	1	203	**135**	36
Lemon Luv (fried pie)	1 serving (85 g)	276	**126**	20
Shrimp, fried	1 serving (116 g)	381	**220**	30
BURGER KING				
Big King Sandwich	1 (226 g)	660	**390**	162
Biscuit				
with Bacon, Egg, & Cheese	1 (171 g)	510	**280**	90
with Sausage	1 (151 g)	590	**360**	117
BK Big Fish Sandwich	1 (252 g)	720	**387**	81
BK Broiler Chicken Sandwich	1 (247 g)	530	**230**	45
Blueberry mini muffins	1 serving	292	**126**	27
Breakfast Buddy with sausage, egg, & cheese	1	255	**144**	54
Broiled Chicken Salad, no dressing	1 (302 g)	190	**90**	45
Cheeseburger	1 (138 g)	380	**170**	81
Deluxe	1	390	**207**	72
Double	1 (210 g)	600	**320**	153
Bacon	1 (218 g)	640	**350**	162
Bacon deluxe	1	584	**342**	144
Chicken Sandwich	1 (229 g)	710	**390**	81
Chicken Tenders	8 pieces (123 g)	350	**200**	63
Croissan'wich				
Bacon, Egg, & Cheese	1	350	**216**	72
Ham, Egg, & Cheese	1	350	**198**	63
Sausage, Egg, & Cheese	1 (176 g)	600	**410**	144

FAST FOODS

FOOD	AMOUNT	CALORIES TOTAL	FAT	SAT-FAT
Danish				
Apple cinnamon	1	390	**117**	27
Cheese	1	406	**144**	45
Dipping Sauces				
Ranch	1 serving (28 g)	171	**162**	27
all other dipping sauces		35–90		
	1 serving (28 g)		**0**	0
French Fries, Medium	1 serving (116 g)	400	**190**	72
French Toast Sticks	1 serving (141 g)	500	**240**	63
Garden Salad	1 (215 g)	100	**45**	27
Hamburger, regular	1 (126 g)	330	**140**	54
Burger Buddies	1 serving	349	**153**	63
Deluxe	1	344	**171**	54
Double Whopper	1 (351 g)	870	**500**	171
with Cheese	1 (375 g)	960	**570**	216
Whopper	1 (270 g)	640	**350**	99
with Cheese	1 (294 g)	730	**410**	144
Whopper Jr.	1 (164 g)	420	**220**	72
with Cheese	1 (177 g)	460	**250**	90
Hash Browns, small	1 serving (75 g)	240	**140**	54
Ocean Catch fish fillet sandwich	1	479	**297**	72
Onion Rings	1 serving (124 g)	310	**130**	18
Pies				
Dutch Apple	1 serving (113 g)	300	**140**	27
Cherry	1	360	**117**	36
Lemon	1	290	**72**	27
Salads				
Chef	1	178	**81**	36
Chunky chicken	1	142	**36**	9
Side	1 (133 g)	60	**25**	18
Scrambled Egg Platter				
croissant, hash browns	1 serving	549	**306**	81
with bacon	1 serving	610	**351**	99
with sausage	1 serving	768	**477**	135
Shakes				
Chocolate, medium	284 g	320	**60**	36
with Syrup	341 g	440	**60**	36

FAST FOODS

FOOD	AMOUNT	CALORIES TOTAL	FAT	SAT-FAT
Shakes (cont.)				
Strawberry, medium, with syrup	341 g	420	**50**	36
Vanilla, medium	284 g	300	**50**	36
Speciality Sandwiches				
Chicken	1	688	**360**	72
Ham & Cheese	1	471	**216**	81
Whaler Sandwich	1	488	**243**	54
with cheese	1	530	**270**	72
CARL'S JR.				
Barbecue Chicken Sandwich	1	310	**54**	14
Breakfast Burrito	1	430	**234**	108
Breakfast Quesadilla	1	300	**126**	54
Carl's Catch Fish Sandwich	1	560	**270**	63
Chicken Club Sandwich	1	550	**261**	72
Cheese Danish	1	400	**198**	45
Cheeseburger				
Western Bacon Cheeseburger	1	870	**315**	144
Double	1	970	**513**	243
Cheesecake, Strawberry Swirl	1 piece (3.5 oz)	300	**153**	81
Chicken Stars	6 pieces	230	**126**	27
Chocolate Cake	1 piece (3 oz)	300	**90**	24
Chocolate Chip Cookie	1	370	**171**	72
Cinnamon Rolls	1 order	420	**117**	36
CrissCut Fries, large	1 serving	550	**306**	81
English Muffin with Margarine	1	230	**90**	14
French Fries, regular	1 serving	370	**180**	63
French Toast Dips	1 order	410	**225**	54
Fudge Moussecake	1 piece	400	**207**	99
"Great Stuff" potato				
Bacon & Cheese	1	630	**261**	63
Broccoli & Cheese	1	530	**198**	45
Plain	1	290	**0**	0
Sour Cream & Chive	1	430	**126**	27
Hamburger	1	200	**72**	36
Big Burger	1	470	**180**	72

FAST FOODS

FOOD	AMOUNT	CALORIES TOTAL	CALORIES FAT	CALORIES SAT-FAT
Hamburger (cont.)				
Famous Big Star	1	610	**342**	99
Super Star	1	820	**477**	180
Hash Brown Nuggets	1 order	270	**153**	36
Hot & Crispy Sandwich	1	400	**198**	45
Hotcakes with margarine	1 serving	510	**216**	45
Muffin				
Blueberry	1	340	**126**	18
Bran	1	370	**117**	18
Onion Rings	1 order	520	**234**	54
Roast Beef Club Sandwich	1	620	**306**	99
Roast Beef Deluxe Sandwich	1	540	**234**	90
Salad-to-Go				
Charbroiled Chicken	1	260	**81**	45
Garden	1	50	**27**	14
Santa Fe Chicken Sandwich	1	530	**270**	63
Scrambled Eggs	1 order (3.6 oz)	160	**99**	36
Shakes, regular	1	350	**63**	36
Shakes				
Chocolate, small	1 (13.5 oz)	390	**63**	45
Strawberry, small	1 (13.5 oz)	400	**63**	45
Vanilla, small	1 (13.5 oz)	330	**72**	45
Sunrise Sandwich	1	370	**189**	54
Teriyaki Chicken Sandwich	1	330	**54**	18
Turkey Club Sandwich	1	530	**207**	54
Zucchini	1 order	380	**207**	54
CHICK-FIL-A				
Chargrilled Chicken Sandwich	1 (150 g)	280	**30**	9
Chargrilled Chicken Club Sandwich	1 (232 g)	390	**110**	45
Cheesecake, plain	1 slice (88 g)	270	**190**	81
with blueberry topping	1 slice (122 g)	350	**173**	NA
with strawberry topping	1 slice (122 g)	343	**173**	NA
Chick-Fil-A-Nuggets				
8-pack	1 serving (110 g)	290	**130**	27
12-pack	1 serving (170 g)	430	**200**	40
Chicken Salad Sandwich	1 (167 g)	320	**40**	18

FAST FOODS

FOOD	AMOUNT	CALORIES TOTAL	CALORIES FAT	CALORIES SAT-FAT
Chicken Sandwich	1 (167 g)	290	**80**	18
Chick-n-Strips	1 serving (119 g)	230	**70**	18
Cole Slaw	1 serving (79 g)	130	**50**	9
Fudge Nut Brownie	1 (74 g)	350	**140**	27
Grilled 'n Lites	2 skewers	97	**18**	NA
Hearty Breast of Chicken Soup	1 serving (215 g)	110	**10**	0
Icedream	1 (127 g)	140	**35**	9
Lemon Pie	1 slice (99 g)	280	**200**	54
Salads				
Carrot & Raisin	1 (76 g)	150	**20**	0
Chargrilled Chicken Garden	1 (397 g)	170	**30**	9
Chicken Salad Plate	1 (468 g)	290	**40**	0
Chick-n-Strips	1 (451 g)	290	**80**	18
Tossed	1 (130 g)	70	**0**	0
Waffle Potato Fries	1 serving (85 g)	290	**90**	36
CHURCH'S FRIED CHICKEN				
Apple pie	1 (3 oz)	300	**171**	NA
Biscuits	1	250	**148**	NA
Cajun Rice	1 order (88 g)	130	**63**	NA
Catfish, fried	3 pieces	201	**108**	NA
Chicken breast fillet sandwich	1	608	**306**	NA
Chicken nuggets				
regular	6 pieces	330	**171**	NA
spicy	6 pieces	312	**153**	NA
Cole slaw	1 serving	83	**63**	NA
Corn on the cob	1 ear	190	**49**	NA
with butter oil	1 ear	237	**81**	NA
Dinner roll	1	83	**18**	NA
Fish fillet sandwich	1	430	**162**	NA
French fries, regular	3 oz	256	**117**	NA
Fried chicken				
Breast	1 serving	278	**153**	NA
Leg	1 serving	147	**81**	NA
Thigh	1 serving	305	**198**	NA
Wing	1 serving	303	**180**	NA
Hush puppies	2 pieces	156	**54**	NA

FAST FOODS

FOOD	AMOUNT	CALORIES TOTAL	CALORIES FAT	CALORIES SAT-FAT
Mashed potatoes & gravy	1 order	90	**30**	NA
Pecan pie	1 serving	367	**180**	NA
DAIRY QUEEN				
Banana Split	1	510	**99**	72
BBQ beef sandwich	1	225	**36**	9
Breaded chicken fillet sandwich	1	430	**180**	36
with Cheese	1	480	**225**	63
Buster Bar	1	460	**261**	81
Butterfinger Blizzard				
regular	1	750	**234**	NA
small	1	520	**162**	NA
Cheese Dog	1	290	**162**	72
Chicken Breast Fillet Sandwich	1	430	**180**	36
with Cheese	1	480	**225**	63
Chicken Strip Basket				
with BBQ Sauce	1 order	810	**333**	81
with Gravy	1 order	860	**378**	99
Chili Dog	1	280	**144**	54
Chili 'n Cheese Dog	1	330	**189**	81
Chocolate Chip Cookie Dough Blizzard				
regular	1	950	**324**	NA
small	1	660	**216**	NA
Chocolate cone				
regular	1	350	**99**	72
small	1	230	**63**	45
Chocolate cone, Dipped				
regular	1	510	**225**	140
small	1	340	**144**	90
Chocolate Dilly Bar	1	210	**117**	54
Chocolate Mint Dilly Bar	1	190	**108**	NA
Chocolate Sandwich Cookie Blizzard				
regular	1	640	**207**	NA
small	1	520	**162**	NA
Double Delight	1 serving	490	**180**	NA
DQ Caramel & Nut Bar	1	260	**117**	NA
DQ Frozen Cake Slice	1	380	**162**	72
DQ Frozen Heart Cake	1/10 cake	270	**81**	NA

FAST FOODS

FOOD	AMOUNT	CALORIES TOTAL	CALORIES FAT	CALORIES SAT-FAT
DQ Frozen Log Cake	⅛ cake	280	**81**	NA
DQ Frozen 8" Round Cake	⅛ cake	340	**108**	NA
DQ Frozen 10" Round Cake	1/12 cake	360	**108**	NA
DQ Frozen Sheet Cake	1/20 cake	350	**108**	NA
DQ Fudge Bar	1	50	**0**	0
DQ Homestyle Burgers				
Bacon Double Cheeseburger	1	610	**324**	162
Cheeseburger	1	340	**153**	72
Deluxe Double Cheeseburger	1	540	**279**	144
Deluxe Double Hamburger	1	440	**198**	90
Double Cheeseburger	1	540	**279**	144
Hamburger	1	290	**108**	45
Ultimate Burger	1	670	**387**	171
DQ Lemon Freez'r	½ cup	80	**0**	0
DQ Nonfat Yogurt	½ cup	100	**0**	0
DQ Sandwich	1	140	**36**	18
DQ Soft Serve				
Chocolate	½ cup	150	**45**	NA
Vanilla	½ cup	140	**41**	NA
DQ Treatzza Pizza				
Heath	⅛ pizza	180	**63**	NA
M&M's	⅛ pizza	190	**63**	NA
Peanut Butter Fudge	⅛ pizza	220	**90**	NA
Strawberry-Banana	⅛ pizza	180	**54**	NA
DQ Vanilla Orange Bar	1	60	**0**	0
Fish Fillet Sandwich	1	370	**144**	32
with Cheese	1	420	**189**	54
Float	1	410	**63**	41
Freeze	1	500	**108**	72
French fries				
large	1 order	390	**162**	36
regular	1 order	300	**126**	27
small	1 order	210	**90**	18
Garden salad, no dressing	1	200	**117**	63
Grilled Chicken Fillet Sandwich	1	310	**90**	23
Heath Blizzard				
regular	1	820	**324**	153
small	1	560	**207**	99

FAST FOODS

FOOD	AMOUNT	CALORIES TOTAL	CALORIES FAT	CALORIES SAT-FAT
Heath Breeze				
regular	1	680	**189**	54
small	1	450	**108**	27
Hot Fudge Brownie Delight	1	710	**261**	126
Malt				
Chocolate				
regular	1	1060	**225**	NA
small	1	880	**198**	NA
Vanilla				
small	1	610	**126**	72
Mr. Misty Float	1	390	**63**	41
Mr. Misty Freeze	1	500	**108**	70
Mr. Misty Kiss	1	70	**0**	0
Mr. Misty				
regular	1	290	**0**	0
small	1	250	**0**	0
Nutty Double Fudge	1	580	**198**	90
Onion Rings, regular	1 order	240	**108**	23
Parfait	1 serving	430	**72**	47
Peanut Buster	1 serving	730	**288**	90
QC Chocolate Big Scoop	1	250	**126**	90
QC Vanilla Big Scoop	1	250	**126**	90
Reese's Peanut Butter Cup Blizzard				
regular	1	790	**297**	NA
small	1	590	**216**	NA
Shakes				
Chocolate				
regular	1	770	**180**	111
small	1	540	**126**	72
Vanilla				
regular	1	600	**144**	90
small	1	520	**126**	72
Starkiss	1	80	**0**	0
Strawberry Blizzard				
regular	1	570	**144**	99
small	1	400	**100**	72

FAST FOODS

FOOD	AMOUNT	CALORIES TOTAL	FAT	SAT-FAT
Strawberry Breeze				
regular	1	460	**9**	5
small	1	320	**5**	0
Strawberry Shortcake	1 order	430	**126**	NA
Strawberry Waffle Cone Sundae	1	350	**108**	45
Sundae				
Chocolate				
regular	1	410	**90**	58
small	1	300	**63**	45
Toffee Dilly Bar with Heath				
Pieces	1	210	**108**	NA
Vanilla cone				
child's	1	140	**36**	27
large	1	410	**108**	72
regular	1	350	**90**	63
small	1	230	**60**	45
Yogurt Cup, regular	1	230	**5**	0
Yogurt Strawberry Sundae, regular	1	300	**5**	0
DOMINO'S PIZZA				
Breadsticks	1	78	**30**	6
Buffalo Wings				
Barbeque	1 order (10 pcs)	501	**219**	59
	1 wing	50	**22**	6
Hot	1 order (10 pcs)	449	**215**	59
	1 wing	45	**22**	6
Cheesy Bread	1 piece	103	**49**	17
Deep Dish Pizza				
6" small pizza				
Cheese only	1 serving (215 g)	595	**247**	95
with Anchovies	1 serving	640	**265**	99
with Bacon	1 serving	677	**310**	117
with Beef	1 serving	639	**283**	110
with Cheddar	1 serving	681	**310**	135
with Extra Cheese	1 serving	654	**287**	118
with Ham	1 serving	612	**253**	97
with Italian Sausage	1 serving	639	**309**	108
with Olives	1 serving	605	**256**	96

FAST FOODS

FOOD	AMOUNT	CALORIES TOTAL	FAT	SAT-FAT
Deep Dish Pizza (cont.)				
with Pepperoni	1 serving	645	**288**	111
all other toppings	1 serving	597	**247**	95
12" medium pizza (8 slices)				
Cheese only	2 slices (180 g)	477	**194**	76
with Anchovies	2 slices	500	**203**	78
with Bacon	2 slices	559	**257**	98
with Beef	2 slices	533	**238**	94
with Cheddar	2 slices	534	**236**	103
with Extra Cheese	2 slices	525	**228**	97
with Ham	2 slices	495	**201**	78
with Italian Sausage	2 slices	532	**233**	92
with Olives	2 slices	490	**204**	77
with Pepperoni	2 slices	552	**255**	100
all other toppings	2 slices	479	**194**	76
14" large pizza (12 slices)				
Cheese only	2 slices (173 g)	455	**178**	70
with Anchovies	2 slices	478	**187**	72
with Bacon	2 slices	530	**219**	90
with Beef	2 slices	499	**214**	85
with Cheddar	2 slices	503	**213**	92
with Extra Cheese	2 slices	500	**210**	90
with Ham	2 slices	472	**184**	72
with Italian Sausage	2 slices	499	**209**	83
with Olives	2 slices	466	**187**	71
with Pepperoni	2 slices	521	**232**	91
all other toppings	2 slices	457	**178**	70
Hand-Tossed Pizza				
12" medium pizza (8 slices)				
Cheese only	2 slices (149 g)	347	**96**	47
with Anchovies	2 slices	370	**105**	49
with Bacon	2 slices	429	**159**	69
with Beef	2 slices	403	**140**	65
with Cheddar	2 slices	404	**138**	74
with Extra Cheese	2 slices	395	**130**	68
with Ham	2 slices	365	**103**	49
with Italian Sausage	2 slices	402	**135**	63
with Olives	2 slices	359	**106**	49

FAST FOODS

FOOD	AMOUNT	CALORIES TOTAL	CALORIES FAT	CALORIES SAT-FAT
Hand-Tossed Pizza (cont.)				
with Pepperoni	2 slices	422	**157**	71
all other toppings	2 slices	349	**96**	47
14" large pizza (12 slices)				
Cheese only	2 slices (137 g)	317	**98**	43
with Anchovies	2 slices	340	**107**	45
with Bacon	2 slices	392	**139**	63
with Beef	2 slices	361	**134**	58
with Cheddar	2 slices	365	**133**	65
with Extra Cheese	2 slices	362	**130**	63
with Ham	2 slices	334	**104**	45
with Italian Sausage	2 slices	361	**129**	56
with Olives	2 slices	329	**108**	44
with Pepperoni	2 slices	383	**152**	64
all other toppings	2 slices	320	**98**	43
Thin Crust Pizza				
12" medium pizza (4 slices)				
Cheese only	1 slice (106 g)	271	**106**	47
with Anchovies	1 slice	294	**115**	49
with Bacon	1 slice	353	**169**	69
with Beef	1 slice	327	**150**	65
with Cheddar	1 slice	328	**148**	74
with Extra Cheese	1 slice	319	**140**	68
with Ham	1 slice	289	**113**	49
with Italian Sausage	1 slice	326	**145**	63
with Olives	1 slice	283	**116**	48
with Pepperoni	1 slice	346	**167**	71
all other toppings	1 slice	273	**106**	47
14" large pizza (6 slices)				
Cheese only	1 slice (99 g)	253	**99**	44
with Anchovies	1 slice	276	**108**	46
with Bacon	1 slice	328	**140**	64
with Beef	1 slice	297	**135**	59
with Cheddar	1 slice	301	**134**	66
with Extra Cheese	1 slice	298	**131**	64
with Ham	1 slice	270	**105**	46
with Italian Sausage	1 slice	297	**130**	57
with Olives	1 slice	264	**109**	45

FAST FOODS

FOOD	AMOUNT	CALORIES TOTAL	CALORIES FAT	CALORIES SAT-FAT
Thin Crust Pizza (cont.)				
with Pepperoni	1 slice	319	**153**	65
all other toppings	1 slice	255	**99**	44
DUNKIN' DONUTS				
Cookies				
Chocolate chunk	1	200	**90**	47
with nuts	1	210	**99**	45
Oatmeal pecan raisin	1	200	**81**	38
Croissant	1	310	**171**	37
Almond	1	420	**234**	46
Chocolate	1	440	**261**	94
Donuts				
Apple Crumb	1 (85 g)	250	**90**	25
Apple filled with cinnamon sugar	1 (79 g)	250	**99**	20
Bavarian filled with chocolate	1	240	**99**	NA
Blueberry filled	1 (68 g)	210	**72**	15
Boston Kreme	1 (79 g)	240	**99**	22
Chocolate Kreme	1 (68 g)	250	**126**	26
Dunkin'	1 (60 g)	240	**126**	26
Jelly filled	1 (68 g)	220	**81**	15
Lemon filled	1 (79 g)	260	**108**	22
Peanut	1 (105 g)	480	**261**	49
Sugared Jelly Stick	1 (85 g)	310	**108**	23
Vanilla Kreme	1 (68 g)	250	**108**	23
Glazed coffee roll	1 (79 g)	280	**108**	21
Glazed cruller	1 (68 g)	260	**99**	20
Glazed French cruller	1 (40 g)	140	**72**	16
Rings				
Buttermilk cake	1 (99 g)	410	**180**	55
Cake, plain	1 (57 g)	262	**162**	33
Chocolate coconut	1 (96 g)	420	**216**	72
Chocolate-frosted cake	1 (65 g)	280	**144**	31
Chocolate-frosted yeast	1 (54 g)	200	**90**	17
Cinnamon cake	1 (62 g)	260	**135**	29
Coconut-coated cake	1 (88 g)	360	**189**	68
Glazed buttermilk	1 (74 g)	290	**126**	29

FAST FOODS

FOOD	AMOUNT	CALORIES TOTAL	CALORIES FAT	CALORIES SAT-FAT
Rings (cont.)				
Glazed chocolate	1 (77 g)	320	**162**	21
Glazed whole-wheat ring	1 (71 g)	280	**135**	28
Glazed yeast ring	1 (54 g)	200	**81**	18
Powdered cake	1 (62 g)	270	**144**	30
Sugared cake	1 (60 g)	270	**162**	34
Vanilla-frosted yeast	1 (57 g)	200	**81**	16
Mini Doughnuts				
Cake	1 (31 g)	100	**54**	11
Chocolate glazed	1 (31 g)	122	**63**	12
Cinnamon cake	1 (26 g)	116	**63**	14
Coconut	1 (34 g)	140	**72**	27
Coffee roll	1 (23 g)	78	**27**	6
Eclair	1 (37 g)	114	**45**	10
Muffins				
Apple 'n spice	1 (96 g)	300	**72**	15
Banana nut	1 (94 g)	310	**90**	18
Blueberry	1 (102 g)	280	**72**	14
Bran with raisins	1 (105 g)	310	**81**	11
Corn	1 (96 g)	340	**108**	15
Cranberry nut	1 (96 g)	290	**81**	16
Oat bran	1 (96 g)	330	**99**	14
Muffins, lowfat				
Apple 'n spice	1 (99 g)	220	**14**	0
Banana	1 (99 g)	240	**14**	0
Blueberry	1 (99 g)	220	**14**	0
Cherry	1 (99 g)	230	**14**	0
Cranberry-orange	1 (99 g)	230	**14**	0
Munchkins				
Butternut cake	1 (17 g)	70	**27**	9
Coconut cake	1 (17 g)	70	**36**	14
Glazed cake	1 (17 g)	60	**27**	5
Glazed chocolate	1 (20 g)	70	**27**	6
Glazed yeast	1 (14 g)	50	**18**	4
Jelly-filled yeast	1 (14 g)	50	**18**	4
Plain cake	1 (14 g)	50	**27**	5
Powdered cake	1 (14 g)	50	**27**	5

FAST FOODS

FOOD	AMOUNT	CALORIES TOTAL	CALORIES FAT	CALORIES SAT-FAT
HARDEE'S				
Apple turnover	1	270	**108**	36
Big Cookie	1	280	**108**	36
Big Country Breakfast				
Bacon	1	820	**441**	117
Country ham	1	670	**342**	81
Ham	1	620	**297**	63
Sausage	1	1000	**594**	180
Big Twin	1	450	**225**	99
Bagel				
Bacon	1	280	**81**	34
Bacon & egg	1	330	**108**	41
Bacon, egg, & cheese	1	375	**144**	68
Egg	1	250	**54**	24
Egg & cheese	1	295	**90**	NA
Plain	1	200	**27**	5
Sausage	1	350	**144**	59
Sausage & egg	1	400	**171**	68
Sausage, egg, & cheese	1	445	**207**	86
Biscuit	1	257	**112**	29
Bacon	1	360	**189**	36
Bacon & egg	1	570	**297**	81
Bacon, egg, & cheese	1	610	**333**	99
Canadian Rise 'n Shine	1	570	**288**	99
Cheese	1	304	**142**	NA
Chicken	1	510	**225**	63
Cinnamon 'n Raisin	1	370	**162**	45
Country ham	1	430	**198**	54
Country ham & egg	1	400	**198**	36
Ham	1	400	**180**	54
Ham & egg	1	370	**171**	36
Ham, egg, & cheese	1	540	**270**	90
Jelly	1	440	**189**	NA
'n Gravy	1	510	**252**	81
Rise 'n Shine	1	390	**189**	54
Sausage	1	510	**279**	90
Sausage & egg	1	630	**360**	99
Steak	1	580	**288**	90

FAST FOODS

FOOD	AMOUNT	CALORIES TOTAL	FAT	SAT-FAT
Biscuit (cont.)				
Steak & egg	1	550	**288**	72
Ultimate omelet	1	570	**297**	99
Breadstick	1	150	**36**	0
Cheeseburger	1	310	**126**	63
Cravin' Bacon	1	690	**414**	144
Quarter-pound Double	1	470	**243**	126
Mesquite Bacon	1	370	**162**	NA
Mushroom 'n Swiss	1	520	**243**	117
The Boss	1	570	**297**	NA
Chicken Fillet Sandwich	1	480	**162**	36
Coleslaw	114 g (4 oz)	240	**180**	27
Combo sub	1	380	**54**	27
Cool Twist cone	1	180	**18**	9
Cool Twist sundae				
hot fudge	1	290	**54**	27
strawberry	1	210	**18**	9
Fisherman's Fillet Sandwich	1	560	**243**	63
French fries				
Crispy Curls	1 order	300	**144**	27
large	1 order	430	**162**	45
medium	1 order	350	**135**	36
small	1 order	240	**90**	27
Fried chicken				
Breast	1	370	**135**	36
Chicken Stix	6 pieces	210	**81**	18
Leg	1	170	**63**	18
Thigh	1	330	**135**	36
Wing	1	200	**72**	18
Frisco Breakfast Ham Sandwich	1	500	**225**	72
Frisco Chicken Sandwich	1	680	**369**	90
Frisco Club Sandwich	1	670	**378**	108
Grilled Chicken Breast Sandwich	1	350	**99**	18
Hamburger	1	270	**99**	45
Big Deluxe	1	530	**270**	117
Frisco Burger	1	760	**450**	162
The Works Burger	1	530	**270**	NA
Ham Sub	1	370	**63**	36

FAST FOODS

FOOD	AMOUNT	CALORIES TOTAL	CALORIES FAT	CALORIES SAT-FAT
Hash Rounds, regular	1 order	230	**126**	27
Hot dog	1	290	**144**	36
Hot Ham 'n Cheese Sandwich	1	350	**117**	54
Marinated Chicken Grill	1	340	**90**	18
Mashed potatoes	114 g (4 oz)	70	**9**	5
Muffin				
Blueberry	1	400	**153**	36
Oatbran raisin	1	410	**144**	27
Pancakes, 3, plain	1 order	280	**18**	9
with 2 strips of bacon	1 order	350	**81**	27
with 1 sausage patty	1 order	430	**144**	54
Roast Beef Sandwich				
big	1	460	**216**	81
regular	1	320	**144**	54
Roast Beef Sub	1	370	**45**	27
Salads				
Chef salad	1	200	**117**	72
Garden, no dressing	1	220	**117**	63
Grilled chicken, no dressing	1	150	**27**	9
Potato salad	1 small	260	**171**	27
Side, no dressing	1	25	**9**	5
Shrimp 'n Pasta Salad	1 serving	362	**261**	NA
Shakes				
Chocolate	1	370	**45**	27
Peach	1	390	**36**	27
Strawberry	1	420	**36**	27
Vanilla	1	350	**45**	27
Turkey Club Sandwich	1	390	**144**	36
Turkey sub	1	390	**63**	36
JACK IN THE BOX				
Apple turnover	1	350	**171**	36
Bacon & Cheddar Potato Wedges	1 order	800	**522**	144
Breakfast Jack	1	280	**108**	45
Cheeseburger	1	320	**135**	54
Bacon bacon	1	710	**405**	135
Double	1	450	**216**	108
Ultimate	1	1030	**711**	234

FAST FOODS

FOOD	AMOUNT	CALORIES TOTAL	FAT	SAT-FAT
Cheesecake	1 piece	310	**162**	81
Chocolate Chip Cookie Dough	1 piece	360	**162**	72
Chicken and mushroom sandwich	1	438	**162**	45
Chicken Caesar Sandwich	1	490	**234**	54
Chicken Fajita Pita	1	280	**81**	27
Chicken Sandwich	1	400	**162**	36
Chicken strips	4 pieces	290	**117**	27
Chicken Supreme Sandwich	1	620	**324**	99
Chicken Teriyaki Bowl	1	670	**36**	9
Chicken wings	6 pieces	846	**396**	96
Country Fried Steak Sandwich	1	450	**225**	63
Crescent				
Sausage	1	670	**432**	171
Supreme	1	570	**324**	135
Curly fries	1 order	360	**180**	45
Dipping sauces				
Buttermilk House	1 oz	130	**117**	45
Tartar	1 oz	150	**135**	9
all others	1 oz	40	**0**	0
Double fudge cake	1 piece	288	**81**	20
Egg rolls	3 pieces	440	**216**	63
Fish Supreme Sandwich	1	590	**306**	54
French fries				
jumbo	1 order	400	**171**	45
regular	1 order	350	**153**	36
small	1 order	220	**99**	23
super scoop	1 order	590	**261**	63
Grilled Chicken Fillet Sandwich	1	430	**171**	45
Guacamole	28 g (1 oz)	50	**36**	5
Hamburger	1	276	**108**	37
Colossus	1	1100	**756**	252
Grilled Sourdough burger	1	670	**387**	144
Ham & Swiss burger	1	638	**351**	NA
Jumbo Jack	1	590	**324**	99
with Cheese	1	680	**396**	144
Mushroom burger	1	477	**243**	NA
Outlaw burger	1	720	**360**	153
Quarter-pound burger	1	510	**243**	90

FAST FOODS

FOOD	AMOUNT	CALORIES TOTAL	FAT	SAT-FAT
Hamburger (cont.)				
Sourdough Jack	1	690	**414**	144
Swiss & bacon burger	1	643	**387**	NA
Hash browns	1 order	170	**108**	27
Jumbo Jack	1	584	**306**	99
with Cheese	1	677	**360**	126
Milk shakes				
Chocolate, regular	1	390	**54**	32
Strawberry, regular	1	330	**63**	36
Vanilla, regular	1	350	**63**	36
Mini chimichangas	4 pieces	571	**252**	77
Moby Jack	1	444	**225**	NA
Old-fashioned Patty Melt	1	713	**414**	133
Monterey Roast Beef Sandwich	1	540	**270**	81
Onion rings	1 order	380	**207**	54
Pancake platter	1	400	**108**	27
Really Big Chicken Sandwich	1	900	**504**	126
Salad				
Chef salad	1	325	**162**	76
Garden chicken	1	200	**81**	36
Pasta seafood salad	1 serving	394	**198**	NA
Side	1	70	**36**	23
Taco salad	1	503	**279**	121
Scrambled egg				
platter	1	560	**288**	78
pocket	1	430	**189**	72
Shakes				
Chocolate	1	330	**63**	39
Oreo Cookie Ice Cream	1	740	**324**	NA
Strawberry	1	320	**63**	39
Vanilla	1	320	**54**	32
Sirloin Steak Sandwich	1	517	**207**	45
Sourdough Breakfast Sandwich	1	380	**180**	63
Spicy Crispy Chicken Sandwich	1	560	**243**	45
Stuffed Jalapeños				
7 pieces	1 order	530	**279**	117
10 pieces	1 order	750	**396**	162

FAST FOODS

FOOD	AMOUNT	CALORIES TOTAL	CALORIES FAT	CALORIES SAT-FAT
Supreme nachos	1 serving	718	**360**	NA
Taco				
regular	1	190	**99**	36
super	1	280	**153**	54
Toasted raviolis	7 pieces	537	**252**	72
Tortilla chips	1 serving	139	**54**	NA
Ultimate Breakfast Sandwich	1	620	**315**	99
KFC (KENTUCKY FRIED CHICKEN)				
BBQ baked beans	1 serving (156 g)	190	**25**	9
Biscuit	1 (56 g)	180	**80**	23
Chicken Little Sandwich	1	169	**90**	18
Chunky Chicken Pot Pie	1 (368 g)	770	**378**	117
Coleslaw	1 serving (142 g)	142	**80**	14
Corn on the cob	1 serving (162 g)	150	**15**	0
Cornbread	1 (56 g)	228	**117**	18
Green beans	1 serving (132 g)	45	**15**	5
Hot Wings	6 pcs (135 g)	471	**297**	72
Kentucky Fried Chicken				
Colonel's Rotisserie Gold Chicken				
Quarter-breast & wing	1 serving (176 g)	335	**168**	49
without skin & wing	1 serving (116 g)	199	**53**	15
Quarter-thigh & leg	1 serving (145 g)	333	**213**	59
without skin	1 serving (116 g)	217	**110**	32
Crispy Strips				
Colonel's	3 (92 g)	261	**142**	33
Spicy Buffalo	3 (120 g)	350	**170**	36
Extra Tasty Crispy Chicken				
Breast	1 (168 g)	470	**250**	**63**
Drumstick	1 (67 g)	190	**100**	**27**
Thigh	1 (118 g)	370	**220**	**54**
Whole Wing	1 (55 g)	200	**120**	**36**
Hot & Spicy Chicken				
Breast	1 (180 g)	530	**310**	**72**
Drumstick	1 (64 g)	190	**100**	**27**
Thigh	1 (107 g)	370	**240**	**63**
Whole Wing	1 (55 g)	210	**130**	36

FAST FOODS

FOOD	AMOUNT	CALORIES TOTAL	CALORIES FAT	CALORIES SAT-FAT
Kentucky Fried Chicken (cont.)				
Original Recipe Chicken				
Breast	1 (153 g)	400	**220**	54
Drumstick	1 (61 g)	140	**80**	18
Thigh	1 (91 g)	250	**160**	41
Whole Wing	1 (47 g)	140	**90**	23
Tender Roast Chicken				
Breast with skin	1 (139 g)	251	**97**	27
without skin	1 (118 g)	169	**39**	11
Drumstick with skin	1 (55 g)	97	**39**	11
without skin	1 (38 g)	67	**22**	6
Thigh with skin	1 (90 g)	207	**126**	34
without skin	1 (59 g)	106	**50**	15
Whole Wing with skin	1 (50 g)	121	**69**	19
Kentucky Nuggets	6 pieces	284	**162**	54
Macaroni & cheese	1 serving (153 g)	180	**70**	27
Mashed potatoes with gravy	1 serving (136 g)	120	**50**	9
Mean Greens	1 serving (152 g)	70	**30**	9
Original Recipe Chicken Sandwich	1 (206 g)	497	**201**	43
Potato Salad	1 serving (160 g)	230	**130**	18
Potato Wedges	1 serving (135 g)	180	**70**	36
Value BBQ Flavored Chicken Sandwich	1 (149 g)	256	**74**	9
LONG JOHN SILVER'S				
Baked Chicken with light herb	1 order	120	**36**	11
dinner	1 order	550	**135**	29
Baked Fish with lemon crumb	1 order (3 pcs)	150	**9**	5
dinner	1 order	570	**108**	19
Baked shrimp	1 order	120	**45**	NA
Batter-dipped chicken	1 piece (57 g)	120	**50**	14
Batter-dipped chicken sandwich, no sauce	1 piece	280	**72**	19
Batter-dipped clams	85 g	300	**150**	36
Batter-dipped fish sandwich, no sauce	1 piece (153 g)	320	**120**	32

FAST FOODS

FOOD	AMOUNT	CALORIES TOTAL	FAT	SAT-FAT
Batter-dipped fish	1 piece (84 g)	170	**100**	24
Batter-dipped shrimp	1 piece (11 g)	35	**20**	5
Battered-fried shrimp	1 piece	47	**27**	6
dinner	1 serving	711	**405**	NA
Breaded clams	1 order	526	**279**	46
Breaded oyster	1 piece	60	**27**	NA
Breaded shrimp	1 order	388	**207**	21
platter	1 order	962	**513**	NA
Cheese sticks	45 g	160	**80**	36
Chicken nuggets dinner	6 pieces	699	**405**	NA
Chicken Planks	2 pieces	240	**108**	29
with fries	2 pieces	490	**234**	51
with fries, coleslaw, 2 hush puppies	3 pieces	890	**396**	86
for kids, incl. fries and hush puppy	2 pieces	560	**261**	57
Chocolate chip cookie	1	230	**81**	51
Chowders				
Clam	1 serving (6.6 oz)	128	**45**	16
Seafood chowder	1 cup	140	**54**	18
Seafood gumbo	1 cup	120	**72**	19
Clams with fries, coleslaw, 2 hush puppies	1 dinner	980	**468**	99
Coleslaw	1 order (96 g)	140	**60**	9
Combination entrees, with fries, slaw, 2 hush puppies				
1 fish + 2 chicken	1 order	950	**441**	99
2 fish + 8 shrimp	1 order	1140	**585**	127
2 fish + 5 shrimp + 1 chicken	1 order	1160	**585**	128
2 fish + 4 shrimp + clams (3 oz)	1 order	1240	**630**	137
Corn cobbette	1 piece (111 g)	140	**70**	14
Crispy fish	1 piece	150	**72**	20
with fries, coleslaw, and 2 hush puppies	3 pieces	980	**450**	102
Fish				
for kids, incl. fries and hush puppy	1 piece	500	**252**	52

FAST FOODS

FOOD	AMOUNT	CALORIES TOTAL	CALORIES FAT	CALORIES SAT-FAT
Fish and chicken				
for kids, incl. fries and hush puppy	1 piece each	620	**306**	67
with fries	1 piece each	550	**288**	61
Fish & Fryes				
2 pieces fish	1 order	651	**324**	72
3 pieces fish	1 order	853	**432**	91
Fish & More, with fries, coleslaw, & 2 hush puppies	2 pieces	890	**432**	91
Fish dinner, fried, 3-piece	1 order	1180	**630**	NA
Fish sandwich, homestyle	1 order	510	**198**	44
Flavorbaked chicken	74 g	110	**30**	9
with rice, green beans, and baked potato	1 piece	448	**68**	NA
with rice and side salad	1 piece	275	**63**	NA
Flavorbaked chicken sandwich	1 (165 g)	290	**90**	18
Flavorbaked fish	1 piece (65 g)	90	**25**	9
with rice, green beans, and baked potato	2 pieces	518	**86**	NA
with rice and side salad	2 pieces	345	**81**	NA
Flavorbaked fish and chicken combo				
with rice, baked potato, and green beans	1 piece each	538	**90**	NA
Flavorbaked fish sandwich	1 (170 g)	320	**120**	63
Fries	1 order (85 g)	250	**130**	23
Green beans	1 order (99 g)	30	**5**	0
Hush puppy	1 piece (23 g)	60	**20**	0
Oatmeal raisin cookie	1	160	**90**	18
Ocean Chef Salad	1	110	**9**	4
Oyster dinner	1 order	789	**405**	NA
Pies				
Apple	1 piece	320	**117**	41
Cherry	1 piece	360	**117**	40
Lemon	1 piece	340	**81**	27
Pineapple cream cheese cake	1 piece (991 g)	310	**162**	81
Popcorn				
chicken	94 g	250	**120**	32

FAST FOODS

FOOD	AMOUNT	CALORIES TOTAL	FAT	SAT-FAT
Popcorn (cont.)				
fish	102 g	290	**130**	36
shrimp	94 g	280	**130**	36
Regular chicken classic wrap	1 (312 g)	730	**320**	63
Rice pilaf	1 order (85 g)	140	**25**	9
Salads, no dressing				
Garden	1	45	**0**	0
Grilled chicken	1	140	**20**	5
Ocean chef	1	130	**20**	0
Side	1	25	**0**	0
Scallop dinner	1 order	747	**405**	NA
Saltines crackers	2	25	**9**	NA
Seafood platter	1 order	976	**522**	NA
Seafood salad	1	380	**279**	46
Shrimp with fries, coleslaw, & 2 hush puppies	10 pieces	840	**423**	87
Tartar sauce	1	50	**45**	9
Ultimate fish sandwich	1 (182 g)	430	**190**	63
Walnut brownie	1	440	**198**	49
MCDONALD'S				
Apple Bran Muffin, lowfat	1	300	**30**	5
Arch Deluxe	1	550	**280**	99
with bacon	1	590	**310**	108
Bacon, Egg, & Cheese Biscuit	1	470	**250**	72
Baked apple pie	1	260	**120**	32
Big Mac	1	560	**280**	90
Biscuit	1	290	**130**	27
Breakfast Burrito	1 order	320	**180**	63
Cheeseburger	1	320	**120**	54
Chicken McNuggets				
4 pieces	1 order	190	**100**	23
6 pieces	1 order	290	**150**	32
9 pieces	1 order	430	**230**	45
Cinnamon roll	1	400	**180**	45
Cookies				
Chocolate chip	1	170	**90**	54
McDonaldland	1 pkg	180	**45**	9

FAST FOODS

FOOD	AMOUNT	CALORIES TOTAL	CALORIES FAT	CALORIES SAT-FAT
Crispy Chicken Deluxe	1	500	**220**	36
Danish				
Apple	1	360	**140**	45
Cheese	1	410	**200**	72
Cinnamon raisin	1	430	**198**	63
Iced cheese	1	390	**198**	54
Raspberry	1	400	**144**	45
Egg McMuffin	1 order	290	**110**	41
English muffin	1 order	140	**20**	0
Fish Fillet Deluxe	1	560	**250**	54
Fries				
small	1 order	210	**90**	14
large	1 order	450	**200**	36
super size	1 order	540	**230**	41
Grilled Chicken Deluxe	1	440	**180**	27
Hamburger	1	130	**70**	14
Hash browns	1 order	144	**81**	26
Hotcakes	1 order	310	**60**	14
with syrup and 2 pats of margarine	1 order	510	**140**	27
McChicken	1	490	**261**	45
McD.L.T.	1 order	680	**396**	133
McLean Deluxe	1	340	**108**	41
with cheese	1	400	**144**	63
Quarter Pounder	1	420	**190**	72
with cheese	1	530	**270**	117
Salads & dressings				
Caesar dressing	1 pkg	160	**130**	27
Croutons	1 pkg	50	**10**	0
Fat-free herb vinaigrette	1 pkg	50	**0**	0
Garden salad	1	35	**0**	0
Grilled Chicken Salad Deluxe	1	120	**10**	0
Ranch dressing	1 pkg	230	**180**	27
Red French Reduced Calorie	1 pkg	160	**70**	9
Sauces				
Barbeque, Sweet 'n Sour, Honey	1 pkg	45–50	**0**	0
Honey mustard	1 pkg	50	**40**	5

FAST FOODS

FOOD	AMOUNT	CALORIES TOTAL	CALORIES FAT	CALORIES SAT-FAT
Sauces (cont.)				
Hot mustard	1 pkg	60	**30**	0
Light mayonnaise	1 pkg	40	**35**	5
Sausage	1	170	**150**	45
Sausage Biscuit	1	470	**280**	81
with egg	1	550	**330**	90
Sausage McMuffin	1	360	**210**	72
with egg	1	440	**260**	90
Scrambled eggs	2	160	**110**	32
Shakes				
Chocolate	1 small	360	**80**	54
Strawberry	1 small	360	**80**	54
Vanilla	1 small	360	**80**	54
Soft serve and cone	1 serving	189	**45**	20
Sundae & toppings				
Hot caramel	1	360	**90**	54
Hot fudge	1	340	**90**	54
Strawberry	1	290	**70**	45
nuts	1 serving	40	**30**	0
Vanilla reduced-fat ice cream cone	1	150	**40**	27
PIZZA HUT				
Bigfoot pizza				
Cheese	2 slices	372	**108**	54
Pepperoni	2 slices	410	**126**	54
Pepperoni, mushroom, & Italian sausage	2 slices	428	**144**	72
Hand-tossed pizza, medium				
Beef	2 slices	520	**162**	72
Cheese	2 slices	470	**126**	72
Ham	2 slices	426	**90**	54
Italian sausage	2 slices	534	**198**	90
Meat Lovers	2 slices	628	**198**	108
Pepperoni	2 slices	476	**144**	72
Pepperoni Lovers	2 slices	612	**252**	108
Pork topping	2 slices	536	**180**	90
Super Supreme	2 slices	592	**234**	90

FAST FOODS

FOOD	AMOUNT	CALORIES TOTAL	CALORIES FAT	CALORIES SAT-FAT
Hand-tossed pizza, medium (cont.)				
Supreme	2 slices	568	**216**	90
Veggie Lovers	2 slices	432	**108**	54
Pan pizza, medium				
Beef	2 slices	572	**234**	90
Cheese	2 slices	522	**198**	90
Ham	2 slices	478	**162**	54
Italian sausage	2 slices	586	**270**	90
Meat Lovers	2 slices	680	**324**	126
Pepperoni	2 slices	530	**216**	72
Pepperoni Lovers	2 slices	664	**306**	126
Pork topping	2 slices	588	**252**	90
Super Supreme	2 slices	646	**306**	108
Supreme	2 slices	622	**270**	108
Veggie Lovers	2 slices	486	**180**	54
Personal pan pizza				
Pepperoni	1 pizza	637	**252**	90
Supreme	1 pizza	722	**306**	108
Thin 'n Crispy pizza, medium				
Beef	2 slices	458	**198**	90
Cheese	2 slices	410	**144**	72
Ham	2 slices	368	**126**	54
Italian sausage	2 slices	472	**216**	90
Meat Lovers	2 slices	576	**234**	108
Pepperoni	2 slices	430	**180**	72
Pepperoni Lovers	2 slices	578	**288**	126
Pork topping	2 slices	474	**216**	90
Super Supreme	2 slices	540	**252**	108
Supreme	2 slices	514	**234**	90
Veggie Lovers	2 slices	372	**126**	54
POPEYE'S CHICKEN				
Apple pie	1 slice	290	**144**	NA
Biscuit	1	250	**135**	NA
Cajun rice	111 g	150	**50**	NA
Cole slaw	114 g	149	**100**	NA
Corn on the cob	1	127	**27**	NA
French fries	1 order	240	**108**	NA

FAST FOODS

FOOD	AMOUNT	CALORIES TOTAL	FAT	SAT-FAT
Fried chicken				
Breast, mild or spicy	1 (105 g)	270	**144**	NA
Leg, mild or spicy	1 (48 g)	120	**66**	NA
Tender, mild or spicy	1 (34 g)	110	**63**	NA
Thigh, mild or spicy	1 (88 g)	300	**204**	NA
Wing, mild or spicy	1 (45 g)	160	**96**	NA
Fried shrimp	1 order (79 g)	250	**148**	NA
Nuggets	1 order	410	**288**	NA
Onion rings	1 order	310	**174**	NA
Potatoes & gravy	108 g	100	**54**	NA
Red beans & rice	167 g	270	**153**	NA
ROY ROGERS				
Baked beans	1 order (142 g)	160	**18**	9
Baked potato	1 (110 g)	130	**9**	0
with margarine	1 (124 g)	240	**117**	18
Big Breakfast Platter with ham	1 (267 g)	710	**351**	99
Big Country Breakfast Platter				
with bacon	1 (217 g)	740	**387**	117
with sausage	1 (274 g)	920	**540**	171
Biscuit	1 (83 g)	390	**189**	54
Bacon & egg	1 (121 g)	470	**234**	72
Bacon	1 (89 g)	420	**207**	63
Cinnamon 'n raisin	1 (80 g)	370	**162**	45
Ham & cheese	1 (127 g)	450	**216**	72
Ham, egg, & cheese	1 (159 g)	500	**243**	90
Sausage & egg	1 (149 g)	560	**315**	99
Sausage	1 (118 g)	510	**279**	90
Biscuits 'n Gravy	1 order (221 g)	510	**252**	81
Cheeseburger	1 (140 g)	393	**198**	108
¼ lb cheeseburger	1 (149 g)	480	**261**	153
Bacon	1 (156 g)	520	**297**	162
Cheesesteak Sandwich	1 (216 g)	580	**324**	126
Chicken soup	1 order (302 g)	225	**63**	18
Chili	1 order (288 g)	295	**135**	63
Coleslaw	1 order (142 g)	295	**225**	36
Fish Sandwich	1 (186 g)	490	**189**	45

FAST FOODS

FOOD	AMOUNT	CALORIES TOTAL	FAT	SAT-FAT
French fries				
large	1 order (173 g)	430	**162**	45
regular	1 order (142 g)	350	**135**	36
Fried chicken				
Breast	1 piece (148 g)	370	**135**	36
Leg	1 piece (69 g)	170	**63**	18
Thigh	1 piece (121 g)	330	**135**	36
Wing	1 piece (66 g)	200	**72**	18
Gold Rush Chicken Sandwich	1 (190 g)	558	**270**	81
Grilled Chicken Sandwich	1 (213 g)	294	**72**	27
Gravy	43 g	20	**5**	5
Hamburger	1 (126 g)	343	**162**	81
¼ lb burger	1 (135 g)	412	**225**	126
Mashed potatoes	1 order (142 g)	92	**5**	5
Nuggets	6 pieces (113 g)	290	**162**	36
Roast beef sandwich	1 (192 g)	329	**90**	27
¼ Roy Roaster				
Dark meat, with skin	1 piece (144 g)	490	**306**	90
without skin	1 piece (111 g)	190	**90**	27
White meat, with skin	1 piece (175 g)	500	**261**	81
without skin	1 piece (134 g)	190	**54**	18
Salads				
Garden	1 (278 g)	110	**45**	27
Grilled Chicken	1 (378 g)	221	**81**	36
Side	1 (140 g)	20	**5**	5
Sourdough ham, egg, & cheese	1 (193 g)	480	**216**	81
Strawberry shortcake	1 order (171 g)	440	**171**	45
Sundaes				
Hot fudge	1 (168 g)	320	**90**	45
Strawberry	1 (158 g)	260	**54**	27
3 pancakes	1 order (137 g)	280	**18**	9
with 1 sausage	1 order (176 g)	430	**144**	54
with 2 bacon	1 order (151 g)	350	**81**	27
Vanilla frozen yogurt cone	1 (118 g)	180	**36**	27
SUBWAY				
Cookies				
Brazil nut & chocolate chips	1	230	**108**	NA

FAST FOODS

FOOD	AMOUNT	CALORIES TOTAL	CALORIES FAT	CALORIES SAT-FAT
Cookies (cont.)				
Chocolate chip	1	210	**90**	NA
Chocolate chip/M&M	1	210	**90**	NA
Chocolate chunk	1	210	**90**	NA
Oatmeal raisin	1	200	**72**	NA
Peanut butter	1	220	**108**	NA
Sugar	1	230	**108**	NA
White chip macadamia nut	1	230	**108**	NA
6-inch cold sandwiches (include meat/poultry/seafood, onions, lettuce, tomatoes, pickles, green peppers, olives)				
BLT				
wheat bread	198 g	327	**95**	27
white bread	191 g	311	**93**	27
Classic Italian BMT				
wheat bread	253 g	460	**194**	63
white bread	246 g	445	**193**	72
Cold-Cut Trio				
wheat bread	253 g	378	**118**	36
white bread	246 g	362	**116**	36
Ham				
wheat bread	239 g	302	**48**	9
white bread	232 g	287	**47**	9
Roast beef				
wheat bread	239 g	303	**42**	9
white bread	232 g	288	**41**	9
Spicy Italian				
wheat bread	239 g	482	**222**	81
white bread	232 g	467	**220**	81
Subway Club				
wheat bread	253 g	312	**45**	9
white bread	246 g	297	**43**	9
Subway Seafood & Crab				
wheat bread	253 g	430	**174**	27
made with light mayo	253 g	347	**89**	18
white bread	246 g	415	**172**	27
made with light mayo	246 g	332	**88**	18

FAST FOODS

FOOD	AMOUNT	CALORIES TOTAL	CALORIES FAT	CALORIES SAT-FAT
6-inch cold sandwiches (cont.)				
Tuna				
wheat bread	253 g	542	**291**	45
made with light mayo	253 g	391	**135**	18
white bread	246 g	527	**289**	45
made with light mayo	246 g	376	**134**	18
Turkey breast				
wheat bread	239 g	289	**36**	9
white bread	232 g	273	**34**	9
Turkey breast & ham				
wheat bread	239 g	295	**42**	9
white bread	232 g	280	**41**	9
Veggie Delite				
wheat bread	182 g	237	**25**	0
white bread	175 g	222	**24**	0
6-inch hot sandwiches (include meat/poultry/seafood, onions, lettuce, tomatoes, pickles, green peppers, olives)				
Chicken taco sub (values include cheese)				
wheat bread	293 g	436	**144**	45
white bread	286 g	421	**143**	45
Meatball				
wheat bread	267 g	419	**145**	54
white bread	260 g	404	**144**	54
Pizza sub (values incl. cheese)				
wheat bread	257 g	383	**196**	81
white bread	250 g	448	**194**	81
Roasted chicken breast				
wheat bread	253 g	348	**54**	9
white bread	246 g	332	**52**	9
Steak & cheese (values incl. cheese)				
wheat bread	264 g	398	**88**	54
white bread	257 g	383	**86**	54
Subway Melt (values incl. cheese)				
wheat bread	258 g	382	**107**	45
white bread	251 g	366	**106**	45
Deli Style Sandwiches (include deli roll, meat/poultry/seafood, onions, lettuce, tomatoes, pickles, green peppers, olives)				
Bologna	171 g	292	**104**	36

FAST FOODS

FOOD	AMOUNT	CALORIES TOTAL	CALORIES FAT	CALORIES SAT-FAT
Deli Style Sandwiches (cont.)				
Ham	171 g	234	**40**	9
Roast beef	180 g	245	**40**	9
Tuna	178 g	354	**161**	27
made with light mayo	178 g	279	**84**	18
Turkey breast	180 g	235	**35**	9
Optional fixin's for sandwiches				
bacon	2 slices (8 g)	45	**36**	9
cheese	2 triangles (11 g)	41	**27**	18
light mayo	1 tsp (5 g)	18	**18**	0
mayo	1 tsp (5 g)	37	**36**	9
oil	1 tsp (5 g)	45	**45**	9
Breads				
Deli style roll	62 g	185	**26**	9
6" wheat bread	77 g	215	**22**	0
12" wheat bread	154 g	430	**45**	9
6" white bread	70 g	200	**21**	0
12" white bread	140 g	400	**42**	9
Salads (values do not include salad dressing)				
BLT	276 g	140	**76**	27
Chicken taco (includes cheese)	370 g	250	**126**	45
Classic Italian BLT	331 g	274	**176**	63
Cold-Cut Trio	330 g	191	**100**	27
Ham	316 g	116	**30**	9
Meatball	345 g	233	**127**	45
Pizza (includes cheese)	335 g	277	**177**	72
Roast beef	316 g	117	**24**	9
Roasted chicken breast	331 g	162	**35**	9
Steak & cheese (includes cheese)	342 g	212	**69**	45
Subway Club	331 g	126	**26**	9
Subway Melt	336 g	195	**89**	36
Subway Seafood & Crab	331 g	244	**155**	27
made with light mayo	331 g	161	**71**	9
Tuna	331 g	356	**272**	45
made with light mayo	331 g	205	**117**	18
Turkey breast	316 g	102	**17**	9
Turkey breast & ham	316 g	109	**24**	9
Veggie Delite	260 g	51	**7**	0

FAST FOODS

FOOD	AMOUNT	CALORIES TOTAL	FAT	SAT-FAT
TACO BELL				
Border Wraps				
Chicken Fajita Wrap	1 order (227 g)	470	**200**	54
Chicken Fajita Wrap Supreme	1 order (255 g)	520	**230**	72
Steak Fajita Wrap	1 order (227 g)	470	**190**	54
Steak Fajita Wrap Supreme	1 order (255 g)	510	**220**	72
Veggie Fajita Wrap	1 order (227 g)	420	**170**	45
Veggie Fajita Wrap Supreme	1 order (255 g)	470	**200**	63
Breakfast				
Breakfast Cheese Quesadilla	1 order (156 g)	380	**190**	81
Breakfast Quesadilla with Bacon	1 order (170 g)	450	**240**	99
Breakfast Quesadilla with Sausage	1 order (170 g)	430	**230**	90
Country Breakfast Burrito	1 order (114 g)	270	**130**	45
Double Bacon & Egg Burrito	1 order (177 g)	480	**250**	81
Fiesta Breakfast Burrito	1 order (99 g)	280	**140**	54
Grande Breakfast Burrito	1 order (177 g)	420	**200**	63
Hash Brown Nuggets	1 order (99 g)	280	**160**	45
Burrito				
7-Layer Burrito	1 order (284 g)	530	**200**	63
Bacon Cheeseburger Burrito	1 order (241 g)	570	**280**	108
Bean Burrito	1 order (199 g)	380	**110**	36
Big Beef Burrito Supreme	1 order (298 g)	520	**210**	90
Big Chicken Burrito Supreme	1 order (255 g)	510	**210**	63
Burrito Supreme	1 order (255 g)	440	**170**	72
Chicken Club Burrito	1 order (227 g)	540	**290**	90
Chili Cheese Burrito	1 order (142 g)	330	**120**	54
Grilled Chicken Burrito	1 order (199 g)	410	**140**	41
Nachos and Sides				
Big Beef Nachos Supreme	1 order (199 g)	450	**220**	72
Choco Taco Ice Cream Dessert	1 order (114 g)	310	**150**	90
Cinnamon Twists	1 order (28 g)	140	**50**	0
Mexican rice	1 order (135 g)	190	**80**	32
Nachos	1 order (99 g)	320	**170**	36
Nachos BellGrande	1 order (312 g)	770	**360**	99
Pintos 'n Cheese	1 order (128 g)	190	**80**	36

FAST FOODS

FOOD	AMOUNT	CALORIES TOTAL	CALORIES FAT	CALORIES SAT-FAT
Sauces and Condiments				
Burger Sauce	14 g	60	**45**	9
Cheddar cheese	7 g	30	**20**	14
Club Sauce	14 g	80	**70**	9
Fajita Sauce	14 g	70	**60**	9
Guacamole	21 g	35	**30**	0
Nacho Cheese Sauce	57 g	120	**90**	23
Sour cream	21 g	40	**35**	23
Three-Cheese Blend	7 g	25	**15**	9
all other sauces are fat-free				
Specialties				
Big Beef MexiMelt	1 order (135 g)	290	**140**	63
Cheese Quesadilla	1 order (121 g)	350	**160**	81
Chicken Quesadilla	1 order (170 g)	410	**190**	90
Mexican Pizza	1 order (220 g)	570	**320**	90
Taco Salad with Salsa	1 order (539 g)	850	**470**	135
without shell	1 order (468 g)	420	**200**	99
Tostada	1 order (177 g)	300	**130**	45
Taco	1 order (78 g)	180	**90**	36
BLT Soft Taco	1 order (128 g)	340	**210**	72
Double Decker Taco	1 order (163 g)	340	**130**	45
Double Decker Taco Supreme	1 order (199 g)	390	**170**	72
Grilled Chicken Soft Taco	1 order (128 g)	240	**110**	32
Grilled Steak Soft Taco	1 order (128 g)	230	**90**	23
Grilled Steak Soft Taco Supreme	1 order (163 g)	290	**130**	45
Soft Taco	1 order (99 g)	220	**90**	41
Soft Taco Supreme	1 order (142 g)	260	**120**	63
Taco Supreme	1 order (114 g)	220	**120**	63
WENDY'S				
Baked potato, plain	1 (284 g)	284	**0**	0
Bacon & cheese	1 (380 g)	530	**160**	36
Broccoli & cheese	1 (411 g)	470	**130**	23
Cheese	1 (383 g)	570	**210**	72
Chili & cheese	1 (439 g)	630	**220**	81
Sour cream & chives	1 (314 g)	380	**60**	36

FAST FOODS

FOOD	AMOUNT	CALORIES TOTAL	CALORIES FAT	CALORIES SAT-FAT
Baked potato (cont.)				
Sour cream	1 pkt (28 g)	60	**50**	32
Whipped margarine	1 pkt (14 g)	60	**60**	14
Big Bacon Classic sandwich	1 (282 g)	580	**270**	108
Breaded chicken	1 fillet (99 g)	230	**100**	23
Sandwich	1 (208 g)	440	**160**	32
Chicken club sandwich	1 (216 g)	470	**180**	36
Cheeseburger				
Double	1	590	**297**	128
Jr. cheeseburger	1 (130 g)	320	**120**	54
Bacon	1 (166 g)	380	**170**	63
Deluxe	1 (180 g)	360	**150**	54
Kid's meal	1 (123 g)	320	**120**	54
Chicken nuggets				
5 piece	75 g	210	**130**	27
4 piece, kid's	60 g	170	**100**	23
Barbeque sauce	1 pkt (28 g)	45	**0**	0
Honey mustard	1 pkt (28 g)	130	**100**	18
Spicy Buffalo Wing sauce	1 pkt (28 g)	25	**10**	0
Sweet & sour sauce	1 pkt (28 g)	50	**0**	0
Chili				
large	1 (340 g)	310	**60**	23
small	1 (227 g)	210	**54**	18
cheddar cheese, shredded	2 tbsp (17 g)	70	**50**	32
Saltine crackers	2 (6 g)	25	**5**	0
Chocolate chip cookie	1 (57 g)	270	**110**	54
Chow mein noodles	¼ cup	74	**36**	5
Cole slaw	½ cup	90	**72**	18
Country fried steak sandwich	1	460	**234**	63
Fish sandwich	1	460	**225**	42
French fries				
biggie	1 order (159 g)	470	**200**	32
medium	1 order (130 g)	390	**170**	27
small	1 order (91 g)	270	**120**	18
Frosty dairy dessert				
large	1 (369 g)	540	**120**	81
medium	1 (298 g)	440	**100**	63
small	1 (227 g)	330	**80**	45

FAST FOODS

FOOD	AMOUNT	CALORIES TOTAL	FAT	SAT-FAT
Fruit-flavored drink	12 fl oz	110	**0**	0
Grilled chicken	1 fillet (82 g)	110	**25**	9
sandwich	1 (189 g)	310	**70**	14
Hamburger	1 (133 g)	360	**150**	54
with everything	1 (219 g)	420	**180**	63
Double	1	520	**243**	96
Jr. hamburger	1 (118 g)	270	**90**	32
Kid's meal	1 (111 g)	270	**90**	32
Hot chocolate	6 fl oz (170 g)	80	**25**	0
Pitas				
Chicken Caesar	1 (237 g)	490	**160**	45
Classic Greek	1 (234 g)	440	**180**	72
Garden ranch chicken	1 (283 g)	480	**160**	36
Garden veggie	1 (257 g)	400	**150**	32
Pita dressings				
Caesar vinaigrette	1 tbsp (17 g)	70	**60**	9
Garden ranch	1 tbsp (16 g)	50	**40**	9
Salads, without dressing				
Caesar side salad	1 (89 g)	100	**35**	14
Deluxe garden	1 (270 g)	110	**50**	9
Grilled chicken Caesar	1 (262 g)	260	**80**	27
Grilled chicken salad	1 (338 g)	200	**70**	14
Side salad	1 (262 g)	60	**25**	0
Taco salad	1 (468 g)	468	**380**	90
Salad dressings (2 tbsp = 1 ladle)				
Blue cheese	2 tbsp (28 g)	180	**170**	32
French	2 tbsp (28 g)	120	**90**	14
fat-free	2 tbsp (28 g)	35	**0**	0
Italian Caesar	2 tbsp (28 g)	150	**140**	23
reduced fat, reduced calorie	2 tbsp (28 g)	40	**30**	0
Hidden Valley ranch	2 tbsp (28 g)	100	**90**	14
reduced fat	2 tbsp (28 g)	60	**50**	9
Salad oil	1 tbsp (14 g)	120	**120**	18
Thousand Island	2 tbsp (28 g)	90	**80**	14
Wine vinegar	1 tbsp (14 g)	0	**0**	0
Soft breadstick	1 (44 g)	130	**30**	5
Spicy chicken	1 fillet (104 g)	210	**80**	14
sandwich	1 (213 g)	410	**130**	23

FAST FOODS

FOOD	AMOUNT	CALORIES TOTAL	FAT	SAT-FAT
Sunflower seeds and raisins	2 tbsp	140	**90**	67
Taco chips	15 (42 g)	210	**100**	14
Turkey ham	¼ cup	35	**18**	4
WHITE CASTLE				
Breakfast sandwich	1	340	**220**	NA
Cheese sticks	1 order	290	**150**	NA
Cheeseburger	1	160	**85**	NA
bacon	1	200	**115**	NA
double	1	285	**165**	NA
Chicken rings	1 order	310	**190**	NA
Chicken sandwich	1	190	**70**	NA
Chocolate shake	1	220	**60**	NA
Fish sandwich	1	160	**60**	NA
French fries	1 order	115	**50**	NA
Hamburger	1	135	**65**	NA
double	1	235	**125**	NA
Onion chips	1 order	180	**80**	NA

FATS AND OILS

FOOD	AMOUNT	CALORIES TOTAL	FAT	SAT-FAT
ANIMAL FATS				
Beef tallow	1 tbsp	116	**116**	58
Butter				
regular	1 pat	36	**36**	23
	1 tbsp	100	**100**	65
	1 stick	813	**813**	515
whipped	1 tsp	23	**23**	12
	1 tbsp	67	**67**	38
Chicken fat	1 tbsp	115	**115**	34
Duck fat	1 tbsp	115	**115**	39
Goose fat	1 tbsp	115	**115**	32
Lard (pork)	1 tbsp	116	**116**	45
Mutton tallow	1 tbsp	116	**116**	55
Turkey fat	1 tbsp	115	**115**	34

FATS AND OILS

FOOD	AMOUNT	CALORIES TOTAL	CALORIES FAT	CALORIES SAT-FAT
BUTTER SUBSTITUTES				
Butter Buds Sprinkles	1 tsp (2 g)	8	**0**	0
Molly McButter	1 tsp (2 g)	5	**0**	0
MARGARINES				
Stick				
Brummel & Brown	1 tbsp (14 g)	90	**90**	18
Fleischmann's	1 tbsp (14 g)	90	**90**	14
Lower Fat	1 tbsp (14 g)	50	**50**	9
I Can't Believe It's Not Butter	1 tbsp (14 g)	90	**90**	18
Imperial	1 tbsp (14 g)	90	**90**	18
Land O Lakes	1 tbsp (14 g)	100	**100**	18
Country Morning Blend	1 tbsp (14 g)	100	**100**	23
Light	1 tbsp (14 g)	50	**50**	27
Blue Bonnet	1 tbsp (14 g)	70	**70**	14
Parkay	1 tbsp (14 g)	90	**90**	14
Promise	1 tbsp (14 g)	90	**90**	23
Shedd's Spread Country Crock				
Churn Style	1 tbsp (14 g)	80	**80**	18
Spreadable Stick	1 tbsp (14 g)	80	**80**	14
Squeeze				
I Can't Believe It's Not Butter	1 tbsp (14 g)	80	**80**	14
Parkay	1 tbsp (14 g)	70	**70**	14
Tub				
Brummel & Brown	1 tbsp (14g)	50	**50**	9
Fleischmann's				
Soft spread	1 tbsp (14 g)	80	**80**	14
Lower Fat	1 tbsp (14 g)	40	**40**	0
I Can't Believe It's Not Butter	1 tbsp (14 g)	90	**90**	18
Light	1 tbsp (14 g)	50	**45**	9
Land O Lakes	1 tbsp (14 g)	100	**100**	18
Spread with sweet cream	1 tbsp (10 g)	80	**80**	18
Parkay				
Soft	1 tbsp (14 g)	90	**90**	18
Spread	1 tbsp (14 g)	60	**60**	14
Promise	1 tbsp (14 g)	80	**80**	18
Buttery Light	1 tbsp (14 g)	50	**50**	9

FATS AND OILS

FOOD	AMOUNT	CALORIES TOTAL	CALORIES FAT	CALORIES SAT-FAT
Promise (cont.)				
Ultra 70% less fat	1 tbsp (14 g)	30	**30**	0
Ultra Fat-Free	1 tbsp (14 g)	5	**0**	0
Shedd's Spread Country Crock	1 tbsp (14 g)	60	**60**	14
Light	1 tbsp (14 g)	50	**50**	9
Churn Style	1 tbsp (14 g)	60	**60**	14
Smart Beat, Super Light	1 tbsp (14 g)	20	**20**	0
SALAD DRESSINGS & SPREADS				
Balsamic & Basil Vinaigrette				
Ken's	2 tbsp	110	**110**	14
Balsamic Vinaigrette				
Newman's Own	2 tbsp	90	**80**	9
Blue cheese				
Chunky (Hellmann's)	2 tbsp	140	**130**	27
Chunky (Ken's)	2 tbsp	140	**140**	27
Chunky (Wish-Bone)	2 tbsp	170	**150**	23
Chunky, Fat-Free (Wish-Bone)	2 tbsp	35	**0**	0
Free (Kraft)	2 tbsp	45	**0**	0
Regular (Kraft)	2 tbsp	130	**120**	23
Safeway Select	2 tbsp	150	**140**	18
Burgundy Basil Vinaigrette (Ken's)	2 tbsp	70	**45**	0
Caesar				
Classic (Kraft)	2 tbsp	110	**90**	18
Classic (Wish-Bone)	2 tbsp	110	**90**	18
Creamy (Hellmann's)	2 tbsp	170	**160**	27
Fat-Free (Hidden Valley)	2 tbsp	30	**0**	0
Free Classic (Kraft)	2 tbsp	45	**0**	0
Gourmet Caesar (Good Seasons)	2 tbsp prepared	150	**140**	0
Italian Free (Kraft)	2 tbsp	25	**0**	0
Lite (Ken's)	2 tbsp	70	**60**	5
Regular (Hellmann's)	2 tbsp	100	**80**	9
Regular (Ken's)	2 tbsp	170	**160**	23
Roasted Garlic (Safeway Select)	2 tbsp	150	**130**	14

FATS AND OILS

FOOD	AMOUNT	CALORIES TOTAL	CALORIES FAT	CALORIES SAT-FAT
California				
Safeway	2 tbsp	110	**80**	9
Catalina				
⅓ less fat (Kraft)	2 tbsp	80	**40**	5
Free (Kraft)	2 tbsp	35	**0**	0
Regular (Kraft)	2 tbsp	130	**100**	14
Coleslaw Dressing				
Hidden Valley	2 tbsp	150	**140**	14
Kraft	2 tbsp	150	**110**	14
Creamy Roasted Garlic				
Fat-Free (Wish-Bone)	2 tbsp	40	**0**	0
French				
Country, Lite (Ken's)	2 tbsp	90	**50**	0
Creamy (Kraft)	2 tbsp	160	**140**	23
Deluxe (Wish-Bone)	2 tbsp	120	**100**	14
Fat-Free	2 tbsp	30	**0**	0
Fat-Free (Hellmann's)	2 tbsp	45	**0**	0
Free (Kraft)	2 tbsp	45	**0**	0
Honey & Bacon (Hidden Valley)	2 tbsp	150	**110**	5
Garlic & Herb (Good Seasons)	2 tbsp prepared	140	**140**	0
Honey Dijon				
Fat-Free (Ken's)	2 tbsp	40	**0**	0
Free (Kraft)	2 tbsp	45	**0**	0
Light (Hidden Valley)	2 tbsp	35	**0**	0
Regular (Kraft)	2 tbsp	140	**120**	23
Honey Mustard				
Lite (Ken's)	2 tbsp	70	**40**	0
Italian				
Creamy (Ken's)	2 tbsp	80	**60**	5
Creamy (Wish-Bone)	2 tbsp	110	**90**	14
Fat-Free (Ken's)	2 tbsp	10	**0**	0
Fat-Free (Wish-Bone)	2 tbsp	10	**0**	0
Free (Kraft)	2 tbsp	20	**0**	0
Herb & Cheese, Fat-Free (Hidden Valley)	2 tbsp	30	**0**	0
Light (Newman's Own)	2 tbsp	45	**35**	0
Lite (Wish-Bone)	2 tbsp	15	**5**	0

FATS AND OILS

FOOD	AMOUNT	CALORIES TOTAL	FAT	SAT-FAT
Italian (cont.)				
Parmesan & Herb (Safeway Select)	2 tbsp	140	**130**	14
Regular (Good Seasons)	2 tbsp prepared	140	**140**	0
Regular (Kraft)	2 tbsp	120	**110**	14
Regular (Safeway)	2 tbsp	90	**70**	9
Regular (Wish-Bone)	2 tbsp	80	**70**	9
Zesty Italian (Good Seasons)	2 tbsp prepared	140	**140**	0
Zesty Italian (Kraft)	2 tbsp	110	**100**	9
Mayonnaise				
Hellman's				
Regular	1 tbsp (15 g)	100	**100**	14
Low-Fat	1 tbsp (15 g)	25	**10**	0
Kraft				
Regular	1 tbsp	100	**100**	18
Free	1 tbsp (15 g)	10	**0**	0
Light	1 tbsp (15 g)	50	**45**	9
Mayonnaise substitute				
Miracle Whip (Kraft)				
Regular	1 tbsp	70	**60**	9
Light	1 tbsp	35	**25**	0
Free	1 tbsp	15	**0**	0
Nacho Cheese Ranch (Hidden Valley)	2 tbsp	130	**120**	NA
Olive Oil & Vinegar				
Newman's Own	2 tbsp	150	**150**	23
Olive Oil Vinaigrette				
Ken's	2 tbsp	60	**50**	5
Oriental				
Wish-Bone	2 tbsp	70	**45**	5
Oriental Sesame (Good Seasons)	2 tbsp prepared	150	**140**	NA
Parmesan, Creamy				
Low-Fat (Hidden Valley)	2 tbsp	30	**0**	0
Romano (Kraft)	2 tbsp	170	**160**	14
Peppercorn Free (Kraft)	2 tbsp	50	**0**	0
Pizza Ranch (Hidden Valley)	2 tbsp	140	**130**	NA

FATS AND OILS

FOOD	AMOUNT	CALORIES TOTAL	CALORIES FAT	CALORIES SAT-FAT
Potato Salad Dressing (Marzetti)	2 tbsp	150	**130**	23
Ranch				
⅓ less fat (Kraft)	2 tbsp	110	**100**	14
Caesar (Kraft)	2 tbsp	110	**90**	18
Cucumber (Kraft)	2 tbsp	140	**130**	18
Cucumber Ranch (Kraft)	2 tbsp	60	**45**	9
Fat-Free (Wish-Bone)	2 tbsp	40	**0**	0
Garlic! (Hidden Valley)	2 tbsp	130	**120**	14
Light (Hidden Valley)	2 tbsp	80	**60**	5
Light (Ken's)	2 tbsp	100	**90**	5
Original (Hidden Valley)	2 tbsp	140	**130**	23
Original with Bacon, fat-free (Hidden Valley)	2 tbsp	50	**0**	0
Original, fat-free (Hidden Valley)	2 tbsp	30	**0**	0
Peppercorn Free (Kraft)	2 tbsp	45	**0**	0
Regular (Kraft)	2 tbsp	170	**170**	27
Regular (Wish-Bone)	2 tbsp	160	**150**	23
Safeway	2 tbsp	140	**140**	9
Super Creamy (Hidden Valley)	2 tbsp	140	**130**	23
Raspberry				
Safeway Select	2 tbsp	130	**100**	9
Raspberry Walnut Vinaigrette, Lite (Ken's)	2 tbsp	80	**50**	0
Red Wine Vinaigrette				
Regular (Wish-Bone)	2 tbsp	90	**45**	5
Fat-Free (Wish-Bone)	2 tbsp	35	**0**	0
Red Wine & Herb				
Fat-Free (Hidden Valley)	2 tbsp	45	**0**	0
Russian				
Wish-Bone	2 tbsp	110	**50**	9
Salsa Zesty Garden (Kraft)	2 tbsp	70	**60**	NA
Sandwich Spread (Kraft)	1 tbsp	50	**35**	5
Reduced fat	1 tbsp	35	**25**	0
Slaw Dressing (Marzetti)	2 tbsp	170	**140**	23
Fat-Free	2 tbsp	45	**0**	0

FATS AND OILS

FOOD	AMOUNT	CALORIES TOTAL	FAT	SAT-FAT
Sun-Dried Tomato & Spices				
(Safeway Select)	2 tbsp	130	**90**	9
Sun-Dried Tomato Vinaigrette,				
Fat-Free (Ken's)	2 tbsp	60	**0**	0
Tangy Tomato Bacon (Kraft)	2 tbsp	130	**100**	9
Taco Ranch (Hidden Valley)	2 tbsp	130	**120**	NA
Thousand Island				
Fat-Free (Wish-Bone)	2 tbsp	35	**0**	0
Free (Kraft)	2 tbsp	40	**0**	0
Regular (Kraft)	2 tbsp	110	**90**	14
Safeway	2 tbsp	120	**100**	14
SHORTENINGS				
Crisco	1 tbsp	110	**110**	27
Flair	1 tbsp	100	**100**	45

FISH AND SHELLFISH

Unless otherwise noted, fish is baked, steamed, or broiled with *no added fat.* If fish is baked in butter or margarine and you are keeping track of total fat, for each teaspoon of butter or margarine you use, add 33 total calories to the total calories listed for each fish and 33 fat calories to the fat calories listed for each fish. If you are keeping track of saturated fat, for each teaspoon of butter you use, add 33 total calories to the total calories listed for each fish and 22 sat-fat calories to the sat-fat calories listed for each fish. For margarine, add 33 total calories to the total calories listed for each fish and 18 sat-fat calories to the sat-fat calories listed for each fish. *See also* **FROZEN, MICROWAVE AND REFRIGERATED FOODS** and **FAST FOODS.**

Remember that most of the following calorie figures are for ***only 1 ounce*** *of seafood!*

FOOD	AMOUNT	CALORIES TOTAL	FAT	SAT-FAT
Abalone				
raw	1 oz	30	**2**	0
cooked, fried	1 oz	54	**17**	4
Anchovy				
raw	1 oz	37	**12**	3

FISH AND SHELLFISH

FOOD	AMOUNT	CALORIES TOTAL	FAT	SAT-FAT
Anchovy (cont.)				
canned in oil, drained	5 anchovies (20 g)	42	**17**	4
	1 oz	60	**25**	6
Bass				
Freshwater, raw	1 oz	32	**9**	2
	1 fillet (79 g)	90	**26**	6
Striped, raw	1 oz	27	**6**	1
	1 fillet (159 g)	154	**33**	7
Bluefish				
raw	1 oz	35	**11**	2
	1 fillet (150 g)	186	**57**	12
Burbot, raw	1 oz	25	**2**	0
	1 fillet (116 g)	104	**8**	2
Butterfish, raw	1 oz	41	**20**	NA
	1 fillet (32 g)	47	**23**	NA
Carp				
raw	1 oz	36	**14**	3
	1 fillet (218 g)	276	**110**	21
cooked, dry heat	1 oz	46	**18**	4
Catfish, channel				
breaded and fried	1 oz	65	**34**	8
	1 fillet (87 g)	199	**104**	26
raw	1 oz	33	**11**	3
	1 fillet (79 g)	92	**30**	7
Caviar, black and red	1 tbsp	40	**26**	15
	1 oz	71	**45**	27
Cisco (lake herring)				
raw	1 oz	28	**5**	1
	1 fillet (79 g)	78	**14**	3
smoked	1 oz	50	**30**	4
Clams				
raw, cherrystones or littlenecks	9 large or 20 small (180 g)	133	**16**	1
	1 oz	22	**3**	0
breaded and fried	1 oz	57	**28**	7
	20 small clams (188 g)	379	**189**	45

FISH AND SHELLFISH

FOOD	AMOUNT	CALORIES TOTAL	CALORIES FAT	CALORIES SAT-FAT
Clams (cont.)				
canned, drained solids	1 oz	42	**5**	0
	½ cup	118	**14**	1
cooked, moist heat	1 oz	42	**5**	0
	20 small clams (90 g)	133	**16**	2
fritters	1 fritter	124	**54**	NA
Cod, Atlantic				
raw	1 oz	23	**2**	0
	1 fillet (231 g)	190	**14**	3
baked	1 oz	30	**2**	0
	1 fillet (180 g)	189	**14**	3
canned	1 oz	30	**2**	1
dried and salted	1 oz	81	**6**	0
Cod, Pacific, raw	1 oz	23	**2**	0
	1 fillet (116 g)	95	**7**	1
Crab				
Alaska King, steamed	1 oz	27	**4**	0
	1 leg (172 g)	129	**18**	2
Alaska King, imitation, made from surimi	1 oz	29	**3**	NA
Blue				
raw	1 oz	25	**3**	0
	1 crab (21 g)	18	**2**	0
cooked, moist heat	1 oz	27	**4**	1
canned	1 oz	28	**3**	0
	½ cup	67	**7**	2
Crab cakes	1 cake	93	**41**	8
	1 oz	44	**19**	4
Chesapeake Bay Deluxe Crab Cakes, frozen	1 oz	65	**41**	NA
Dungeness, raw	1 oz	24	**2**	0
Nutri Sea Crab Sticks	1 oz	29	**3**	NA
Nutri Sea King Crab	1 oz	31	**3**	NA
Sea Legs Crabmeat Salad Style	1 oz	27	**3**	NA

FISH AND SHELLFISH

FOOD	AMOUNT	CALORIES TOTAL	CALORIES FAT	CALORIES SAT-FAT
Crayfish				
raw	1 oz	25	**3**	0
	8 crayfish (27 g)	24	**2**	0
steamed	1 oz	32	**3**	0
Croaker, Atlantic				
raw	1 oz	30	**8**	3
	1 fillet (79 g)	83	**22**	8
breaded and fried	1 oz	63	**32**	9
	1 fillet (87 g)	192	**99**	27
Cusk, raw	1 oz	25	**2**	NA
Cuttlefish, raw	1 oz	22	**2**	0
Dolphinfish, raw	1 oz	24	**2**	0
	1 fillet (204 g)	174	**13**	3
Drum, freshwater, raw	1 oz	34	**13**	3
	1 fillet (198 g)	236	**88**	20
Eel				
raw	1 oz	52	**30**	6
baked	1 oz	67	**38**	8
	1 fillet (159 g)	375	**214**	43
Flatfish (flounder or sole)				
raw	1 oz	26	**3**	1
	1 fillet (163 g)	149	**17**	4
baked or steamed	1 oz	33	**4**	1
	1 fillet (127 g)	148	**17**	4
Gefilte fish	1 piece	35	**7**	2
	1 oz	24	**4**	1
Grouper				
raw	1 oz	26	**3**	1
	1 fillet (259 g)	238	**24**	5
baked or steamed	1 oz	33	**3**	1
	1 fillet (202 g)	238	**24**	5
Haddock				
raw	1 oz	25	**2**	0
	1 fillet (193 g)	168	**12**	2
baked or steamed	1 oz	32	**2**	0
	1 fillet (150 g)	168	**12**	2
smoked	1 oz	33	**2**	0

FISH AND SHELLFISH

FOOD	AMOUNT	CALORIES TOTAL	CALORIES FAT	CALORIES SAT-FAT
Halibut, Atlantic and Pacific				
baked or steamed	1 oz	40	**7**	1
	½ fillet (159 g)	223	**42**	6
Herring, Atlantic				
raw	1 oz	45	**23**	5
	1 fillet (184 g)	291	**150**	34
baked or steamed	1 oz	57	**30**	7
canned	1 oz	59	**35**	7
in tomato sauce	1 herring (37 g)	97	**52**	10
pickled	1 oz	65	**39**	6
	1 herring	112	**68**	9
	1 piece (15 g)	33	**21**	3
smoked, kippered	1 oz	60	**33**	7
	1 fillet (40 g)	87	**45**	10
Herring, Pacific, raw	1 oz	55	**35**	8
Lobster, Northern				
raw	1 oz	26	**2**	0
	1 lobster (150 g)	136	**12**	NA
cooked, moist heat	1 oz	28	**2**	0
	1 cup	142	**8**	1
Newburg (with butter, eggs, sherry, cream)	1 cup	485	**239**	160
Salad (with mayonnaise)	½ cup or 4 oz	286	**149**	NA
Lox (smoked salmon)	1 oz	33	**11**	2
Mackerel, Atlantic				
raw	1 oz	58	**35**	8
	1 fillet (112 g)	229	**140**	33
baked or steamed	1 oz	74	**45**	11
	1 fillet (88 g)	231	**141**	33
Mackerel, Jack, canned	1 cup	296	**108**	30
Mackerel, King, raw	1 oz	30	**9**	1
	½ fillet (198 g)	207	**36**	6
Mackerel, Pacific and Jack, raw	1 oz	44	**20**	6
	1 fillet (225 g)	353	**160**	46

FISH AND SHELLFISH

FOOD	AMOUNT	CALORIES TOTAL	CALORIES FAT	CALORIES SAT-FAT
Mackerel, Spanish				
raw	1 oz	39	**16**	5
	1 fillet (187 g)	260	**106**	31
baked or steamed	1 oz	45	**16**	5
	1 fillet (146 g)	230	**83**	24
Milkfish, raw	1 oz	42	**17**	NA
Monkfish, raw	1 oz	21	**4**	NA
Mullet, striped				
raw	1 oz	33	**10**	3
	1 fillet (119 g)	139	**41**	12
baked	1 oz	42	**12**	4
	1 fillet (93 g)	139	**41**	12
Mussels, Blue				
raw	1 oz	24	**6**	1
	1 cup	129	**30**	6
Ocean perch, Atlantic				
raw	1 oz	27	**4**	1
	1 fillet (64 g)	60	**9**	1
baked	1 oz	34	**5**	1
	1 fillet (50 g)	60	**9**	1
breaded and fried	1 fillet	185	**99**	NA
Octopus, raw	1 oz	23	**3**	1
Oysters, Eastern				
raw	6 medium (84 g)	58	**19**	5
	1 cup	170	**55**	14
breaded and fried	1 oz	56	**32**	8
	6 medium (88 g)	173	**100**	25
canned	1 oz	19	**6**	2
	½ cup	85	**28**	7
steamed	1 oz	39	**13**	3
	6 medium (42 g)	58	**19**	5
stew (2 parts milk, 1 part oyster)	1 cup	233	**139**	80
Oyster, Pacific, raw	1 oz	23	**6**	1
	1 medium (50 g)	41	**10**	2

FISH AND SHELLFISH

FOOD	AMOUNT	CALORIES TOTAL	CALORIES FAT	CALORIES SAT-FAT
Pike, Northern				
raw	1 oz	25	**2**	0
	½ fillet (198 g)	175	**12**	2
baked	1 oz	32	**2**	0
	½ fillet (155 g)	176	**12**	2
Pike, Walleye, raw	1 oz	26	**3**	1
	1 fillet (159 g)	147	**17**	4
Pollock, Atlantic, raw	1 oz	26	**2**	0
	½ fillet (193 g)	177	**17**	2
Pollock, Walleye				
raw	1 oz	23	**2**	0
	1 fillet (77 g)	62	**6**	1
baked	1 oz	32	**3**	1
	1 fillet (60 g)	68	**6**	1
Pompano, Florida				
raw	1 oz	47	**24**	9
	1 fillet (112 g)	184	**95**	35
baked	1 oz	60	**31**	11
	1 fillet (88 g)	185	**96**	36
Pout, ocean, raw	1 oz	22	**2**	1
	½ fillet (176 g)	140	**14**	5
Rockfish, Pacific				
raw	1 oz	27	**4**	1
	1 fillet (191 g)	180	**27**	6
baked	1 oz	34	**5**	1
	1 fillet (149 g)	180	**27**	6
Roughy, Orange, raw	1 oz	36	**18**	0
Sablefish				
raw	1 oz	55	**39**	8
	½ fillet (193 g)	377	**266**	56
smoked	1 oz	72	**51**	10
Salmon, Atlantic, raw	1 oz	40	**16**	3
Salmon, Chinook				
raw	1 oz	51	**27**	6
smoked	1 oz	33	**11**	2
Salmon, Chum				
raw	1 oz	34	**10**	2
canned	1 oz	40	**14**	4

FISH AND SHELLFISH

FOOD	AMOUNT	CALORIES TOTAL	CALORIES FAT	CALORIES SAT-FAT
Salmon, Coho				
raw	1 oz	41	**15**	3
Salmon, Pink				
raw	1 oz	33	**9**	1
canned	1 oz	39	**15**	4
Salmon, Sockeye				
raw	1 oz	48	**22**	4
canned, drained	1 oz	40	**14**	4
Salmon, smoked (lox)	1 oz	33	**11**	2
Sardines, Atlantic, canned in oil, drained	1 oz	59	**29**	3
	2 sardines (24 g)	50	**25**	3
	1 can (3¼ oz)	192	**95**	13
Sardines, Pacific, canned in tomato sauce, drained	1 oz	51	**31**	8
	1 sardine (38 g)	68	**41**	11
Scallops				
raw	1 oz	25	**2**	0
	2 large or 5 small (30 g)	26	**2**	0
breaded, fried	1 oz	61	**28**	6
	2 large (31 g)	67	**31**	7
steamed	1 oz	32	**4**	1
Scup, raw	1 oz	30	**7**	NA
	1 fillet (64 g)	67	**16**	NA
Sea bass				
raw	1 oz	27	**5**	1
	1 fillet (129 g)	125	**23**	6
baked	1 oz	35	**7**	2
	1 fillet (101 g)	125	**23**	6
Sea trout, raw	1 oz	29	**9**	3
	1 fillet (238 g)	248	**77**	22
Shad				
raw	1 oz	56	**35**	NA
	1 fillet (184 g)	362	**228**	NA
baked	1 oz	57	**29**	NA

FISH AND SHELLFISH

FOOD	AMOUNT	CALORIES TOTAL	FAT	SAT-FAT
Shark				
raw	1 oz	37	**11**	2
batter-dipped and fried	1 oz	65	**35**	8
Sheepshead				
raw	1 oz	31	**6**	2
	1 fillet (238 g)	257	**52**	13
baked	1 oz	36	**4**	1
	1 fillet (186 g)	234	**27**	6
Shrimp				
raw	1 oz	30	**4**	1
	4 large (28 g)	30	**4**	1
breaded, fried	1 oz	69	**31**	5
	4 large (30 g)	73	**33**	6
canned	1 oz	34	**5**	1
	½ cup	77	**11**	2
Cocktail (Sau-Sea)	½ cup	90	**0**	0
steamed	1 oz	28	**3**	1
	4 large (22 g)	22	**2**	1
Smelt, Rainbow				
raw	1 oz	28	**6**	1
baked	1 oz	35	**8**	2
Snapper				
raw	1 oz	28	**3**	1
	1 fillet (218 g)	217	**26**	6
baked	1 oz	36	**4**	1
	1 fillet (170 g)	217	**26**	6
Sole (*see* Flatfish)				
Spiny Lobster, raw	1 oz	32	**4**	1
	1 lobster (209 g)	233	**28**	4
Spot, raw	1 oz	35	**12**	4
	1 fillet (64 g)	79	**28**	8
Squid (calamari)				
raw	1 oz	26	**4**	1
fried	1 oz	50	**19**	5
Sturgeon				
raw	1 oz	30	**10**	2

FISH AND SHELLFISH

FOOD	AMOUNT	CALORIES TOTAL	CALORIES FAT	CALORIES SAT-FAT
Sturgeon (cont.)				
baked	1 oz	38	**13**	3
smoked	1 oz	48	**11**	3
Sucker, White, raw	1 oz	26	**6**	1
	1 fillet (159 g)	147	**33**	6
Sunfish, Pumpkinseed, raw	1 oz	25	**2**	0
	1 fillet (48 g)	43	**3**	0
Surimi	1 oz	28	**2**	0
Swordfish				
raw	1 oz	34	**10**	3
baked	1 oz	44	**13**	4
Tilefish				
raw	1 oz	27	**6**	1
	½ fillet (193 g)	184	**40**	8
baked	1 oz	42	**12**	2
	½ fillet (150 g)	220	**63**	12
Trout, Rainbow				
raw	1 oz	33	**9**	2
	1 fillet (79 g)	93	**24**	5
baked	1 oz	43	**11**	2
	1 fillet (62 g)	94	**24**	5
Tuna				
raw	1 oz	41	**12**	3
baked	1 oz	52	**16**	4
canned, drained				
solid white in water	1 oz	37	**6**	2
chunk light in oil, drained	1 oz	56	**21**	4
Tuna salad	½ cup	190	**85**	14
Turbot, European, raw	1 oz	27	**8**	NA
	½ fillet (204 g)	194	**54**	NA
Whelk				
raw	1 oz	39	**1**	0
steamed	1 oz	78	**2**	0

FISH AND SHELLFISH

FOOD	AMOUNT	CALORIES TOTAL	CALORIES FAT	CALORIES SAT-FAT
Whitefish				
raw	1 oz	38	**15**	2
	1 fillet (198 g)	266	**104**	16
smoked	1 oz	30	**2**	1
Whiting				
raw	1 oz	26	**3**	1
	1 fillet (92 g)	83	**11**	2
baked	1 oz	33	**4**	1
	1 fillet (72 g)	83	**11**	2
Wolffish, Atlantic, raw	1 oz	27	**6**	1
	½ fillet (153 g)	147	**33**	5
Yellowtail, raw	1 oz	41	**13**	NA
	½ fillet (187 g)	273	**88**	NA

FROZEN, MICROWAVE, AND REFRIGERATED FOODS

FOOD	AMOUNT	CALORIES TOTAL	CALORIES FAT	CALORIES SAT-FAT
APPETIZERS				
Cocktail Beef Franks (Cohen's)	7 (89 g)	320	**240**	81
Eggrolls				
Beef Steak Teriyaki (Lo-An)	1 (78 g)	140	**35**	9
Chicken & Shrimp (Lo-An)	1 (78 g)	140	**35**	9
Chicken Munchers mini eggrolls (La Choy)	6 eggrolls (85 g)	210	**80**	23
Egg Roll Bites (Matlaw's)	2 pieces (28 g)	45	**5**	0
Lobster (Lo-An)	1 (78 g)	150	**35**	9
Pork (Chung's)	2 (168 g)	400	**180**	45
Shrimp Munchers mini eggrolls (La Choy)	6 eggrolls (85 g)	190	**60**	14
Shrimp (Lo-An)	1 (78 g)	150	**45**	9
Vegetable (Barney's)	3 (81 g)	140	**35**	14
White Meat Chicken (Chung's)	2 (168 g)	340	**100**	14
White Meat Chicken (Lo-An)	1 (78 g)	140	**35**	NA
Potato Knishes	1 knish (210 g)	436	**150**	36

FROZEN, MICROWAVE, AND REFRIGERATED FOODS

FOOD	AMOUNT	CALORIES TOTAL	CALORIES FAT	CALORIES SAT-FAT
Puffs				
Mushroom Puffs (Mother's)	2 pieces (66 g)	190	**100**	18
Potato Puff (Manischewitz)	5 pieces (92 g)	340	**180**	45
Spinach & Potato Puff (Manischewitz)	5 pieces (92 g)	320	**180**	45
Spinach Puffs (Mother's)	2 pieces (66 g)	200	**100**	18
Quiche				
petite quiche appetizers	6 (128 g)	370	**200**	90
Turnover				
Beef (Manischewitz)	4 pieces (85 g)	300	**190**	54
BREADS				
New York Brand				
Texas Garlic Toast	1 slice (40 g)	170	**90**	18
with cheese	1 slice (48 g)	190	**100**	36
Pepperidge Farm				
Five Cheese & Garlic Cheese Bread	2 ¼" slices (56 g)	200	**90**	41
Garlic Toast	1 slice (40 g)	160	**90**	14
Mozzarella Garlic Cheese Bread	2 ¼" slices (56 g)	200	**90**	45
Sourdough Garlic Bread	2 ½" slices (50 g)	170	**60**	9
BREAKFAST FOODS				
Blintzes				
Empire Kosher				
Apple	2 (124 g)	220	**50**	14
Cherry	2 (124 g)	200	**35**	9
Golden				
Blueberry	1 (62 g)	90	**9**	0
Cherry	1 (62 g)	95	**9**	0
Potato	1 (62 g)	90	**36**	9
Ratner's				
Blueberry Cheese	1 (61 g)	100	**5**	0
Cheese	1 (61 g)	90	**5**	0
jumbo	1 (94.5 g)	133	**6**	5
Cherry	1 (61 g)	100	**5**	0
Potato	1 (61 g)	110	**30**	18

FROZEN, MICROWAVE, AND REFRIGERATED FOODS

FOOD	AMOUNT	CALORIES TOTAL	CALORIES FAT	CALORIES SAT-FAT
Breakfast Burritos				
Bacon (Great Starts, Swanson)	1 pkg (99 g)	250	**100**	36
Breakfast Sandwich				
Great Starts (Swanson)				
Egg, Canadian Bacon & Cheese on a Muffin	1 pkg (116 g)	290	**140**	54
French Toast Sticks with Syrup	1 pkg (120 g)	320	**90**	45
Pancakes with Bacon	1 pkg (128 g)	400	**180**	63
Pancakes with Sausage	1 pkg (170 g)	490	**230**	99
Sausage, Egg & Cheese on a Biscuit	1 pkg (156 g)	460	**250**	99
Scrambled Eggs & Bacon with Home-Fried Potatoes	1 pkg (149 g)	290	**170**	81
Scrambled Eggs & Sausage with Hashed Brown Potatoes	1 pkg (177 g)	360	**230**	90
Morningstar Farms				
Breakfast Links	2 links (45 g)	60	**20**	5
Breakfast Patties	1 patty (38 g)	70	**25**	5
Breakfast Strips	2 strips (16 g)	60	**40**	5
Grillers	1 patty (64 g)	140	**60**	5
Weight Watchers Smart Ones				
English Muffin Sandwich	1 (113 g)	210	**50**	18
Handy Ham & Cheese Omelet	1 (113 g)	220	**45**	23
Croissants				
Original (Sara Lee)	1 (43 g)	170	**70**	NA
French Toast, Frozen				
Aunt Jemima, all types	2 (118 g)	240	**60**	18
Breakfast Blast Mini Sticks (Swanson)	1 pkg (120 g)	310	**130**	NA
Downyflake				
Cinnamon Swirl	2 (113 g)	270	**50**	NA
Plain	2 (113 g)	260	**60**	NA

FROZEN, MICROWAVE, AND REFRIGERATED FOODS

FOOD	AMOUNT	CALORIES TOTAL	CALORIES FAT	CALORIES SAT-FAT
French Toast, Frozen (cont.)				
Great Starts (Swanson)				
Cinnamon Swirl French Toast with Sausage	1 pkg (156 g)	440	**250**	108
French Toast Sticks	1 pkg (120 g)	320	**90**	NA
French Toast with Sausage	1 pkg (156 g)	410	**230**	81
Muffins				
Blueberry (Sara Lee)	1 muffin (64 g)	220	**100**	NA
Pancakes, Frozen				
Aunt Jemima				
Buttermilk Pancake Batter	½ cup batter (4 4" pancakes)	260	**30**	9
Low-fat	3 (97 g)	150	**15**	0
Great Starts (Swanson)				
6 Silver Dollar Pancakes with Sausage	1 pkg (106 g)	340	**160**	81
Hungry Jack (Pillsbury)				
Blueberry	3 (116 g)	270	**35**	9
Buttermilk	3 (116 g)	270	**40**	9
Toaster Strudel				
Apple	1 pastry (54 g)	200	**80**	14
Brown Sugar Cinnamon	1 pastry (54 g)	190	**70**	14
Cherry	1 pastry (54 g)	190	**70**	14
Cream Cheese & Blueberry	1 pastry (54 g)	200	**90**	27
Strawberry	1 pastry (54 g)	200	**80**	18
Strawberry Kiwi	1 pastry (54 g)	190	**70**	18
Tropical Wave	1 pastry (54 g)	190	**70**	14
Wildberry	1 pastry (54 g)	190	**70**	18
Waffles, Frozen				
Aunt Jemima				
Blueberry	2 waffles (72 g)	190	**50**	14
Buttermilk	2 waffles (72 g)	200	**50**	14
Homestyle	2 waffles (72 g)	200	**50**	14
Low-fat	2 waffles (74 g)	160	**15**	5
Oatmeal	2 waffles (84 g)	200	**70**	NA
Original	2 waffles (72 g)	200	**60**	0

FROZEN, MICROWAVE, AND REFRIGERATED FOODS

FOOD	AMOUNT	CALORIES TOTAL	CALORIES FAT	CALORIES SAT-FAT
Waffles, Frozen (cont.)				
Belgian Chef Belgian Waffles	2 waffles (75 g)	170	**25**	5
Breakfast Blast 5 Waffle Sticks	1 (78 g)	330	**150**	NA
Downyflake				
Homestyle and Buttermilk	2 waffles (68 g)	170	**35**	NA
Eggo (Kellogg's)				
Blueberry	2 waffles (78 g)	220	**80**	14
Buttermilk, Strawberry, or Homestyle	2 waffles (78 g)	220	**70**	14
Common Sense Oat Bran	2 waffles (78 g)	200	**60**	NA
Fat free	2 waffles (58 g)	120	**0**	0
Minis (Homestyle)	3 sets of 4 waffles (93 g)	260	**80**	18
Special K	2 waffles (58 g)	120	**0**	0
Nutrigrain Eggo				
Apple Cinnamon	2 waffles (78 g)	220	**70**	14
Banana Bread	2 waffles (78 g)	200	**60**	9
Cinnamon Twist	3 sets of 4 waffles (92 g)	290	**90**	18
Multi-bran	2 waffles (78 g)	180	**50**	9
Nut & Honey	2 waffles (78 g)	240	**90**	18
Whole Wheat	2 waffles (78 g)	190	**60**	9
Hungry Jack (Pillsbury)				
Apple Cinnamon	2 waffles (71 g)	200	**50**	18
Buttermilk	2 waffles (68 g)	190	**50**	18
Homestyle	2 waffles (68 g)	180	**50**	18
Mini Funfetti	3 sets of 4 waffles (93 g)	260	**70**	23
DISHES OR DINNERS				
Amy's				
Burritos				
cheese	1 burrito (170 g)	280	**70**	23
non-dairy	1 burrito (170 g)	250	**95**	9
California Veggie Burger	1 burger (72 g)	100	**25**	0
Enchiladas				

FROZEN, MICROWAVE, AND REFRIGERATED FOODS

FOOD	AMOUNT	CALORIES TOTAL	CALORIES FAT	CALORIES SAT-FAT
Amy's (cont.)				
Black Bean, Vegetable	1 enchilada (135 g)	130	**40**	0
Cheese	1 enchilada (135 g)	210	**80**	23
Spanish Rice & Beans	1 meal (284 g)	250	**70**	9
Lasagna				
Tofu Vegetable	1 lasagna (269 g)	300	**90**	9
Vegetable	1 lasagna (269 g)	300	**90**	36
Macaroni & Cheese	1 entree (255 g)	390	**130**	72
Mexican Tamale Pie	1 pie (227 g)	220	**30**	0
Pot Pies				
Beef	1 pie (198 g)	400	**210**	99
Broccoli	1 pie (213 g)	430	**190**	90
Vegetable	1 pie (213 g)	360	**160**	99
Shepherd's Pie	1 pie (227 g)	160	**35**	0
Veggie Loaf	1 entree (284 g)	260	**50**	5
Authentic Chinese Entrees				
Kung Pao Chicken	1 cup (199 g)	250	**40**	9
Moo Goo Gai Pan	1 cup (199 g)	260	**70**	18
Sweet & Sour Pork	1 cup (199 g)	250	**40**	9
Banquet				
Boneless Pork Riblet Meal	1 meal (283 g)	400	**170**	72
Brown Gravy & Salisbury Steaks	1 patty with gravy (132 g)	230	**150**	72
Chicken Fried Beef Steak Meal	1 meal (283 g)	420	**210**	63
Chicken Nugget Meal	1 meal (191 g)	430	**210**	72
Chicken Pot Pie	1 pie (198 g)	380	**200**	81
Country Fried Chicken	3 oz (84 g)	270	**160**	45
Creamy Broccoli, Chicken, & Cheese Meal	1 cup (219 g)	280	**130**	63
Fish Stick Meal	1 meal (187 g)	300	**120**	32
Fried Chicken	3 oz (84 g)	280	**160**	45
Fried Chicken Meal	1 meal (255 g)	470	**240**	81
Fried Chicken (The Hearty One)	416 g	870	**500**	117
Grilled Chicken Meal	1 meal (280 g)	330	**120**	27
Homestyle Gravy & Sliced Turkey	2 slices with gravy (134 g)	120	**70**	27

FROZEN, MICROWAVE, AND REFRIGERATED FOODS

FOOD	AMOUNT	CALORIES TOTAL	CALORIES FAT	CALORIES SAT-FAT
Banquet (cont.)				
Lasagna with Meat Sauce	1 cup (224 g)	350	**100**	45
Potato, Ham, & Broccoli au Gratin	⅔ cup (140 g)	220	**120**	45
Salisbury Steak Meal	1 meal (269 g)	380	**220**	81
Salisbury Steak Meal (The Hearty One)	1 meal (467 g)	780	**490**	189
Southern Fried Chicken	3 oz (84 g)	280	**160**	45
Turkey (mostly white meat) Meal	1 meal (262 g)	280	**90**	23
Turkey (mostly white meat) Meal (The Hearty One)	1 meal (481 g)	630	**290**	72
Turkey Pot Pie	1 pie (198 g)	370	**180**	72
Birds Eye Easy Recipe				
Asian Stir Fry	2¼ cups (253 g)			
	as packaged	230	**15**	0
	as prepared	330	**90**	18
Original Stir Fry	2¼ cups (270 g)			
	as packaged	210	**35**	5
	as prepared	300	**100**	25
Primavera	1¾ cups (215 g)			
	as packaged	180	**45**	9
	as prepared	250	**80**	14
Southwestern	1¾ cups (233 g)			
	as packaged	200	**50**	9
	as prepared	270	**70**	14
Teriyaki Stir Fry	2 cups (250 g)			
	as packaged	210	**25**	5
	as prepared	290	**60**	9
Birds Eye Pasta Secrets				
Italian Pesto	2⅓ cups (181 g) frozen or 1 cup cooked	240	**80**	18
Primavera	2⅓ cups (189 g) frozen or 1 cup cooked	230	**90**	27
Three Cheese	2 cups (173 g) froz. or 1 cup cooked	230	**70**	23

FROZEN, MICROWAVE, AND REFRIGERATED FOODS

FOOD	AMOUNT	CALORIES TOTAL	FAT	SAT-FAT
Birds Eye Pasta Secrets (cont.)				
White Cheddar	2 cups (180 g) froz. or 1 cup cooked	240	**90**	23
Zesty Garlic	2 cups (167 g) froz. or 1 cup cooked	240	**90**	23
Boca Burger (vegan)	1 burger (71 g)	84	**0**	0
Budget Gourmet				
Beef Pepper Steak with Rice	1 entree (255 g)	270	**70**	23
Cheese Manicotti with Meat Sauce	1 entree (255 g)	350	**170**	81
Escalloped Noodles & Turkey	1 entree (226 g)	320	**150**	63
Fettuccini Alfredo with Four Cheeses	1 entree (226 g)	320	**110**	45
Italian Sausage Lasagna	1 entree (283 g)	410	**190**	81
Macaroni & Cheese (side dish)	1 pkg (163 g)	260	**100**	54
Rice Pilaf with green beans (side dish)	1 pkg (141 g)	220	**90**	27
Rigatoni in Cream Sauce with Broccoli & White Chicken	1 entree (226 g)	250	**60**	32
Roast Beef Supreme	1 entree (255 g)	310	**120**	63
Spinach au Gratin (side dish)	1 pkg (141 g)	160	**110**	63
Stir Fry Rice & Vegetables	1 entree (226 g)	350	**140**	54
Spicy Szechuan Vegetables & Chicken	1 entree (226 g)	290	**80**	27
Swedish Meatballs	1 entree (283 g)	550	**310**	162
Three Cheese Lasagna	1 entree (255 g)	330	**110**	72
Budget Gourmet Low Fat				
Angel Hair Pasta	1 entree (226 g)	230	**45**	14
Chicken Oriental & Vegetables	1 entree (255 g)	290	**70**	23
Chinese Style Vegetables & Chicken	1 entree (226 g)	250	**60**	23
Fettuccini Primavera in Herb Sauce with Chicken	1 entree (255 g)	280	**70**	32

FROZEN, MICROWAVE, AND REFRIGERATED FOODS

FOOD	AMOUNT	CALORIES TOTAL	CALORIES FAT	CALORIES SAT-FAT
Budget Gourmet Low Fat (cont.)				
Glazed Turkey	1 entree (255 g)	250	**45**	18
Italian Style Vegetables & Chicken	1 entree (226 g)	250	**60**	23
Linguini with Clams & Shrimp	1 entree (255 g)	280	**70**	45
Orange Glazed Chicken Breast	1 entree (255 g)	280	**30**	9
Pasta in Wine & Mushroom Sauce with Chicken	1 entree (255 g)	280	**60**	18
Penne Pasta with Chunky Tomatoes & Italian Sausage in Sauce	1 entree (226 g)	270	**45**	14
Rigatoni in Cream Sauce with Broccoli & White Chicken	1 entree (226 g)	250	**60**	32
Roast Chicken with Herb Gravy	1 meal (283 g)	260	**70**	32
Spaghetti Marinara	1 entree (226 g)	260	**50**	9
Ziti Parmesano	1 entree (226 g)	260	**60**	18
Butterball Chicken Requests				
Italian Style Herb	1 piece (99 g)	190	**60**	18
Original	1 piece (99 g)	180	**60**	18
Parmesan	1 piece (99 g)	200	**60**	27
Celentano				
Broccoli Stuffed Shells	4 shells (280 g)	190	**35**	9
Cheese Ravioli	6 ravioli (182 g)	360	**35**	18
Eggplant Parmigiana	½ tray (196 g)	320	**190**	45
Eggplant Rollettes	1 tray (280 g)	330	**130**	36
Low Fat Lasagne	1 tray (280 g)	260	**25**	9
Manicotti	2 pieces (196 g)	310	**130**	63
Stuffed Shells	3 shells (196 g)	300	**130**	63
Dinty Moore				
Beef Stew	1 cup (213 g)	190	**90**	45
Chicken and Dumplings	1 cup (213 g)	200	**50**	18
Corned Beef Hash	1 cup (213 g)	350	**200**	81
Don Miguel				
Bean and Cheese Burrito	1 (198 g)	420	**120**	NA

FROZEN, MICROWAVE, AND REFRIGERATED FOODS

FOOD	AMOUNT	CALORIES TOTAL	CALORIES FAT	CALORIES SAT-FAT
Don Miguel (cont.)				
Bean and Cheese Chimichanga	1 (198 g)	470	**160**	NA
Beef and Cheese Burrito	1 (198 g)	390	**100**	NA
Chicken and Cheese Burrito	1 (198 g)	410	**130**	NA
Chicken Burrito	1 (198 g)	360	**70**	NA
Empire Kosher				
Breaded Mushrooms	7 pieces (81 g)	90	**5**	0
Breaded Zucchini	7 pieces (83 g)	100	**5**	0
Chicken Fat	1 tbsp (14 g)	120	**120**	36
Chicken Pie	1 pie (227 g)	440	**190**	45
Chicken Nuggets	5 nuggets (85 g)	180	**80**	14
Chicken Stix	4 stix (88 g)	180	**90**	18
Ground Turkey	4 oz (112 g)	150	**70**	18
Potato Pancakes (Latkes)	1 piece (56 g)	100	**40**	5
Golden				
Potato Pancakes (Latkes)	1 (38 g)	71	**27**	0
Gorton's				
Butterfly Shrimp	20 (91 g)	240	**120**	27
Crunchy Golden Breaded Fish Fillets	1 fillet (108 g)	250	**130**	36
Crunchy Golden Fish Sticks	6 sticks (104 g)	250	**120**	32
Garlic & Herb Breaded Fish Fillets	1 fillet (104 g)	220	**100**	27
Lemon Pepper Battered Fish Fillets	1 fillet (104 g)	270	**160**	45
Parmesan Fish Sticks	2 fillets (104 g)	260	**140**	36
Southern Fried Fish Fillets	1 fillet (104 g)	230	**130**	36
Gorton's Grilled Fillets				
Italian Herb	1 fillet (108 g)	130	**50**	9
all other flavors	1 fillet (108 g)	120	**50**	9
Gorton's Homestyle Baked Fillets				
Au Gratin	1 fillet (131 g)	130	**45**	18
Garlic Butter Crumb	1 fillet (131 g)	170	**80**	14
Primavera	1 fillet (131 g)	120	**45**	23

FROZEN, MICROWAVE, AND REFRIGERATED FOODS

FOOD	AMOUNT	CALORIES TOTAL	CALORIES FAT	CALORIES SAT-FAT
Green Giant Create a Meal! Meal Starter				
Beefy Noodle Flavor	1¾ cups (178 g) as packaged	170	**15**	0
	1¼ cups as prep.	350	**130**	45
Cheesy Pasta & Vegetable	1¾ cups (176 g) as packaged	230	**90**	54
	1¼ cups as prep.	440	**210**	108
Country Chicken Noodle Flavor with Pasta	1¼ cups (176 g) as packaged	130	**5**	0
	1¼ cups as prep.	280	**80**	18
Fajita Style	⅓ package (168 g)	140	**60**	41
	2 fajitas prep.	430	**140**	54
Hearty Vegetable Stew	1¼ cups (186 g) as packaged	130	**5**	0
	1¼ cups as prep.	280	**80**	18
Homestyle Stew	1¼ cups (189 g) as packaged	140	**25**	5
	1 cup as prep.	340	**140**	54
Lo Mein Stir Fry	2⅓ cups (212 g) as packaged	170	**10**	0
	1¼ cups as prep.	320	**60**	14
Oven Roasted Garlic Herb	1¾ cups (184 g) as packaged	160	**0**	0
	1¾ cups as prep.	360	**80**	14
Oven Roasted Homestyle Pot Roast	1¾ cups (191 g) as packaged	150	**5**	0
	2 cups as prep.	370	**120**	27
Oven Roasted Savory Onion	1¾ cups (185 g) as packaged	130	**10**	0
	1¾ cups as prep.	340	**120**	27
Skillet Lasagna	1¾ cups (188 g) as packaged	160	**5**	0
	1¼ cups as prep.	350	**120**	45
Szechuan Stir Fry	1¾ cups (200 g) as packaged	150	**45**	5
	1¼ cups as prep.	310	**130**	27

FROZEN, MICROWAVE, AND REFRIGERATED FOODS

FOOD	AMOUNT	CALORIES TOTAL	CALORIES FAT	CALORIES SAT-FAT
Green Giant Create a Meal! Meal Starter (cont.)				
Teriyaki Stir Fry	1¾ cups (199 g) as packaged	100	**5**	0
	1¼ cups as prep.	230	**50**	9
Green Giant Pasta Secrets				
Alfredo	2 cups (160 g) froz. or 1 cup cooked	260	**90**	27
Garlic Seasoning	2 cups (188 g) froz. or 1 cup cooked	250	**90**	45
Lasagna Style	2 cups (188 g) froz. or 1 cup cooked	260	**90**	27
White Cheddar Sauce	2 cups (181 g) froz. or 1 cup cooked	270	**80**	23
Healthy Choice				
Beef Macaroni	1 meal (240 g)	220	**35**	18
Beef Pepper Steak Oriental	1 meal (269 g)	250	**35**	18
Beef Stroganoff	1 meal (311 g)	310	**60**	27
Beef Tips, Traditional	1 meal (318 g)	260	**50**	27
Breast of Turkey, Traditional	1 meal (298 g)	290	**40**	18
Cacciatore Chicken	1 meal (354 g)	270	**35**	9
Charbroiled Beef Patty	1 meal (311 g)	280	**50**	27
Chicken & Vegetables Marsala	1 meal (326 g)	240	**35**	18
Chicken Breast Con Queso Burrito	1 meal (299 g)	350	**50**	23
Chicken Enchilada Suprema	1 meal (320 g)	300	**60**	27
Chicken Fettuccini Alfredo	1 meal (240 g)	280	**60**	23
Chicken Parmigiana	1 meal (326 g)	330	**70**	27
Chicken Teriyaki	1 meal (311 g)	270	**50**	27
Country Breaded Chicken	1 meal (290 g)	350	**80**	18
Country Glazed Chicken	1 meal (240 g)	230	**35**	14
Country Herb Chicken	1 meal (344 g)	320	**50**	23
Country Inn Roast Turkey	1 meal (283 g)	250	**50**	18
Country Roast Turkey with Mushrooms	1 meal (240 g)	220	**35**	9
Fiesta Chicken Fajitas	1 meal (198 g)	260	**35**	9
Ginger Chicken Hunan	1 meal (357 g)	380	**45**	9
Grilled Chicken Sonoma	1 meal (255 g)	230	**35**	9

FROZEN, MICROWAVE, AND REFRIGERATED FOODS

FOOD	AMOUNT	CALORIES TOTAL	CALORIES FAT	CALORIES SAT-FAT
Healthy Choice (cont.)				
Grilled Chicken with Mashed Potatoes	1 meal (226 g)	170	**30**	14
Grilled Glazed Pork Patty	1 meal (272 g)	300	**60**	18
Honey Mustard Chicken	1 meal (269 g)	270	**35**	14
Lemon Pepper Fish	1 meal (303 g)	320	**60**	18
Macaroni & Cheese	1 meal (255 g)	320	**60**	23
Meatloaf, Traditional	1 meal (340 g)	320	**45**	23
Mesquite Chicken BBQ	1 meal (298 g)	310	**45**	18
Pasta Shells Marinara	1 meal (340 g)	390	**70**	32
Penne Pasta & Roasted Tomato Sauce	1 meal (226 g)	230	**45**	9
Roasted Chicken	1 meal (311 g)	230	**45**	23
Salisbury Steak, Traditional	1 meal (326 g)	330	**60**	27
Sesame Chicken	1 meal (276 g)	240	**25**	5
Sesame Chicken Shanghai	1 meal (340 g)	300	**45**	9
Shrimp & Vegetables Maria	1 meal (354 g)	290	**45**	18
Southwestern Grilled Chicken	1 meal (289 g)	230	**60**	27
Spaghetti & Sauce with Seasoned Beef	1 meal (283 g)	280	**50**	18
Yankee Pot Roast	1 meal (311 g)	290	**60**	27
Zucchini Lasagna	1 meal (383 g)	330	**15**	9
Hormel				
Beef Stew	1 cup (213 g)	190	**90**	36
Chili, no beans	1 cup (209 g)	220	**50**	23
Macaroni & Cheese	1 cup (213 g)	270	**100**	NA
Noodles & Chicken (Hearty Helpings)	1 cup (298 g)	250	**100**	NA
Inland Valley MunchSkin Meals				
Cheese & Potato Potato Skins Kit	2 topped potato skins (113 g)	250	**140**	63
Kid Cuisine				
Circus Show Corn Dog	1 meal (249 g)	450	**140**	45
Cosmic Chicken Nuggets	1 meal (257 g)	500	**230**	90
Game Time Taco Roll-Ups	1 meal (208 g)	420	**160**	63
High Flying Fried Chicken	1 meal (286 g)	440	**180**	81
Magical Macaroni & Cheese	1 meal (300 g)	440	**110**	45

FROZEN, MICROWAVE, AND REFRIGERATED FOODS

FOOD	AMOUNT	CALORIES TOTAL	CALORIES FAT	CALORIES SAT-FAT
Kid Cuisine (cont.)				
Pirate Pizza with Cheese	1 meal (226 g)	430	**100**	45
Wave Rider Waffle Sticks	1 meal (187 g)	380	**70**	18
Kids Fun Feast (Swanson)				
Chillin' Cheese Pizza	1 pkg (224 g)	350	**80**	41
Chompin' Chicken Drumlets	1 pkg (255 g)	490	**220**	81
Frazzlin' Fried Chicken	1 pkg (312 g)	660	**320**	NA
Frenzied Fish Sticks	1 pkg (198 g)	370	**130**	45
Munchin' Mini Tacos	1 pkg (204 g)	390	**140**	54
Razzlin' Rings	1 pkg (340 g)	390	**100**	45
Roarin' Ravioli	1 pkg (312 g)	440	**110**	41
Kid's Kitchen (Hormel)				
Beans & Weiners	1 cup (220 g)	310	**110**	NA
Beefy Mac	1 cup (213 g)	190	**50**	NA
Kosherific				
Fish Sticks	6 (113 g)	280	**130**	23
Lean Cuisine				
Baked Chicken	1 pkg (244 g)	230	**35**	14
Baked Fish	1 pkg (255 g)	270	**50**	18
Beef Peppercorn	1 pkg (248 g)	220	**60**	18
Beef Portabello	1 pkg (255 g)	220	**60**	32
Beef Pot Roast	1 pkg (255 g)	210	**60**	18
Cheese Cannelloni	1 pkg (258 g)	230	**35**	18
Cheese Lasagna with Chicken Breast Scaloppini	1 pkg (283 g)	290	**70**	18
Cheese Ravioli	1 pkg (240 g)	270	**60**	27
Chicken a l'Orange	1 pkg (255 g)	250	**15**	5
Chicken & Vegetables	1 pkg (297 g)	250	**45**	9
Chicken Carbonara	1 pkg (255 g)	280	**60**	18
Chicken Chow Mein	1 pkg (255 g)	220	**35**	9
Chicken Enchilada Suiza	1 pkg (255 g)	280	**40**	14
Chicken Fettuccine	1 pkg (262 g)	280	**50**	18
Chicken in Wine Sauce	1 pkg (230 g)	210	**50**	18
Chicken Lasagna	1 pkg (283 g)	270	**70**	27
Chicken Medallions with Cheese Sauce	1 pkg (265 g)	260	**70**	27
Chicken Mediterranean	1 pkg (297 g)	270	**30**	9
Chicken Parmesan	1 pkg (308 g)	220	**40**	14

FROZEN, MICROWAVE, AND REFRIGERATED FOODS

FOOD	AMOUNT	CALORIES TOTAL	FAT	SAT-FAT
Lean Cuisine (cont.)				
Chicken in Peanut Sauce	1 pkg (255 g)	290	**50**	9
Chicken Piccata	1 pkg (255 g)	270	**60**	18
Chicken Pie	1 pkg (269 g)	290	**80**	23
Chicken with Basil Cream Sauce	1 pkg (240 g)	270	**60**	18
Classic Cheese Lasagna	1 pkg (326 g)	270	**35**	23
Fettuccini Alfredo	1 pkg (262 g)	300	**60**	23
Fiesta Chicken	1 pkg (240 g)	250	**45**	5
Glazed Chicken	1 pkg (240 g)	240	**50**	9
Glazed Turkey Tenderloins	1 pkg (255 g)	240	**40**	9
Grilled Chicken & Penne Pasta	1 pkg (265 g)	260	**70**	27
Grilled Chicken Salsa	1 pkg (251 g)	270	**60**	23
Herb Roasted Chicken	1 pkg (226 g)	210	**40**	9
Homestyle Turkey	1 pkg (265 g)	230	**40**	9
Honey Roasted Chicken	1 pkg (240 g)	290	**50**	18
Honey Roasted Pork	1 pkg (269 g)	250	**50**	23
Lasagna with Meat Sauce	1 pkg (297 g)	290	**60**	32
Macaroni and Cheese	1 pkg (283 g)	290	**60**	36
Meatloaf	1 pkg (265 g)	250	**60**	27
Oriental-Style Dumplings	1 pkg (255 g)	300	**50**	14
Oven Roasted Beef	1 pkg (262 g)	260	**70**	27
Roasted Turkey Breast	1 pkg (276 g)	270	**20**	5
Salisbury Steak	1 pkg (269 g)	280	**70**	36
Southern Beef Tips	1 pkg (248 g)	290	**50**	18
Spaghetti with Meat Sauce	1 pkg (326 g)	290	**45**	14
Shrimp & Angel Hair Pasta	1 pkg (283 g)	290	**50**	9
Swedish Meatballs	1 pkg (258 g)	290	**50**	23
Three-Bean Chili	1 pkg (283 g)	250	**50**	18
Vegetable Eggroll	1 pkg (255 g)	340	**60**	18
Lean Cuisine Great for Lunch				
Alfredo Pasta Primavera	1 pkg (283 g)	290	**60**	27
Broccoli & Cheddar Cheese Sauce over Baked Potato	1 pkg (290 g)	250	**80**	NA
Cheese Lasagna Casserole	1 pkg (283 g)	270	**50**	27
Mandarin Chicken	1 pkg (255 g)	250	**35**	5

FROZEN, MICROWAVE, AND REFRIGERATED FOODS

FOOD	AMOUNT	CALORIES TOTAL	CALORIES FAT	CALORIES SAT-FAT
Lean Cuisine Great for Lunch (cont.)				
Penne Pasta with Tomato Basil Sauce	1 pkg (283 g)	270	**30**	9
Roasted Potatoes with Broccoli & Cheddar Cheese Sauce	1 pkg (290 g)	260	**50**	32
Teriyaki Stir-Fry	1 pkg (283 g)	290	**30**	5
Lean Cuisine Hearty Portions				
Cheese & Spinach Manicotti	1 pkg (439 g)	340	**60**	23
Chicken & Barbecue Sauce	1 pkg (393 g)	380	**50**	14
Chicken Florentine	1 pkg (395 g)	420	**80**	27
Grilled Beef Patty & Gravy with Whipped Potatoes	1 pkg (439 g)	370	**80**	36
Grilled Chicken & Penne Pasta	1 pkg (396 g)	380	**60**	23
Lasagna	1 pkg (425 g)	440	**80**	36
Roasted Chicken with Mushrooms	1 pkg (354 g)	380	**60**	9
Roasted Turkey Breast	1 pkg (396 g)	290	**45**	18
Lo-An Eggrolls (*see* APPETIZERS: Eggrolls)				
Mama Lucia				
Italian Style Meatballs	3 (84g)	270	**190**	81
Marie Callender's				
Beef Tips in Mushroom Sauce	1 meal (385 g)	430	**170**	63
Breaded Chicken Parmigiana Dinner	1 dinner (454 g)	620	**250**	72
Chicken (White Meat) & Broccoli Pot Pie	1 cup (290 g)	710	**440**	117
Chicken Cordon Bleu	1 meal (368 g)	590	**230**	72
Chicken Pot Pie	1 cup (241 g)	520	**290**	72
Chili & Cornbread	1 cup + 1.5 oz cornbread (297 g)	350	**120**	54
Chunky Chicken & Noodles	1 meal (368 g)	520	**270**	99
Country Fried Chicken & Gravy	1 meal (454 g)	620	**270**	81
Fettuccini Primavera with Tortellini	1 cup (234 g)	430	**240**	108

FROZEN, MICROWAVE, AND REFRIGERATED FOODS

FOOD	AMOUNT	CALORIES TOTAL	CALORIES FAT	CALORIES SAT-FAT
Marie Callender's (cont.)				
Fettuccini with Broccoli & Chicken in Alfredo Sauce	1 cup (221 g)	410	**220**	90
Grilled Chicken in Mushroom Sauce	1 dinner (397 g)	480	**140**	54
Grilled Turkey Breast Strips & Rice Pilaf	1 meal (333 g)	310	**90**	32
Herb Roasted Chicken	1 dinner (397 g)	670	**380**	135
Lasagna with Meat Sauce	1 cup (246 g)	370	**170**	81
Macaroni & Cheese	1 meal (382 g)	510	**160**	81
Meatloaf & Gravy with Mashed Potatoes	1 meal (397 g)	540	**270**	108
Sirloin Salisbury Steak & Gravy	1 meal (397 g)	550	**220**	99
Spaghetti & Meat Sauce	1 cup + 2 oz garlic bread (290 g)	380	**120**	36
Stuffed Pasta Trio	1 meal (297 g)	380	**160**	81
Swedish Meatballs	1 meal (354 g)	520	**240**	90
Turkey Pot Pie	1 cup (280 g)	600	**320**	81
Turkey with Gravy & Dressing	1 meal (397 g)	500	**170**	81
Matlaw's				
Stuffed Clams	1 clam (71 g)	120	**50**	NA
Michelina's				
Chicken Italiano with Parmesan Cheese	1 pkg (213 g)	250	**60**	18
Chili-Mac	1 container (227 g)	270	**80**	27
Gravy with Egg Noodles and Swedish Meatballs	1 pkg (284 g)	360	**110**	41
Fettuccine Alfredo	1 pkg (269 g)	410	**150**	72
Four Cheese Lasagna	1 pkg (227 g)	290	**70**	36
Lasagna with Meat Sauce	1 pkg (255 g)	300	**70**	32
Linguini with Clams	1 pkg (255 g)	310	**30**	5
Macaroni & Beef	1 container (227 g)	250	**60**	23
Macaroni & Cheese	1 container (227 g)	340	**120**	54
Penne Pollo	1 pkg (255 g)	320	**80**	36
Pepper Steak & Rice	1 pkg (241 g)	270	**40**	14

FROZEN, MICROWAVE, AND REFRIGERATED FOODS

FOOD	AMOUNT	CALORIES TOTAL	CALORIES FAT	CALORIES SAT-FAT
Michelina's (cont.)				
Rigatoni Pomodoro	1 container (227 g)	220	**20**	0
Salisbury Steak & Gravy	1 pkg (241 g)	300	**160**	63
Spaghetti Bolognese	1 pkg (255 g)	290	**35**	9
Wheels & Cheese	1 container (227 g)	300	**80**	41
Michelina's Lean 'n Tasty				
Black Bean Chili	1 pkg (284 g)	400	**45**	9
Fettuccine with Creamy Pesto Sauce	1 pkg (241 g)	250	**60**	32
Glazed Chicken	1 pkg (227 g)	280	**30**	5
Gravy with Egg Noodles & Swedish Meat Balls	1 pkg (284 g)	300	**60**	14
Honey Barbecue Sauce with Chicken & Rice	1 pkg (241 g)	300	**20**	0
Mac & Beef	1 pkg (227 g)	230	**50**	18
Macaroni & Cheese	1 pkg (227 g)	270	**60**	32
Penne Arrabiata	1 pkg (255 g)	230	**20**	0
Penne Pasta with Mushrooms	1 pkg (227 g)	250	**50**	32
Spaghetti & Meatballs with Tomato Sauce	1 pkg (255 g)	290	**60**	23
Spaghetti with Onions, Green Peppers & Mushrooms	1 pkg (255 g)	280	**50**	9
Teriyaki Chicken with Rice	1 pkg (241 g)	290	**25**	5
Mrs. Budd's				
Chicken à la King	1 cup (238 g)	420	**80**	23
Chicken Supreme	1 cup (238 g)	430	**100**	27
White Meat Chicken Pie				
Fancy Vegetables	1 cup (227 g)	310	**140**	45
Original Recipe	1 cup (227 g)	330	**150**	45
Mrs. Paul's				
Deviled Crabs	1 cake (82 g)	170	**60**	14
Eggplant Parmigiana	½ cup (118 g)	190	**100**	27
Grilled Salmon				
Creamy Dill	1 fillet (92 g)	90	**25**	9
Honey Mustard	1 fillet (92 g)	90	**15**	5

FROZEN, MICROWAVE, AND REFRIGERATED FOODS

FOOD	AMOUNT	CALORIES TOTAL	CALORIES FAT	CALORIES SAT-FAT
Mrs. Paul's (cont.)				
Grilled Tuna				
Barbecue	1 fillet (92 g)	100	**5**	0
Sesame Teriyaki	1 fillet (92 g)	110	**15**	0
Mrs. Paul's Dream Kitchen				
Crispy Crunchy Fish Fillets	2 fillets (106 g)	240	**110**	27
Crispy Crunchy Fish Sticks	6 sticks (95 g)	210	**90**	23
Mrs. Paul's Grilled Fillets				
Garlic Butter	1 fillet (104 g)	130	**50**	9
Lemon Pepper	1 fillet (104 g)	130	**50**	9
Mrs. Paul's Premium Fillets				
Flounder	1 fillet (80 g)	170	**70**	23
Haddock	1 fillet (120 g)	230	**100**	23
Nancy's Quiche				
Quiche Florentine	1 quiche (170 g)	440	**230**	108
Quiche Lorraine	1 quiche (170 g)	470	**240**	108
Petite Quiche Appetizers all types	6 (128 g)	370	**200**	90
Natural Touch				
Garden Veggie Pattie	1 patty (67 g)	100	**25**	5
Patio Burritos				
Bean & Cheese	1 burrito (142 g)	300	**80**	41
Beef & Bean	1 burrito (142 g)	320	**110**	41
Chicken	1 burrito (142 g)	290	**70**	27
Patio Dinners				
Cheese Enchilada Dinner	1 meal (340 g)	370	**110**	45
Fiesta Dinner	1 meal (340 g)	350	**100**	45
Mexican Style Dinner	1 meal (376 g)	470	**170**	72
Quiche St. Jacques				
Quiche Jardiniere	1/11 of 9" diameter quiche (57 g)	160	**100**	45
Quiche Lorraine	1/11 of 9" diameter quiche (57 g)	170	**110**	54
Quiche Maximilian	1/11 of 9" diameter quiche (57 g)	160	**110**	45
Quiche Provençal	1/11 of 9" diameter quiche (57 g)	160	**100**	45

FROZEN, MICROWAVE, AND REFRIGERATED FOODS

FOOD	AMOUNT	CALORIES TOTAL	CALORIES FAT	CALORIES SAT-FAT
Ratner's				
Potato Pancakes (Latkes)	1 pancake (43 g)	110	**60**	27
Rice Gourmet				
Broccoli Chicken Rice Bowl	1 bowl (312 g)	500	**220**	72
Teriyaki Style Rice Bowl (beef)	1 bowl (312 g)	370	**60**	14
Teriyaki Style Rice Bowl (chicken)	1 bowl (312 g)	420	**45**	9
Safeway Select				
Beef Pot Pie	1 cup (213 g)	570	**270**	63
Chicken & Broccoli Pot Pie	1 cup (241 g)	640	**310**	90
Chicken Pot Pie	1 cup (241 g)	630	**350**	81
Turkey Pot Pie	1 cup (241 g)	610	**310**	90
Safeway Select Gourmet Club				
(10 lowfat) Chicken Fillets	2 pieces (142 g)	120	**0**	0
Boneless Pork Shoulder Country Style Ribs	84 g	160	**70**	23
Cheese & Broccoli Potatoes	1 potato (147 g)	260	**90**	41
Cheese Cannelloni	1 piece (189 g)	210	**60**	27
Chicken Breasts	1 fillet (113 g)	225	**85**	18
Chicken Nuggets	6 nuggets (112 g)	290	**180**	45
Chicken Strips	4 strips (122 g)	230	**120**	27
Deluxe Beef Shepherd's Pie	1 cup (227 g)	360	**180**	81
Deluxe Beef Steak Pot Pie	1 cup (227 g)	400	**170**	45
Deluxe Chicken Pot Pie	1 cup (227 g)	480	**250**	63
Extra Lean Roadhouse Beef Patties	1 patty (113 g)	130	**40**	18
Extra Lean Steakhouse Beef Patties	1 patty (113 g)	130	**45**	18
Four Stuffed Baked Potatoes	1 potato (147 g)	280	**100**	45
Lean Turkey with Beans & Salsa	1 cup (235 g)	180	**50**	9
Lowfat Tuna Noodle Casserole	1 cup (235 g)	320	**25**	9
Lowfat Turkey Lasagna	1 cup (227 g)	300	**30**	9
Meat Lasagna	1 cup (227 g)	300	**90**	45
Six Vegetable Lasagna	1 cup (227 g)	310	**150**	99

FROZEN, MICROWAVE, AND REFRIGERATED FOODS

FOOD	AMOUNT	CALORIES TOTAL	CALORIES FAT	CALORIES SAT-FAT
Safeway Select Gourmet Club (cont.)				
St. Louis Style Pork Spareribs	2 ribs (125 g)	370	**210**	72
Veal Cannelloni	1 piece (189 g)	250	**100**	63
Sea Pak				
Clam Strips	1 pkg (141 g)	410	**200**	36
Jumbo Butterfly Shrimp	4 shrimp (85 g)	200	**80**	9
Popcorn Shrimp	15 shrimp (85 g)	210	**110**	18
Spare the Rib				
Pork, no bones	5 oz (140 g)	380	**260**	72
Stouffer's				
Baked Chicken Breast (Homestyle)	1 pkg (251 g)	260	**100**	54
Breaded Pork Cutlet (Homestyle)	1 pkg (283 g)	420	**200**	90
Cheddar Pasta with Beef & Tomatoes	1 pkg (311 g)	500	**210**	90
Cheese Ravioli	1 pkg (301 g)	380	**120**	54
Chicken à la King	1 pkg (269 g)	350	**110**	36
Chicken & Dumplings (Homestyle)	1 pkg (283 g)	280	**70**	32
Chicken Breast in Barbecue Sauce (Homestyle)	1 pkg (283 g)	510	**210**	108
Chicken Breast with Mushroom Gravy (Homestyle)	1 pkg (283 g)	360	**130**	63
Chicken Parmigiana (Homestyle)	1 pkg (340 g)	460	**140**	36
Chicken Pie	1 pkg (283 g)	540	**290**	90
	1 cup (250 g)	500	**280**	90
Chili with Beans	1 pkg (248 g)	290	**90**	36
Chunky Beef & Tomatoes (Homestyle)	1 pkg (283 g)	280	**80**	32
Country Style Biscuit	1 pkg (255 g)	510	**260**	72
Creamed Chipped Beef	½ cup (125 g)	160	**100**	27
Escalloped Apples	⅔ cup (158 g)	190	**25**	9
Escalloped Chicken and Noodles	1 pkg (283 g)	430	**240**	45

FROZEN, MICROWAVE, AND REFRIGERATED FOODS

FOOD	AMOUNT	CALORIES TOTAL	CALORIES FAT	CALORIES SAT-FAT
Stouffer's (cont.)				
Fettuccini Alfredo	1 pkg (283 g)	520	**250**	144
Fish Filet (Homestyle)	1 pkg (255 g)	430	**190**	45
Five Cheese Lasagna	1 pkg (304 g)	360	**120**	63
Fried Chicken Breast (Homestyle)	1 pkg (251 g)	400	**150**	54
Green Pepper Steak	1 pkg (297 g)	320	**80**	27
Lasagna Bake with Meat Sauce	1 pkg (290 g)	370	**100**	45
Lasagna with Meat & Sauce	1 cup (215 g)	260	**90**	36
Lasagna with Tomato Sauce & Italian Sausage	1 pkg (308 g)	340	**100**	45
Macaroni and Beef	1 pkg (326 g)	340	**80**	36
Macaroni and Cheese	1 cup (225 g)	320	**140**	63
Macaroni and Cheese with Broccoli	1 pkg (297 g)	360	**150**	72
Meatloaf (Homestyle)	1 pkg (279 g)	390	**180**	99
Roast Turkey Breast (Homestyle)	1 pkg (272 g)	310	**110**	54
Salisbury Steak (Homestyle)	1 pkg (272 g)	380	**160**	72
Sliced Beef Brisket (Homestyle)	1 pkg (283 g)	370	**150**	90
Spaghetti with Meatballs	1 pkg (357 g)	440	**130**	45
Spaghetti with Meat Sauce	1 pkg (272 g)	350	**100**	36
Stuffed Peppers	1 pepper & sauce (220 g)	180	**70**	9
Swedish Meatballs with Pasta	1 pkg (290 g)	480	**210**	81
Tuna Noodle Casserole	1 pkg (283 g)	340	**130**	41
Turkey Tetrazzini	1 pkg (283 g)	360	**150**	63
Veal Parmigiana (Homestyle)	1 pkg (329 g)	430	**150**	45
Vegetable Lasagna	1 pkg (297 g)	440	**180**	54
Welsh Rarebit	¼ cup (62 g)	120	**80**	36
Stouffer's Family Style Favorites				
Lasagna with Meat Sauce	1 cup (215 g)	270	**80**	41
Macaroni & Cheese	1 cup (255 g)	380	**150**	72

FROZEN, MICROWAVE, AND REFRIGERATED FOODS

FOOD	AMOUNT	CALORIES TOTAL	FAT	SAT-FAT
Stouffer's Hearty Portions				
Beef Pot Roast	1 pkg (453 g)	430	**160**	63
Country Fried Beef Steak	1 pkg (453 g)	750	**370**	117
Pork with Roasted Potatoes	1 pkg (453 g)	570	**140**	45
Roast Beef	1 pkg (453 g)	430	**160**	54
Salisbury Steak	1 pkg (453 g)	630	**260**	117
Sliced Turkey Breast	1 pkg (453 g)	570	**240**	63
Swanson Classic Mac & More				
Macaroni & Cheese	1 pkg (170 g)	240	**80**	36
Swanson Dinners				
Beef Pot Pie	1 pkg (198 g)	415	**210**	81
Boneless Pork Rib	1 pkg (298 g)	470	**170**	63
Chicken Nuggets	1 pkg (284 g)	590	**230**	63
Chicken Pot Pie	1 pie (198 g)	410	**200**	81
Country Fried Beef Steak	1 pkg (305 g)	460	**200**	90
Fish 'n Chips	1 pkg (284 g)	490	**180**	36
Fried Chicken				
Dark Portions	1 pkg (312 g)	580	**270**	90
White Portions	1 pkg (312 g)	630	**280**	72
Meatloaf	1 pkg (305 g)	380	**130**	54
Salisbury Steak	1 pkg (312 g)	340	**140**	54
Sirloin Beef Tips	1 pkg (447 g)	450	**140**	NA
Turkey (Mostly White Meat)	1 pkg (333 g)	320	**70**	27
Turkey Pot Pie	1 pie (198 g)	400	**190**	72
Veal Parmigiana	1 pkg (319 g)	390	**160**	72
Yankee Pot Roast	1 pkg (326 g)	250	**40**	14
Swanson Hungry Man				
Beef Pot Pie	1 pkg (397 g)	660	**299**	117
Boneless Chicken	1 pkg (489 g)	630	**200**	54
Boneless Pork Rib	1 pkg (400 g)	770	**340**	117
Chicken Pot Pie	1 pkg (397 g)	650	**320**	126
Country Fried Beef Steak	1 pkg (454 g)	660	**300**	126
Fisherman's Platter	1 pkg (368 g)	650	**230**	54
Fried Chicken (Mostly White Meat)	1 pkg (439 g)	800	**350**	99
Salisbury Steak	1 pkg (461 g)	610	**310**	153
Traditional Pot Roast	1 pkg (454 g)	360	**50**	18

FROZEN, MICROWAVE, AND REFRIGERATED FOODS

FOOD	AMOUNT	CALORIES TOTAL	CALORIES FAT	CALORIES SAT-FAT
Swanson Hungry Man (cont.)				
Turkey (Mostly White Meat)	1 pkg (475 g)	510	**130**	36
Turkey Pot Pie	1 pkg (397 g)	650	**310**	117
Taj Gourmet				
Bean Masala	1 pkg (342 g)	330	**50**	5
Chicken Korma	1 pkg (312 g)	330	**80**	18
Chicken Tikka Masala	1 pkg (312 g)	330	**80**	27
Dal Bahaar	1 pkg (342 g)	330	**60**	5
Palak Paneer	1 pkg (342 g)	320	**60**	27
Vegetable Korma	1 pkg (342 g)	300	**50**	9
Tyson				
Beef Fajitas	3½ fajitas (357 g)	480	**120**	36
Van De Kamp's				
Baked Breaded Fish Fillets	2 fillets (99 g)	150	**25**	5
Battered Fish Fillets	1 fillet (75 g)	180	**100**	14
Breaded Fish Fillets	2 fillets (99 g)	280	**170**	27
Breaded Fish Sticks	6 sticks (114 g)	290	**150**	23
Garlic & Herb Baked Breaded Fish Fillets	1 fillet (78 g)	150	**45**	9
Lemon Pepper Baked Breaded Fish Fillets	1 fillet (78 g)	140	**45**	9
Weight Watchers Smart Ones				
Angel Hair Pasta	1 entree (255 g)	180	**20**	0
Bowtie Pasta & Mushrooms Marsala	1 entree (273 g)	270	**60**	41
Broccoli and Cheese Baked Potato	1 entree (283 g)	250	**50**	36
Chicken Fettuccini	1 entree (283 g)	300	**60**	18
Creamy Rigatoni with Broccoli & Chicken	1 entree (255 g)	230	**20**	5
Fiesta Chicken	1 entree (241 g)	210	**20**	5
Grilled Salisbury Steak	1 entree (241 g)	250	**80**	36
Homestyle Macaroni & Cheese	1 entree (255 g)	290	**60**	27
Hunan Style Rice & Vegetables	1 entree (293 g)	280	**70**	18
Lasagna Florentine	1 entree (283 g)	200	**20**	0
Lemon Herb Chicken Piccata	1 entree (241 g)	200	**20**	5

FROZEN, MICROWAVE, AND REFRIGERATED FOODS

FOOD	AMOUNT	CALORIES TOTAL	CALORIES FAT	CALORIES SAT-FAT
Weight Watchers Smart Ones (cont.)				
Macaroni and Cheese	1 entree (255 g)	220	**20**	5
Pasta & Spinach Romano	1 entree (294 g)	260	**70**	27
Pasta with Tomato Basil Sauce	1 entree (272 g)	260	**60**	23
Penne Pasta with Sundried Tomatoes	1 entree (283 g)	280	**70**	45
Penne Pollo	1 entree (283 g)	290	**50**	27
Ravioli Florentine	1 entree (241 g)	220	**20**	5
Roast Turkey Medallions	1 entree (241 g)	180	**20**	5
Santa Fe Style Rice & Beans	1 entree (283 g)	290	**70**	36
Shrimp Marinara	1 entree (255 g)	180	**20**	5
Swedish Meatballs	1 entree (255 g)	300	**90**	36
Tuna Noodle Casserole	1 entree (269 g)	270	**60**	23
Yu Sing (Michelina's)				
Chicken Fried Rice	1 pkg (227 g)	360	**70**	14
Chicken Lo Mein	1 meal (227 g)	230	**40**	NA
Oriental Beef & Peppers	1 pkg (227 g)	290	**60**	23
Shrimp Fried Rice	1 pkg (227 g)	360	**50**	5
Shrimp Lo Mein	1 pkg (227 g)	220	**15**	0
Sweet & Sour Chicken	1 pkg (241 g)	340	**35**	9

PIZZAS, FROZEN

FOOD	AMOUNT	CALORIES TOTAL	CALORIES FAT	CALORIES SAT-FAT
Amy's Pizza				
Cheese	⅓ pizza (123 g)	310	**100**	36
Spinach	⅓ pizza (132 g)	320	**100**	36
Vegetable, no cheese	⅓ pizza (113 g)	270	**70**	9
Celeste Pizza-for-One				
Cheese	1 pizza (170 g)	420	**180**	90
Deluxe	1 pizza (202 g)	470	**220**	81
Four Cheese				
Original	1 pizza (174 g)	480	**240**	99
Zesty	1 pizza (174 g)	470	**210**	108
Pepperoni	1 pizza (170 g)	470	**240**	81
Suprema	1 pizza (221 g)	530	**260**	90
Vegetable	1 pizza (187 g)	420	**180**	63

FROZEN, MICROWAVE, AND REFRIGERATED FOODS

FOOD	AMOUNT	CALORIES TOTAL	CALORIES FAT	CALORIES SAT-FAT
Di Giorno				
Four Cheese				
small	⅓ pizza (114 g)	280	**90**	45
large	⅙ pizza (139 g)	330	**100**	54
Pepperoni				
small	⅓ pizza (120 g)	320	**120**	54
large	⅙ pizza (148 g)	390	**150**	63
Spinach, Mushroom & Garlic	⅓ pizza (125 g)	270	**80**	36
Supreme	⅙ pizza (165 g)	400	**150**	63
Fox De Luxe				
Sausage & Pepperoni	1 pizza (184 g)	460	**160**	45
Healthy Choice French Bread Pizza				
Cheese	1 pizza (170 g)	340	**45**	14
Pepperoni	1 pizza (170 g)	340	**45**	14
Supreme	1 pizza (180 g)	330	**45**	14
Vegetable	1 pizza (170 g)	280	**35**	14
Hot Pockets				
Pizza Mini's				
Double Cheese	6 pieces (85 g)	240	**80**	36
Pepperoni	6 pieces (85 g)	250	**100**	36
Sausage & Pepperoni	6 pieces (85 g)	230	**80**	27
Pizza snacks				
Pepperoni	6 pieces (85 g)	220	**70**	31
Toaster Breaks Pizza				
Double Cheese	1 piece (60 g)	190	**80**	27
Pepperoni	1 piece (60 g)	200	**90**	31
Jeno's Crisp 'n Tasty Pizza				
Chompin' Cheese	1 pizza (195 g)	460	**170**	54
Crazy Combination (Sausage & Pepperoni)	1 pizza (198 g)	520	**250**	63
Power Pepperoni	1 pizza (192 g)	510	**240**	54
Lean Cuisine French Bread Pizza				
Deluxe	1 pkg (173 g)	300	**50**	23
Pepperoni	1 pkg (148 g)	310	**60**	27
Mama Celeste				
Four Cheese	⅙ pizza (140 g)	340	**90**	41
Pepperoni	⅙ pizza (142 g)	380	**140**	63
Supreme	⅙ pizza (157 g)	380	**150**	63

FROZEN, MICROWAVE, AND REFRIGERATED FOODS

FOOD	AMOUNT	CALORIES TOTAL	CALORIES FAT	CALORIES SAT-FAT
McCain Ellio's				
Cheese	1 slice (75 g)	160	**45**	18
Ore Ida Bagel Bites				
Cheese & Pepperoni	4 pieces (88 g)	200	**60**	31
Cheese, Sausage & Pepperoni	4 pieces (88 g)	190	**60**	32
Three Cheese	4 pieces (88 g)	190	**50**	31
Red Baron Deep Dish Singles Pizzas				
Cheese	1 pizza (168 g)	500	**230**	90
Pepperoni	1 pizza (168 g)	540	**280**	99
Supreme	1 pizza (168 g)	490	**240**	90
Safeway Select Gourmet Club				
French Bread Pepperoni Pizza	1 pizza (85 g)	230	**90**	36
Stouffer's French Bread Pizza				
Cheese	1 pizza (147 g)	370	**150**	54
Deluxe	1 pizza (175 g)	420	**160**	63
Extra Cheese	1 pizza (167 g)	400	**140**	63
Five Cheese	1 pizza (147 g)	370	**150**	54
Grilled Vegetable	1 pizza (165 g)	350	**100**	45
Pepperoni	1 pizza (159 g)	390	**140**	54
Pepperoni & Mushroom	1 pizza (174 g)	440	**170**	63
Sausage & Pepperoni	1 pizza (177 g)	470	**200**	72
Three Meat (sausage, pepperoni, bacon)	1 pizza (177 g)	470	**200**	72
Tombstone				
Light				
Vegetable	⅕ pizza (131 g)	240	**60**	23
Original				
Pepperoni	¼ pizza (152 g)	400	**190**	81
Supreme	⅕ pizza (130 g)	320	**140**	63
Thin Crust				
Four Meat Combination	¼ pizza (143 g)	380	**200**	90
Pepperoni	¼ pizza (138 g)	400	**230**	99
Tombstone for One Deep Dish Pizza				
Cheese	1 pizza (177 g)	470	**190**	108
Pepperoni	1 pizza (177 g)	510	**240**	117

FROZEN, MICROWAVE, AND REFRIGERATED FOODS

FOOD	AMOUNT	CALORIES TOTAL	CALORIES FAT	CALORIES SAT-FAT
Totino's Party Pizza				
Pepperoni	½ pizza (145 g)	380	**190**	45
Supreme	½ pizza (155 g)	390	**190**	41
Totino's Pizza Rolls				
Cheese	6 rolls (85 g)	210	**70**	27
Combination (Sausage & Pepperoni)	6 rolls (85 g)	220	**100**	27
Pepperoni	6 rolls (85 g)	240	**110**	27
Pepperoni Supreme	6 rolls (85 g)	230	**100**	23
Supreme	6 rolls (85 g)	220	**90**	18
Wolfgang Puck's Pizza				
Artichoke Heart	½ pizza (140 g)	340	**150**	54
Barbecue Chicken	½ pizza (150 g)	340	**100**	45
Four Cheese	½ pizza (130 g)	360	**130**	54
Pepperoni Mushroom	½ pizza (155 g)	390	**130**	54
Zucchini & Tomato	½ pizza (146 g)	290	**100**	45
POULTRY				
Perdue				
Breaded Chicken Breast Cutlets				
Homestyle	84 g	120	**15**	0
Italian Style	84 g	130	**25**	9
Breaded Chicken Breast Nuggets	5 pieces (93 g)	200	**110**	27
Breaded Chicken Breast Tenderloins	84 g	160	**60**	23
Swift Premium Turkey Roast				
Boneless White Turkey	5 oz (40 g)	160	**80**	27
White & Dark Turkey	5 oz (40 g)	190	**110**	41
Tyson				
Barbecue Style Chicken Wings	3 pieces (91 g)	200	**110**	31
Breaded Chicken Breast Fillets	2 fillets (81 g)	180	**70**	14
Breaded Chicken Breast Patties	1 patty (73 g)	190	**110**	27
fat free	1 patty (74 g)	80	**0**	0

FROZEN, MICROWAVE, AND REFRIGERATED FOODS

FOOD	AMOUNT	CALORIES TOTAL	CALORIES FAT	CALORIES SAT-FAT
Tyson (cont.)				
Breaded Chicken Breast Tenders	3 tenders (92 g)	100	**0**	0
Chicken Fajitas	3½ fajitas (374 g)	460	**100**	27
Hot 'n Spicy Chicken Wings	4 pcs (96 g)	220	**130**	31
Mandarin Sesame Wraps	1½ wraps (416 g)	630	**140**	31
Wamplers				
100% Pure Ground Chicken	4 oz (112 g)	220	**120**	36
100% Pure Ground Turkey	4 oz (112 g)	210	**130**	27
100% Pure Turkey Burgers	1 burger (112 g)	210	**130**	27
Turkey Mignons	1 mignon (142 g)	230	**110**	45
SANDWICHES				
Amy's in a Pocket Sandwich				
Spinach Feta	1 (128 g)	200	**60**	27
Vegetable Pie	1 (142 g)	230	**60**	5
Banquet Hot Sandwich Toppers				
Creamed Chipped Beef	1 bag (113 g)	120	**50**	23
Gravy & Salisbury Steak	1 bag (141 g)	210	**140**	63
Gravy & Sliced Beef	1 bag (113 g)	70	**20**	9
Gravy & Sliced Turkey	1 bag (141 g)	160	**100**	36
Bob Evans Farms				
Sausage & Biscuit	2 sandwiches (112 g)	370	**210**	63
Chef America				
Croissant Pockets				
Chicken, Broccoli & Cheddar	1 piece (128 g)	290	**80**	36
Egg, Sausage & Cheese	1 piece (128 g)	340	**130**	54
Ham & Cheese	1 piece (128 g)	320	**110**	54
Pepperoni Pizza	1 piece (128 g)	360	**140**	63
Turkey & Ham with Swiss	1 piece (128 g)	290	**90**	36
Hot Pockets				
Beef & Cheddar	1 piece (128 g)	350	**150**	81
Beef with Barbecue	1 piece (128 g)	340	**110**	45
Chicken & Cheddar with Broccoli	1 piece (128 g)	300	**90**	45
Meatballs with Mozzarella	1 piece (128 g)	320	**100**	54

FROZEN, MICROWAVE, AND REFRIGERATED FOODS

FOOD	AMOUNT	CALORIES TOTAL	CALORIES FAT	CALORIES SAT-FAT
Chef America (cont.)				
Pepperoni & Sausage Pizza	1 piece (128 g)	330	**120**	54
Pepperoni Pizza	1 piece (128 g)	360	**140**	63
Lean Pockets				
Chicken Broccoli Supreme	1 piece (128 g)	260	**60**	27
Chicken Fajita	1 piece (128 g)	270	**60**	23
Chicken Parmesan	1 piece (128 g)	280	**60**	23
Ham & Cheddar	1 piece (128 g)	280	**70**	27
Pepperoni Pizza Deluxe (reduced fat)	1 piece (128 g)	270	**60**	23
Philly Steak & Cheese	1 piece (128 g)	260	**60**	27
Turkey, Broccoli & Cheese	1 piece (128 g)	250	**60**	27
Toaster Breaks Melts				
Ham & Cheese	1 piece (61 g)	180	**80**	23
Veggie Pockets				
Bar-B-Q Style	1 pocket (127 g)	290	**80**	5
Broccoli & Cheddar Style	1 pocket (127 g)	290	**80**	5
Greek Style	1 pocket (127 g)	250	**80**	5
Indian Style	1 pocket (127 g)	260	**80**	5
Oriental Style	1 pocket (127 g)	250	**80**	5
Pizza Style	1 pocket (127 g)	270	**80**	5
Tex-Mex Style	1 pocket (127 g)	280	**80**	5
Veggie Burger	1 sandwich (71 g)	130	**10**	0
Gardenburger Veggie Patties				
Fire Roasted Vegetable	1 patty (71 g)	120	**25**	14
Gardenburger with cheese	1 patty (71 g)	110	**25**	14
Savory Mushroom	1 patty (71 g)	120	**25**	14
The Original	1 patty (71 g)	130	**25**	9
Veggie Medley	1 patty (71 g)	100	**0**	0
Green Giant Harvest Burgers				
Original flavor	1 patty (90 g)	140	**35**	14
Healthy Choice Hearty Handfuls				
Chicken & Broccoli	1 hearty handful (173 g)	320	**50**	14
Chicken & Mushrooms	1 hearty handful (173 g)	310	**45**	14

FROZEN, MICROWAVE, AND REFRIGERATED FOODS

FOOD	AMOUNT	CALORIES TOTAL	CALORIES FAT	CALORIES SAT-FAT
Healthy Choice Hearty Handfuls (cont.)				
Garlic Chicken	1 hearty handful (173 g)	330	**45**	14
Ham & Cheese	1 hearty handful (173 g)	320	**45**	14
Italian Style Meatball	1 hearty handful (173 g)	320	**45**	14
Philly Beef Steak	1 hearty handful (173 g)	290	**45**	14
Jimmie Dean				
Bacon, Egg, & Cheese on a Biscuit	1 sandwich (102 g)	300	**150**	54
Sausage, Egg, & Cheese on a Biscuit	1 sandwich (128 g)	390	**240**	90
Sausage Biscuits	2 sandwiches (113 g)	390	**220**	72
Morningstar Farms				
Better 'n Burgers	1 patty (78 g)	70	**0**	0
Chik Nuggets	4 nuggets (86 g)	160	**40**	5
Chik Patties	1 patty (71 g)	150	**50**	9
Garden Veggie Patties	1 patty (67 g)	100	**25**	5
Spicy Black Bean	1 patty (78 g)	110	**10**	0
Veggie Dogs	1 link (57 g)	80	**5**	0
Quaker Maid				
Italian Meatball Sandwich	1 (170 g)	460	**220**	90
Philly Cheese Steak	1 sandwich (170 g)	400	**80**	36
Pure Beef Sandwich Steak	1 steak (56 g)	170	**130**	54
Safeway Stuffed Sandwiches				
Barbecue Beef	1 piece (127 g)	310	**120**	23
Ham & Cheese	1 piece (128 g)	320	**140**	54
Lean Turkey, Broccoli & Cheddar	1 piece (128 g)	240	**60**	18
Pepperoni Pizza	1 piece (128 g)	370	**170**	63
Steak-umm				
100% All Beef Sandwich Steaks	1 steak, raw (57 g)	190	**150**	63
Reduced Fat Real Beef Sandwich Steaks	1 steak, raw (57 g)	110	**63**	18

FROZEN, MICROWAVE, AND REFRIGERATED FOODS

FOOD	AMOUNT	CALORIES TOTAL	FAT	SAT-FAT
Steak-umm (cont.)				
Sandwich to Go Cheese Steak	1 sandwich (133 g)	290	**120**	54
Tennessee Pride				
Sausage & Buttermilk Biscuit	2 sandwiches (91 g)	320	**180**	63
Sausage & Egg with Cheese Biscuit	2 sandwiches (125 g)	400	**230**	81
Weight Watchers Sandwiches				
English Muffin with Ham & Cheese	1 sandwich (113 g)	210	**50**	27
White Castle				
Cheeseburger	2 sandwiches (104 g)	310	**160**	81
Hamburger	2 sandwiches (90 g)	270	**130**	54
VEGETABLES AND SIDE DISHES (*see also* **VEGETABLES**)				
Birds Eye Side Orders				
Bavarian Style Vegetables	1 cup (156 g)	150	**70**	36
French Green Beans with Toasted Almonds	¾ cup (116 g)	80	**35**	0
New England Style Vegetables	1 pkg (255 g)	260	**130**	45
Peas & Pearl Onions	⅔ cup (121 g)	90	**5**	0
Budget Gourmet				
Cheddared Potatoes & Broccoli	1 pkg (141 g)	160	**70**	54
Oriental Rice with Vegetables	1 pkg (148 g)	200	**100**	41
Rice Pilaf with Green Beans	1 pkg (141 g)	220	**90**	27
Spinach au Gratin	1 pkg (141 g)	160	**110**	63
Green Giant				
Alfredo Vegetables	¾ cup (109 g)	80	**25**	14
Broccoli & Cheese	⅔ cup (112 g)	70	**25**	9
Cheesy Rice & Broccoli	1 pkg (283 g)	300	**45**	14
Creamed Spinach	½ cup (109 g)	80	**25**	14
Cut Leaf Spinach in Butter Sauce	½ cup (98 g)	40	**15**	9
Green Bean Casserole	⅔ cup (109 g)	90	**45**	9

FROZEN, MICROWAVE, AND REFRIGERATED FOODS

FOOD	AMOUNT	CALORIES TOTAL	FAT	SAT-FAT
Green Giant (cont.)				
Rice Pilaf	1 pkg (283 g)	260	**25**	14
Shoepeg White Corn & Butter	¾ cup (112 g)	120	**25**	14
Southwestern Style Corn & Roasted Red Peppers	¾ cup (99 g)	90	**10**	0
Teriyaki Vegetables	1¼ cup (110 g)	100	**60**	9
Vegetable Medley & Butter	¾ cup (103 g)	60	**20**	14
White & Wild Rice	1 pkg (283 g)	280	**50**	9
Ore Ida				
Country Style Hash Browns	1¼ cup (91 g)	80	**0**	0
Country Style Potato Wedges	13 pieces (84 g)	120	**40**	5
Country Style Steak Fries	8 pieces (84 g)	110	**30**	0
Crispers	17 fries (84 g)	220	**110**	18
Golden Crinkles	14 pieces (84 g)	120	**30**	9
Golden Fries	16 pieces (84 g)	120	**35**	5
Golden Patties Shredded Potatoes	1 patty (71 g)	140	**70**	14
Mashed Potatoes	⅔ cup (69 g)	90	**20**	0
Mini Tater Tots	19 pieces (84 g)	180	**90**	18
Onion Ringers	6 rings (88 g)	210	**100**	14
Onion Rings	4 pieces (84 g)	220	**100**	18
Oven Chips	7 pieces (84 g)	180	**70**	9
Potato Wedges with skins	8 pieces (84 g)	110	**25**	5
Shoestrings	44 pieces (84 g)	150	**50**	9
Tater Tots	9 pieces (86 g)	150	**60**	14
Toaster Hash Browns	2 patties (99 g)	190	**90**	18
Topped Baked Potatoes, Broccoli & Cheese	½ baker (159 g)	160	**35**	18
Twice Baked Potatoes				
Cheddar cheese	1 piece (141 g)	190	**70**	18
Sour cream & chives	1 piece (141 g)	190	**60**	14
Safeway Select Side Dish				
BBQ Beans	½ cup	200	**15**	5
Cheddar Cheese & Bacon Mashed Potatoes	⅔ cup	190	**90**	54

FROZEN, MICROWAVE, AND REFRIGERATED FOODS

FOOD	AMOUNT	CALORIES TOTAL	CALORIES FAT	CALORIES SAT-FAT
Safeway Select Side Dish (cont.)				
Creamed Spinach	½ cup	130	**90**	54
Roasted Garlic Mashed				
Potatoes	⅔ cup	110	**40**	27
Stouffer's				
Creamed Spinach	½ cup (125 g)	180	**100**	63
Corn Souffle	½ cup (125 g)	170	**60**	18
Spinach Soufflé	½ cup (118 g)	140	**80**	18

FRUITS AND FRUIT JUICES

FOOD	AMOUNT	CALORIES TOTAL	CALORIES FAT	CALORIES SAT-FAT
Apple				
fresh	1	81	**0**	0
cooked, boiled	½ cup slices	46	**0**	0
canned, sweetened	½ cup slices	68	**0**	0
Apple butter	1 tbsp	35	**0**	0
Apple juice	1 cup	116	**0**	0
Applesauce				
unsweetened	½ cup	53	**0**	0
sweetened	½ cup	97	**0**	0
Apricot	3	51	**0**	0
dried, uncooked	10 halves	83	**0**	0
	1 cup halves	310	**0**	0
dried, cooked	1 cup halves	211	**0**	0
Avocado, fresh				
California	1	306	**270**	41
Florida			**24**	
	1	339	**3**	48
Banana, fresh	1	105	**5**	2
Blackberries, fresh	½ cup	37	**0**	0
Blueberries, fresh	1 cup	82	**0**	0
Boysenberries, canned, heavy				
syrup	½ cup	113	**0**	0
Cantaloupe, fresh	½ fruit	94	**0**	0

FRUITS AND FRUIT JUICES

FOOD	AMOUNT	CALORIES TOTAL	CALORIES FAT	CALORIES SAT-FAT
	1 cup cubes	57	**0**	0
Cherries, sour, red, fresh	1 cup with pits	51	**0**	0
canned, light syrup	½ cup	94	**0**	0
canned, heavy syrup	½ cup	116	**0**	0
Cherries, sweet, fresh	10	49	**0**	0
	1 cup	104	**0**	0
canned, light syrup	½ cup	85	**0**	0
canned, heavy syrup	½ cup	107	**0**	0
Cranberries, fresh	1 cup whole	46	**0**	0
dried cranberries	⅓ cup	130	**0**	0
Cranberry juice cocktail	1 cup	147	**0**	0
Cranberry sauce	½ cup	209	**0**	0
Dates	5–6 pitted dates	120	**0**	0
	1 cup chopped	489	**7**	3
Figs, fresh	1 medium	37	**0**	0
	1 large	47	**0**	0
dried, uncooked	2	150	**0**	0
	1 cup	508	**21**	4
dried, cooked	1 cup	279	**11**	2
Fruit cocktail, canned				
juice pack	1 cup	113	**0**	0
light syrup pack	1 cup	110	**0**	0
heavy syrup pack	1 cup	186	**0**	0
Grapefruit, fresh	½ fruit	38	**0**	0
	1 cup sections	74	**0**	0
Grapefruit juice				
fresh	4 oz	47	**0**	0
canned, unsweetened	4 oz	47	**0**	0
canned, sweetened	4 oz	57	**0**	0
Grapes, fresh	10	15	**0**	0
	1 cup	58	**0**	0
Grape juice, canned or bottled	1 cup	155	**0**	0
Guava, fresh	1	45	**0**	0
Honeydew melon	⅒ melon	46	**0**	0
	1 cup cubes	60	**0**	0
Kiwi, fresh	1 medium	46	**0**	0
	1 large	55	**0**	0
Kumquat, fresh	1	12	**0**	0

FRUITS AND FRUIT JUICES

FOOD	AMOUNT	CALORIES TOTAL	CALORIES FAT	CALORIES SAT-FAT
Lemon, fresh	1 medium	17	**0**	0
	1 large	25	**0**	0
Lemon juice	1 tbsp	4	**0**	0
	1 cup	60	**0**	0
Lime, fresh	1	20	**0**	0
Lime juice	1 tbsp	4	**0**	0
	1 cup	66	**0**	0
Mango, fresh	1	135	**0**	0
	1 cup slices	108	**0**	0
Mixed fruit, dried	11 oz package	712	**13**	0
Mulberries, fresh	10	7	**0**	0
	1 cup	61	**0**	0
Nectarine, fresh	1	67	**0**	0
	1 cup slices	68	**0**	0
Orange, fresh	1	60	**0**	0
Orange juice	juice from 1 fruit	39	**0**	0
	1 cup	111	**0**	0
Papaya, fresh	1	117	**0**	0
	1 cup cubes	54	**0**	0
Peaches, fresh	1	37	**0**	0
canned in juice	1 cup halves	109	**0**	0
canned in light syrup	1 cup halves	136	**0**	0
canned in heavy syrup	1 cup halves	190	**0**	0
dried, uncooked	10 halves	311	**9**	1
	1 cup halves	383	**11**	1
dried, cooked	1 cup halves	198	**6**	1
Pears, fresh	1	98	**0**	0
canned in juice	1 cup halves	123	**0**	0
canned in light syrup	1 cup halves	144	**0**	0
canned in heavy syrup	1 cup halves	188	**0**	0
dried, uncooked	10 halves	459	**10**	1
	1 cup halves	472	**10**	1
dried, cooked	1 cup halves	325	**7**	0
Pineapple, fresh	1 slice (¾" thick)	42	**0**	0
	1 cup diced pieces	77	**0**	0
canned in juice	1 cup chunks	150	**0**	0
canned in light syrup	1 cup	131	**0**	0
canned in heavy syrup	1 cup	199	**0**	0

FRUITS AND FRUIT JUICES

FOOD	AMOUNT	CALORIES TOTAL	FAT	SAT-FAT
Pineapple juice	1 cup	139	**0**	0
Plantain, fresh	1	218	**6**	0
cooked	1 cup slices	179	**3**	0
Plums, fresh	1	36	**0**	0
	1 cup slices	91	**0**	0
Prunes				
canned in heavy syrup	5 fruits	90	**0**	0
	1 cup	245	**0**	0
dried, uncooked	10 fruits	201	**0**	0
	1 cup	385	**7**	0
dried, cooked	1 cup	227	**4**	0
Prune juice	1 cup	181	**0**	0
Raisins, seedless	1 cup packed	494	**7**	0
Raspberries, fresh	1 cup	61	**0**	0
Rhubarb, fresh	1 cup diced pcs	26	**0**	0
Strawberries, fresh	1 cup	45	**0**	0
canned in heavy syrup	1 cup	234	**0**	0
Tangerine, fresh	1	37	**0**	0
	1 cup sections	86	**0**	0
Watermelon, fresh	1 cup diced pcs	50	**0**	0

GRAINS AND PASTA

FOOD	AMOUNT	CALORIES TOTAL	FAT	SAT-FAT
BREAD				
Bagels				
Freshly baked, grocery				
Blueberry	1 (114 g)	280	**10**	0
Bran	1 (114 g)	260	**10**	0
Cinnamon Raisin	1 (114 g)	280	**10**	0
Combination	1 (114 g)	280	**15**	0
Garlic	1 (114 g)	270	**10**	0
Honey Wheat	1 (114 g)	260	**10**	0
Oat Bran	1 (114 g)	270	**10**	0

GRAINS AND PASTA

FOOD	AMOUNT	CALORIES TOTAL	CALORIES FAT	CALORIES SAT-FAT
Bagels (cont.)				
Onion	1 (114 g)	270	**10**	0
Plain	1 (114 g)	270	**10**	0
Poppy Seed	1 (114 g)	280	**20**	0
Pumpernickel	1 (114 g)	260	**10**	0
Raisin Bran	1 (114 g)	260	**10**	0
Rye	1 (114 g)	270	**10**	0
Sesame Seed	1 (114 g)	280	**20**	0
Sourdough	1 (114 g)	280	**10**	0
Frozen				
Lenders				
Bagelettes	2 (25 g)	140	**10**	0
Big 'n Crusty				
Cinnamon Raisin	1 (85 g)	230	**15**	0
Plain or Onion	1 (85 g)	230	**10**	0
Blueberry Swirl or Cinnamon Swirl	1 (57 g)	160	**5**	0
Onion or Egg	1 (57 g)	160	**15**	0
Plain	1 (57 g)	150	**5**	0
Soft	1 (71 g)	210	**30**	0
Sara Lee				
Cinnamon & Raisin	1 (79 g)	220	**5**	0
Oat bran or poppy seed	1 (79 g)	210	**10**	0
Plain	1 (79 g)	210	**0**	0
Bialy	1 (110 g)	270	**15**	0
Biscuits (*see also* **FAST FOODS**)				
Beaten Biscuits	2" biscuit	98	**60**	24
Pillsbury Ready-to-Bake				
Big Country				
Butter Tastin'	1 biscuit (34 g)	100	**35**	9
Buttermilk	1 biscuit (34 g)	100	**35**	9
Grands!				
Golden Wheat	1 biscuit (61 g)	200	**70**	18
Butter Tastin' Biscuits	1 biscuit (61 g)	200	**90**	23
Buttermilk Biscuits	1 biscuit (61 g)	200	**90**	27
Golden Corn	1 biscuit (61 g)	210	**90**	23

GRAINS AND PASTA

FOOD	AMOUNT	CALORIES TOTAL	CALORIES FAT	CALORIES SAT-FAT
Biscuits (cont.)				
Flaky Biscuits	1 biscuit (61 g)	200	**80**	18
Southern Style Biscuits	1 biscuit (61 g)	200	**90**	23
Biscuit Baking Mix (Jiffy)	¼ cup mix (32 g)	130	**40**	9
	baked	190	**60**	20
Bisquick All-Purpose Baking Mix (Betty Crocker)				
Biscuits, Pancakes	⅓ cup mix (40 g)	170	**50**	14
reduced fat	⅓ cup mix (40 g)	150	**25**	5
Buttermilk Biscuit Mix (Washington)	5 tbsp mix (40 g)	160	**45**	14
Hungry Jack				
Butter Tastin' or Buttermilk	1 biscuit (34 g)	100	**40**	9
Bran'nola	1 slice (38 g)	100	**15**	0
Country Oat	1 slice (38 g)	90	**25**	5
Bread Crumbs				
Cracker Meal (OTC)	¼ cup (28 g)	110	**0**	0
Italian Style	¼ cup (28 g)	110	**15**	0
Plain	¼ cup (28 g)	100	**15**	0
Kellogg's Corn Flake Crumbs	2 tbsp (11 g)	40	**0**	0
Oven Fry				
Extra Crispy Chicken	⅛ packet (15 g) (coats 1 piece)	60	**10**	0
Extra Crispy Pork	⅛ pkt (15 g) (coats 1 chop)	60	**10**	0
Shake 'n Bake				
Barbecue Chicken Glaze	⅛ packet (12 g) (coats 1–2 pieces)	45	**10**	0
Buffalo Wings	1/10 pkg (10 g) (coats 2 pieces)	40	**5**	0
Honey Mustard Chicken Glaze	¼ packet (25 g) (coats 2 pieces)	100	**20**	9
Hot & Spicy Chicken or Pork coating	⅛ pkg (10 g) (coats 1 piece)	40	**10**	0
Pork Original Recipe	⅛ packet (11 g) (coats 1 chop)	40	**0**	0

GRAINS AND PASTA

FOOD	AMOUNT	CALORIES TOTAL	CALORIES FAT	CALORIES SAT-FAT
Bread Crumbs (cont.)				
Tangy Honey Glaze	¼ packet (25 g) (coats 2 pieces)	90	**15**	5
Bread Mixes				
Bread Machine				
Dromedary				
Country White	½ inch (50 g)	140	**10**	5
Italian Herb	½ inch (50 g)	140	**20**	14
Sourdough	½ inch (50 g)	140	**15**	9
Stoneground Wheat	½ inch (50 g)	140	**15**	9
Pillsbury				
Cracked Wheat	1/12 pkg (36 g)	130	**20**	0
Crusty White	1/12 pkg (36 g)	130	**15**	0
Brown Bread Raisin, New England Style (B & M)	½" slice (56 g)	130	**5**	0
	baked with water & 2 eggs	70	**35**	14
Hot Roll Mix (Pillsbury)	1 pan roll (28 g)	100	**10**	0
Quick Mix (Pillsbury)				
Apple Cinnamon	1/12 loaf (37 g mix)	140	**10**	0
Banana	1/12 loaf (33 g mix)	120	**10**	0
Spoon Bread Mix (Washington)	2 tbsp mix (13 g)	60	**30**	9
Bread Sticks				
(Stella D'Oro)	1 stick (9 g)	40	**10**	0
Grissini-Style fat free (Stella D'Oro)	3 sticks (15 g)	60	**0**	0
Italian (Angonoa's)	3 sticks (15 g)	60	**10**	0
Mini Sesame (Angonoa's)	24 sticks (28 g)	130	**35**	5
Sesame (Stella D'Oro)	1 stick (11 g)	50	**20**	0
Sesame	4 sticks (28 g)	130	**35**	5
Soft (Bread Du Jour)	1 stick (53 g)	130	**10**	0
Soft (Pillsbury)	1 stick (39 g)	110	**25**	5
Soft, Cheese	1 stick (28 g)	80	**20**	9
Bread Stick Mixes				
Pillsbury				
Breadsticks	1 stick (39 g)	110	**20**	0
Garlic Breadsticks	2 sticks (60 g)	180	**60**	14

GRAINS AND PASTA

FOOD	AMOUNT	CALORIES TOTAL	CALORIES FAT	CALORIES SAT-FAT
Bread Stuffing				
Arnold				
Cornbread	2 cups (67 g)	250	**35**	9
Seasoned	2 cups (67 g)	250	**30**	5
Pepperidge Farm Top of Stove Stuffing Mix				
Apple & Raisin	½ cup (36 g)	140	**15**	0
Corn Bread	¾ cup (43 g)	170	**20**	0
Country Garden Herb	½ cup (34 g)	150	**45**	9
Herb Seasoned Stuffing	¾ cup (43 g)	170	**15**	0
Cubed	¾ cup (37 g)	140	**15**	0
Honey Pecan	½ cup (34 g)	140	**45**	5
Pepperidge Farm Crunchy Croutons				
Cheese & Garlic	9 croutons (7 g)	35	**15**	0
Onion & Garlic	9 croutons (7 g)	30	**15**	0
Seasoned	9 croutons (7 g)	35	**15**	0
Ritz Stuffing Mix	⅔ cup (38 g)	200	**80**	18
Stove Top Stuffing Mix (Kraft)				
Cornbread	⅙ box (28 g)	110	**5**	0
	½ cup prep.	170	**80**	18
for Chicken	½ cup dry (28 g)	120	**25**	0
	½ cup prep.	170	**80**	14
lower sodium	⅙ box (28 g)	110	**10**	0
	½ cup prep.	180	**80**	14
for Pork	⅙ box (28 g)	110	**10**	0
	½ cup prep.	170	**80**	18
for Turkey	⅙ box (28 g)	110	**10**	0
	½ cup prep.	170	**80**	18
Savory Herbs	⅙ box (28 g)	110	**10**	0
	½ cup prep.	170	**80**	14
Wild Rice & Mushroom	⅔ cup (37 g)	170	**50**	14
Challah	1" slice (50 g)	160	**35**	9
Cocktail				
Rye or pumpernickel	3 slices (31 g)	80	**10**	0
Corn Bread	⅛ of 8" diameter	225	**77**	22
	3" x 4.5"	325	**100**	35
	2" x 2"	104	**57**	28
Crackling Corn Bread	⅛ of 8" diameter	263	**150**	55

GRAINS AND PASTA

FOOD	AMOUNT	CALORIES TOTAL	CALORIES FAT	CALORIES SAT-FAT
Corn Bread Mixes				
Buttermilk Corn Bread Mix (Washington)	3 tbsp (25 g)	90	**15**	5
	baked with whole milk & 1 egg	110	**30**	14
Corn Bread mix (Aunt Jemima)	⅓ cup mix (35 g)	140	**35**	9
	as prepared	160	**45**	11
Corn Bread Mix (Marie Callender's)	¼ cup mix (66 g)	150	**30**	0
Cornbread Twists (Pillsbury)	1 twist (41 g)	130	**50**	14
Cracked wheat				
Pepperidge Farm	1 slice (25 g)	70	**10**	0
Croissant				
Almond Filled	1 (98 g)	430	**280**	36
Apple Filled	1 (110 g)	290	**150**	27
Butter	1 (55 g)	200	**100**	63
Butter (Vie-de-France)	1 (56 g)	230	**110**	72
Cheese Filled	1 (104 g)	380	**250**	81
Cherry Filled	1 (110 g)	330	**140**	27
Chocolate Filled	1 (91 g)	350	**180**	36
Margarine	1 (70 g)	260	**140**	27
Croissant Mixes				
Pillsbury				
Original Crescent	1 roll (28 g)	110	**50**	14
reduced fat	1 roll (28 g)	100	**40**	9
Croutons				
Pepperidge Farm	2 tbsp (7 g)	30	**10**	0
English muffins				
Thomas	1 (57 g)	120	**10**	0
Bran'nola (Arnolds)	1 (66 g)	130	**15**	0
Raisin	1 (61 g)	140	**10**	0
Raisin (Sun Maid)	1 (68 g)	160	**10**	0
Flatbread (New York)				
low fat				
all flavors	1 piece (11 g)	50	**15**	5

GRAINS AND PASTA

FOOD	AMOUNT	CALORIES TOTAL	CALORIES FAT	CALORIES SAT-FAT
Flatbread (cont.)				
fat free				
Roasted Garlic	1 piece (11 g)	40	**0**	0
Mini's				
Low fat	3 pieces (13 g)	58	**17**	0
Fat free	3 pieces (13 g)	45	**0**	0
French bread	1 slice (2")	130	**0**	0
French loaf (Bread Du Jour)	3" slice (56 g)	130	**10**	0
Crusty French Loaf (Pillsbury)	⅕ loaf (62 g)	150	**10**	0
French Loaf Mix (Pillsbury)	⅕ pkg (62 g)	150	**20**	9
French toast (*see also* **FROZEN, MICROWAVE, AND REFRIGERATED FOODS**)				
Garlic bread	3½" slice (56 g)	190	**80**	14
with cheese	3½" slice (56 g)	180	**60**	14
Italian Bread	1¼" slice (46 g)	130	**10**	0
Pepperidge Farm	2" slice (57 g)	150	**20**	9
Italian Bread Shell (Boboli)	⅕ shell (57 g)	150	**30**	9
Italian Olive Bread	2 slices (50 g)	150	**30**	5
Multi-grain	1" slice (50 g)	160	**20**	NA
Oatmeal (Pepperidge Farm)	1 slice (25 g)	60	**10**	0
Soft (Pepperidge Farm)	1 slice (25 g)	60	**5**	NA
Pita				
White	1 (45 g)	110	**5**	0
Whole-wheat	1 (45 g)	110	**5**	0
Pizza Crusts				
Boboli				
Thin Pizza Crust	⅕ crust (57 g)	160	**35**	14
2 Pizza Crusts	½ crust (57 g)	150	**30**	9
Popover Mix, New England Style (Washington)	¼ cup mix (27 g)	110	**25**	9
	baked with water & 2 eggs	140	**40**	14
Pumpernickel	1 slice (32 g)	80	**10**	0
Raisin bread (Pepperidge Farm)	1 slice (28 g)	80	**10**	0

GRAINS AND PASTA

FOOD	AMOUNT	CALORIES TOTAL	CALORIES FAT	CALORIES SAT-FAT
Rolls				
Cinnamon rolls (Pillsbury)	1 roll (40 g)	140	**45**	14
Club (Pepperidge Farm)	1 (47 g)	120	**10**	5
Cornbread Twists (Pillsbury)	1 twist (41 g)	130	**50**	14
Dinner rolls				
Arnold	2 (38 g)	110	**25**	5
Country Style (Pepperidge Farm)	3 (57 g)	150	**30**	9
Crescent (Pillsbury)	2 (57 g)	200	**100**	23
Parker House (Pepperidge Farm)	3 (53 g)	150	**40**	14
Party (Pepperidge Farm)	5 (53 g)	170	**40**	14
Egg Twist	1 (47 g)	180	**35**	14
French (Pepperidge Farm)	½ roll (71 g)	180	**20**	5
Seven Grain (Pepperidge Farm)	1 (38 g)	80	**20**	0
Hamburger	1 (43 g)	130	**20**	9
Potato roll (Martins)	1 (53 g)	150	**20**	0
Hard rolls	1 (57 g)	160	**10**	5
Hoagie with sesame seeds (Pepperidge Farm)	1 (69 g)	200	**40**	23
Hot dog	1 (39 g)	110	**20**	5
Italian rolls (Bread Du Jour)	1 (35 g)	80	**5**	0
Oat Bran	1 roll (52 g)	140	**20**	5
Pumpernickel	1 roll (57 g)	160	**20**	0
Sub rolls (8") (La Parisienne)	⅗ roll (50 g)	130	**0**	0
Wheat	1 roll (57 g)	190	**25**	5
White	1 (6" in diam)	150	**10**	0
Whole-wheat	1 (6" in diam)	180	**10**	0
Roll mixes (refrigerated)				
Pillsbury				
Cinnamon Raisin Rolls with Icing	1 roll (49 g)	170	**50**	14
Cinnamon Rolls with Icing	1 roll (41 g)	150	**45**	14
reduced fat	1 roll (44 g)	140	**30**	9

GRAINS AND PASTA

FOOD	AMOUNT	CALORIES TOTAL	CALORIES FAT	CALORIES SAT-FAT
Roll mixes (cont.)				
Dinner Rolls	1 roll (40 g)	110	**20**	0
Grands!				
Cinnamon rolls with cream cheese icing	1 roll (99 g)	330	**100**	27
Cinnamon rolls with icing	1 roll (99 g)	320	**90**	23
reduced fat	1 roll (99 g)	300	**60**	18
Orange Sweet Rolls with Icing	1 roll (49 g)	170	**60**	14
Rye				
Jewish	1 slice (32 g)	80	**10**	0
Seeded	1 slice (22 g)	55	**9**	5
Onion	1 slice (32 g)	80	**10**	0
Sunflower seed	1 slice (50 g)	120	**7**	0
Seven grain (Dimplemeier)	1 slice (50 g)	120	**18**	0
Sicilian bread	¼ loaf (57 g)	150	**15**	0
Sourdough	1" slice (50 g)	130	**10**	0
Sourdough Boule (La Parisienne)	1 slice (57 g)	120	**10**	0
Spoon Bread	1 slice	325	**185**	100
Tortillas				
Corn	1 tortilla (28 g)	60	**0**	0
Flour	1 tortilla (35 g)	110	**25**	0
White	1 slice (25-27 g)	65-80	8-15	0
Light White	1 slice (22 g)	40	**5**	0
Very Thin White (Pepperidge Farm)	1 slice (15 g)	37	**5**	0
Whole-wheat				
Pepperidge Farm	1 slice (25 g)	60	**10**	0
Pepperidge Farm Soft	1 slice (25 g)	60	**5**	0
Stroehmann	1 slice (36 g)	80	**15**	0
Wonder	1 slice (34 g)	80	**15**	0
BREAKFAST CEREALS, COLD				
100% Bran	⅓ cup (29 g)	80	**5**	0
All-Bran, original	½ cup (31 g)	80	**10**	0
All-Bran, Extra Fiber	½ cup (26 g)	50	**10**	0

GRAINS AND PASTA

FOOD	AMOUNT	CALORIES TOTAL	CALORIES FAT	CALORIES SAT-FAT
Alpha-Bits				
Frosted	1 cup (32 g)	130	**10**	0
Marshmallow	1 cup (29 g)	120	**10**	0
Apple Jacks	1 cup (33 g)	120	**0**	0
Amaranth	¾ cup (28 g)	100	**0**	0
Banana Nut Crunch	1 cup (59 g)	250	**50**	9
Basic 4	1 cup (55 g)	200	**25**	0
Berry Berry Kix	¾ cup (30 g)	120	**10**	0
Blueberry Morning	1¼ cups (55 g)	210	**20**	5
Bran, 100% Natural	¾ cup (49 g)	160	**0**	0
Bran, Health Fiber	¾ cup (28 g)	100	**0**	0
Cap'n Crunch	¾ cup (27 g)	110	**15**	5
Crunch Berries	¾ cup (26 g)	100	**15**	5
Peanut Butter Crunch	¾ cup (27 g)	110	**25**	5
Cheerios	1 cup (30 g)	110	**15**	0
Apple Cinnamon	¾ cup (30 g)	120	**15**	0
Frosted	1 cup (30 g)	120	**0**	0
Honey Nut	1 cup (30 g)	120	**10**	0
Multi-grain Plus	1 cup (30 g)	110	**10**	0
Cinnamon Grahams	¾ cup (30 g)	120	**10**	0
Cinnamon Toast Crunch	¾ cup (30 g)	130	**30**	5
Cocoa Frosted Flakes	¾ cup (31 g)	120	**0**	0
Cocoa Krispies	¾ cup (30 g)	120	**10**	5
Cocoa Pebbles	¾ cup (29 g)	120	**10**	9
Cocoa Puffs	1 cup (30 g)	120	**10**	0
Complete Bran Flakes	¾ cup (29 g)	90	**5**	0
Complete Oat Bran Flakes	¾ cup (30 g)	110	**10**	0
Complete Wheat Bran Flakes	¾ cup (29 g)	90	**5**	0
Cookie Crisp Chocolate Chip	1 cup (30 g)	120	**10**	0
Corn Chex	1 cup (30 g)	110	**0**	0
Corn Flakes	1 cup (28 g)	100	**0**	0
Corn Flakes Honey Crunch	¾ cup (30 g)	110	**10**	0
Corn Pops	1 cup (31 g)	120	**0**	0
Count Chocula	1 cup (30 g)	120	**10**	0
Cracklin' Oat Bran	¾ cup (49 g)	190	**50**	0
Crispix	1 cup (29 g)	110	**0**	0
Crispy Wheaties 'n Raisins	1 cup (55 g)	190	**5**	0

GRAINS AND PASTA

FOOD	AMOUNT	CALORIES TOTAL	CALORIES FAT	CALORIES SAT-FAT
Crunchy Corn Bran	¾ cup (27 g)	90	**10**	0
Double Chex	1¼ cups (30 g)	120	**0**	0
Fiber One	½ cup (30 g)	60	**10**	0
French Toast Crunch	¾ cup (30 g)	120	**10**	0
Froot Loops	1 cup (32 g)	120	**10**	5
Frosted Flakes	¾ cup (31 g)	120	**0**	0
Frosted Mini-Wheats	5 biscuits (51 g)	180	**10**	0
Bite-size	1 cup (59 g)	200	**10**	0
Fruit & Fiber Peaches, Raisins & Almonds	1 cup (55 g)	210	**25**	5
Fruity Pebbles	¾ cup (27 g)	110	**10**	0
Golden Crisp	¾ cup (27 g)	110	**0**	0
Golden Grahams	¾ cup (30 g)	120	**10**	0
Granola, low-fat (Kellogg's)				
with raisins	⅔ cup (60 g)	220	**30**	9
without raisins	½ cup (49 g)	190	**25**	5
98% Fat Free	⅔ cup (55 g)	180	**10**	0
Grape Nuts	½ cup (58 g)	200	**10**	0
Grape Nuts Flakes	¾ cup (29 g)	100	**10**	0
Great Grains				
Crunchy Pecan	⅔ cup (53 g)	220	**60**	9
Raisin, Date, Pecan	⅔ cup (54 g)	210	**45**	5
Healthy Choice				
Almond Crunch with Raisins	1 cup (58 g)	210	**20**	0
Multigrain Raisins, Crunchy Oat Clusters & Almonds	1 cup (54 g)	210	**20**	0
Multigrain Squares	1 cup (54 g)	190	**10**	0
Toasted Brown Sugar Squares	1 cup (54 g)	190	**10**	0
Honey Bunches of Oats				
Honey Roasted	¾ cup (30 g)	120	**15**	5
with Almonds	¾ cup (31 g)	130	**30**	5
Honeycomb	1⅓ cups (29 g)	110	**5**	0
Honey Nut Clusters	1 cup (55 g)	210	**20**	0
Just Right				
Crunchy Nugget	1 cup (55 g)	210	**15**	0
Fruit & Nut	1 cup (60 g)	220	**20**	0

GRAINS AND PASTA

FOOD	AMOUNT	CALORIES TOTAL	FAT	SAT-FAT
Kix	1⅓ cups (30 g)	120	**5**	0
Life	¾ cup (32 g)	120	**15**	0
Cinnamon	¾ cup (32 g)	120	**10**	0
Lucky Charms	1 cup (30 g)	120	**10**	0
Muesli				
Cranberry with almonds or walnuts	¾ cup (58 g)	220	**25**	0
Raspberry with almonds	¾ cup (58 g)	220	**25**	0
Swiss (Familia)				
Granola No Added Sugar	½ cup (57 g)	200	**30**	5
Granola Original Recipe	½ cup (60 g)	210	**30**	5
Granola Original Recipe Lowfat	⅔ cup (55 g)	180	**15**	0
Puffed Wheat	½ cup (47 g)	170	**45**	9
Müeslix				
Apple & Almond Crunch	¾ cup (57 g)	200	**45**	9
Raisin & Almond Crunch	⅔ cup (55 g)	200	**25**	0
Multi Bran Chex	1 cup (58 g)	200	**15**	0
Natural Bran Flakes	⅔ cup (28 g)	90	**5**	0
100% Natural	½ cup (51 g)	230	**80**	32
Oats, Honey & Raisins Low fat	⅔ cup (55 g)	210	**27**	9
Nutri-Grain Almond Raisin	1¼ cups (49 g)	180	**25**	0
Oatmeal Crisp				
with Almonds	1 cup (55 g)	220	**45**	5
with Raisins	1 cup (55 g)	210	**25**	0
Oatmeal Squares	1 cup (56 g)	220	**25**	5
Oat Bran O's	1¼ cup (28 g)	100	**0**	0
Oat Bran Flakes	1¼ cup (28 g)	100	**0**	0
Product 19	1 cup (30 g)	100	**0**	0
Puffed Rice	1 cup (14 g)	50	**0**	0
Puffed Wheat	1¼ cups (15 g)	50	**0**	0
Puffins	¾ cup (27 g)	90	**10**	0
Raisin Bran	1 cup (61 g)	200	**15**	0
Raisin Bran Flakes	1¼ cup (55 g)	190	**0**	0
Raisin Nut Bran	¾ cup (55 g)	200	**35**	5

GRAINS AND PASTA

FOOD	AMOUNT	CALORIES TOTAL	CALORIES FAT	CALORIES SAT-FAT
Raisin Squares	¾ cup (53 g)	180	**14**	0
Reese's Peanut Butter Puffs	¾ cup (30 g)	130	**25**	5
Rice Chex	1¼ cups (31 g)	120	**0**	0
Rice Krispies	1¼ cups (33 g)	120	**0**	0
Razzle-Dazzle	¾ cup (28 g)	110	**0**	0
Shredded Wheat (Barbara's)	2 biscuits (40 g)	140	**10**	0
Shredded Wheat (Post)	2 biscuits (46 g)	160	**5**	0
Frosted Shredded Wheat	1 cup (52 g)	190	**10**	0
Honey Nut Shredded Wheat	1 cup (52 g)	200	**15**	0
Spoon Size Shredded Wheat	1 cup (49 g)	170	**5**	0
Wheat 'N Bran Shredded Wheat	1¼ cups (59 g)	200	**5**	0
Smacks	¾ cup (27 g)	10	**5**	0
Smart Starts	1 cup (50 g)	180	**5**	0
Special K	1 cup (31 g)	110	**0**	0
10 Bran Cereal	¾ cup (27 g)	100	**0**	0
Toasted Oatmeal				
Honey Nut	1 cup (49 g)	190	**25**	5
With Clusters	1 cup (49 g)	190	**25**	5
Total	¾ cup (30 g)	110	**10**	0
Corn Flakes	1⅓ cups (30 g)	110	**0**	0
Raisin Bran	1 cup (55 g)	180	**10**	0
Whole Grain	¾ cup (30 g)	110	**10**	0
Triples	1 cup (30 g)	120	**10**	0
Trix	1 cup (30 g)	120	**15**	0
Waffle Crisp	1 cup (30 g)	130	**25**	0
Wheat Chex	1 cup (50 g)	180	**10**	0
Wheaties	1 cup (30 g)	110	**10**	0
Honey Gold	¾ cup (30 g)	110	**0**	0
BREAKFAST CEREALS, HOT				
Alpen	⅔ cup (55 g)	200	**25**	0
Cream of Rice	1 oz dry	100	**0**	0
Cream of Wheat	1 pkt (35 g)	100	**0**	0
	1 cup cooked	120	**0**	0

GRAINS AND PASTA

FOOD	AMOUNT	CALORIES TOTAL	CALORIES FAT	CALORIES SAT-FAT
Cream of Wheat (cont.)				
Flavored	1 pkt (35 g)	130	**0**	0
regular, quick, instant, cooked	3 tbsp (33 g)	120	**0**	0
Farina (Pillsbury)	3 tbsp (28 g)	100	**0**	0
Grits (Quaker)	¼ cup (41 g)	140	**5**	0
Instant (Quaker)				
Country bacon	1 pkt (28 g)	100	**5**	0
Original	1 pkt (28 g)	100	**0**	0
Real butter	1 pkt (28 g)	100	**15**	5
Real cheddar cheese	1 pkt (28 g)	100	**15**	5
Red Eye Gravy & Country Ham	1 pkt (28 g)	10	**5**	0
Malt-O-Meal, cooked	1 cup	110	**0**	0
Maypo	1 oz dry	100	**9**	NA
Oatmeal				
Quaker	½ cup (40 g)	150	**25**	5
Instant	½ cup (40 g)	150	**25**	5
Irish (John McCann's)	⅓ cup (60 g)	230	**40**	9
Oat Bran				
Quaker	½ cup dry (40 g)	150	**30**	5
Mother's Oat Bran	½ cup dry (40 g)	150	**25**	9
Quinoa	¼ cup	159	**22**	2
Ralston 100% Wheat	½ cup (38 g)	130	**10**	0
Wheatena	⅓ cup (41 g)	150	**5**	0
Whole Wheat (Mother's)	½ cup (40 g)	130	**5**	0
CEREAL GRAINS (*see also* **FLOUR**)				
Barley, pearled				
uncooked	½ cup	352	**10**	0
cooked	½ cup	97	**3**	0
Buckwheat	½ cup	292	**26**	6
Buckwheat groats, roasted				
uncooked	½ cup	283	**20**	4
cooked	½ cup	91	**5**	0
Kasha	¼ cup (45 g)	170	**15**	NA

GRAINS AND PASTA

FOOD	AMOUNT	CALORIES TOTAL	CALORIES FAT	CALORIES SAT-FAT
Bulgur				
uncooked	½ cup	239	**8**	0
cooked	½ cup	76	**2**	0
Corn Grits				
uncooked	½ cup	290	**8**	0
cooked	1 cup	146	**5**	0
Couscous				
uncooked	½ cup	346	**5**	0
cooked (no added fat)	½ cup	101	**1**	0
Farina				
uncooked	½ cup	325	**4**	0
cooked	1 cup	116	**2**	0
Hominy, canned	½ cup	57	**6**	0
Millet				
uncooked	½ cup	378	**38**	7
cooked	½ cup	143	**11**	2
Oat bran				
uncooked	½ cup	116	**30**	6
cooked	1 cup	87	**17**	3
Oatmeal				
uncooked	⅓ cup	104	**15**	3
cooked	1 cup	145	**22**	4
Quinoa	¼ cup (42 g)	159	**18**	0
Rice				
Brown				
uncooked	1 cup	684	**49**	10
cooked	½ cup	109	**8**	0
White, long-grain				
uncooked	1 cup	676	**11**	3
cooked	½ cup	131	**3**	0
White, long-grain, parboiled				
uncooked	1 cup	686	**9**	2
cooked	½ cup	100	**2**	0
White, glutinous				
uncooked	1 cup	685	**9**	2
cooked	½ cup	116	**2**	0

GRAINS AND PASTA

FOOD	AMOUNT	CALORIES TOTAL	CALORIES FAT	CALORIES SAT-FAT
Rice (cont.)				
White, Instant				
uncooked	1 cup	360	**2**	0
cooked	½ cup	80	**1**	0
Wild				
uncooked	1 cup	571	**16**	2
cooked	½ cup	83	**3**	0
Rice bran, crude	⅓ cup	88	**53**	10
Rye	½ cup	282	**18**	2
Wheat bran	½ cup	65	**12**	2
	1 tbsp	8	**1**	0
Wheat germ, toasted	1 tbsp	27	**7**	0
	½ cup	216	**54**	9
FLOUR				
Arrowroot flour	1 tbsp	29	**0**	0
Bisquick, original (Betty Crocker)	⅓ cup (40 g)	170	**50**	14
reduced fat	⅓ cup (40 g)	140	**25**	5
Buckwheat flour	½ cup	201	**17**	5
Cake flour	½ cup	195	**4**	0
Corn flour				
whole grain	½ cup	209	**20**	3
masa	⅓ cup	139	**13**	2
Cornmeal, yellow	½ cup	210	**10**	1
Cornstarch	1 tbsp	31	**0**	0
Pastry flour	⅓ cup (30 g)	100	**5**	0
Potato starch	1 tbsp (10 g)	30	**0**	0
Rice flour				
brown	½ cup	287	**20**	4
white	½ cup	289	**10**	3
Rye flour				
dark	½ cup	207	**15**	2
medium	½ cup	181	**8**	1
Semolina	½ cup	303	**8**	1
Soy flour	½ cup	179	**78**	12

GRAINS AND PASTA

FOOD	AMOUNT	CALORIES TOTAL	CALORIES FAT	CALORIES SAT-FAT
White flour	½ cup	200	**0**	0
All-purpose flour	½ cup	226	**5**	0
Bread flour	½ cup	249	**10**	0
Whole-wheat flour	½ cup	203	**10**	2
PASTAS (*see also* **FROZEN, MICROWAVE, AND REFRIGERATED FOODS)**				
All dry types, cooked				
Macaroni, noodles, spaghetti, shells, etc.	2 oz dry	211	**8**	0
	1 cup cooked	197	**8**	0
Barley egg	¼ cup (55 g)	220	**25**	NA
Cellophane (Chinese)	1 cup dry	174	**7**	0
Chinese Style Noodles	57 g dry	200	**5**	0
Egg noodles	2 oz dry	217	**22**	5
	1 cup cooked	212	**21**	5
Whole-wheat	2 oz dry	198	**7**	0
	1 cup cooked	174	**7**	0
Chow Mein noodles				
Goodman's	⅔ cup (26 g)	120	**40**	NA
La Choy	½ cup (28 g)	140	**60**	9
Chung King	⅓ cup (28 g)	140	**65**	NA
Soba noodles (Japanese)	2 oz dry	192	**4**	0
	1 cup cooked	113	**0**	0
Somen noodles (Japanese)	2 oz dry	203	**4**	0
	1 cup cooked	230	**3**	0
Oriental noodles, miscellaneous				
Chinese Style	57 g	200	**5**	0
Japanese Style	55 g	200	**15**	0
Oriental Style	55 g	195	**1**	0
Rice Stick	¼ cup (55 g)	200	**5**	0
Refrigerated pasta				
Celentano				
Broccoli Stuffed Shells	3 shells (280 g)	190	**35**	NA
Contadina				
Angel hair	1¼ cup (80 g)	230	**20**	9
Fettuccine	1¼ cup (83 g)	240	**20**	9
Linguine	1¼ cup (85 g)	240	**20**	9
Ravioli				
Chicken & Herb	1¼ cups (113 g)	370	**120**	36

GRAINS AND PASTA

FOOD	AMOUNT	CALORIES TOTAL	CALORIES FAT	CALORIES SAT-FAT
Refrigerated pasta (cont.)				
Four Cheese	1 cup (88 g)	230	**35**	18
Garden Vegetable	1 cup (90 g)	250	**45**	18
Spinach Fettuccine	1¼ cups (89 g)	250	**30**	14
Tortelloni				
Cheese & Herb	1 cup (111 g)	320	**80**	45
Chicken & Prosciutto	1 cup (109 g)	360	**110**	41
Garlic & Cheese	1 cup (104 g)	280	**40**	23
Mushroom & Cheese	1 cup (104 g)	290	**50**	18
Sausage & Bell Pepper	1 cup	330	**90**	NA
Sun-dried Tomato	1 cup (100 g)	320	**90**	23
Sweet Italian Sausage	1 cup (107 g)	320	**70**	27
3 Cheese	¾ cup (85 g)	250	**45**	27
DiGiorno				
Angel hair	⅕ pkg (56 g)	160	**10**	0
Fettuccine	¼ pkg (70 g)	200	**15**	0
Four Cheese Ravioli	1 cup (110 g)	350	**140**	81
Linguine	¼ pkg (70 g)	200	**15**	0
Mozzarella Garlic Tortelloni	1 cup (99 g)	300	**80**	45
Three Cheese Tortellini	¾ cup (81 g)	250	**60**	31
Refrigerated Pasta Sauces: *see* **SAUCES, GRAVIES, AND DIPS**				
Mixed Pasta Dishes				
Fettuccine Alfredo	1 cup	880	**610**	377
Chef Boyardee				
Beef Ravioli	1 cup (253 g)	240	**50**	23
Beefaroni	1 cup (249 g)	260	**60**	27
Cheese Ravioli	1 cup (246 g)	220	**30**	14
Lasagna	1 cup (249 g)	270	**70**	27
Mini Ravioli	1 cup (252 g)	240	**60**	27
Spaghetti & Meatballs	1 cup (257 g)	270	**90**	45
Tortellini	1 cup (258 g)	230	**210**	0
Chef Boyardee 99% Fat Free				
Beef Ravioli	1 cup (244 g)	210	**10**	0
Cheese Ravioli	1 cup (251 g)	240	**25**	9
Franco-American				
Garfield				
Beef Ravioli	1 cup (252 g)	300	**90**	36

GRAINS AND PASTA

FOOD	AMOUNT	CALORIES TOTAL	CALORIES FAT	CALORIES SAT-FAT
Franco-American (cont.)				
Life with Louie				
Pasta with Meatballs	1 cup (252 g)	260	**100**	45
Pasta with Tomato & Cheese Sauce	1 cup (252 g)	190	**20**	5
RavioliOs				
Beef Ravioli	1 cup (252 g)	230	**30**	14
Sonic the Hedgehog				
Pasta with Meatballs	1 cup (252 g)	280	**100**	36
Spaghetti in Tomato Sauce	1 cup (252 g)	210	**20**	9
SpaghettiOs				
Pasta with Meatballs	1 cup (252 g)	260	**100**	45
Pasta with Sliced Franks	1 cup (252 g)	250	**100**	45
Pasta with Tomato & Cheese Sauce	1 cup (252 g)	190	**20**	5
Superiore				
Beef Ravioli	1 cup (259 g)	280	**80**	36
Spaghetti & Meatballs	1 cup (252 g)	270	**90**	45
TeddyOs				
Pasta with Tomato & Cheese Sauce	1 cup (252 g)	190	**20**	5
Where's Waldo				
Pasta in Tomato Sauce	1 cup (252 g)	190	**20**	5
Pasta with Meatballs	1 cup (252 g)	260	**100**	45
Hamburger Helper (Betty Crocker)				
Cheddar Cheese Melt	¾ cup mix (42 g)	150	**15**	5
	1 cup prepared	310	**100**	36
Cheeseburger Macaroni	⅓ cup mix (45 g)	180	**40**	14
	1 cup prepared	360	**140**	54
Chili Macaroni	⅓ cup mix (41 g)	140	**10**	0
	1 cup prepared	290	**90**	36
Italian Herb	½ cup mix (40 g)	130	**10**	0
	1 cup prepared	270	**90**	36
Lasagne	⅔ cup mix (41 g)	140	**5**	0
	1 cup prepared	280	**90**	36
Ravioli	½ cup mix (41 g)	140	**5**	0
	1 cup prepared	280	**90**	36

GRAINS AND PASTA

FOOD	AMOUNT	CALORIES TOTAL	CALORIES FAT	CALORIES SAT-FAT
Hamburger Helper (cont.)				
Southwestern Beef	⅓ cup mix (44 g)	150	**10**	0
	1 cup prepared	300	**90**	36
Stroganoff	⅔ cup mix (41 g)	160	**25**	9
	1 cup prepared	320	**110**	45
Supreme Topping				
Italian Parmesan	⅓ cup mix (44 g)	160	**15**	0
	1 cup prepared	300	**100**	36
Cheesy Hashbrowns	½ cup mix (48 g)	170	**15**	5
	1 cup prepared	400	**170**	54
Hormel				
Italian Style Lasagna	1 bowl (284 g)	350	**72**	NA
Macaroni & Cheese	1 cup (213 g)	270	**100**	NA
Noodles & Chicken (Hearty Helpings)	1 cup (298 g)	250	**100**	NA
Spaghetti	1 bowl (284 g)	240	**23**	NA
Kid's Kitchen (Hormel)				
Beefy Mac	1 cup (213 g)	190	**50**	23
Cheezy Mac 'n Beef	1 cup (213 g)	260	**60**	27
Cheezy Mac 'n Cheese	1 cup (213 g)	260	**100**	54
Mini Beef Ravioli	1 cup (213 g)	240	**60**	27
Noodle Rings & Chicken	1 cup (213 g)	150	**35**	14
Spaghetti & Mini Meatballs	1 cup (213 g)	220	**60**	36
Spaghetti Rings & Franks	1 cup (213 g)	240	**80**	31
Spaghetti Rings with Meatballs	1 cup (213 g)	230	**60**	27
Kraft				
Deluxe Macaroni & Cheese Dinner				
Four Cheese	98 g mix	320	**90**	63
Light	98 g mix	290	**40**	23
Original	98 g mix	320	**90**	54
Macaroni & Cheese Dinner				
The Cheesiest, Cheesy Alfredo, Spirals, Super Heroes, Rugrats, Bugs Bunny, ABC's, 123	70 g mix	260	**25**	9
	1 cup prepared	410	**170**	41

GRAINS AND PASTA

FOOD	AMOUNT	CALORIES TOTAL	CALORIES FAT	CALORIES SAT-FAT
Kraft (cont.)				
Thick 'n Creamy	70 g mix	260	**25**	9
	1 cup prepared	420	**170**	45
Three Cheese	70 g mix	260	**25**	9
	1 cup prepared	410	**160**	41
White Cheddar	70 g mix	260	**25**	9
	1 cup prepared	410	**160**	41
Velveeta Shells & Cheese	112 g mix	360	**120**	72
Lipton				
Noodles & Sauce				
Alfredo	⅔ cup (62 g)	250	**60**	36
	1 cup prepared	330	**130**	54
Beef Flavor	⅔ cup (60 g)	230	**30**	9
	1 cup prepared	280	**90**	23
Butter & Herb	⅔ cup (62 g)	250	**60**	31
	1 cup prepared	300	**110**	41
Chicken Broccoli	⅔ cup (59 g)	230	**35**	14
	1 cup prepared	310	**100**	27
Chicken Flavor	⅔ cup (60 g)	240	**40**	18
	1 cup prepared	290	**100**	27
Creamy Chicken	⅔ cup (59 g)	240	**50**	27
	1 cup prepared	320	**110**	43
Parmesan	⅔ cup (60 g)	250	**70**	36
	1 cup prepared	330	**140**	54
Stroganoff	⅔ cup (56 g)	220	**35**	18
	1 cup prepared	300	**100**	36
Pasta & Sauce				
Cheddar Broccoli	⅔ cup (68 g)	260	**35**	14
	1 cup prepared	340	**100**	30
Creamy Garlic	⅔ cup (69 g)	270	**50**	23
	1 cup prepared	350	**120**	40
Mild Cheddar Cheese	¾ cup (59 g)	210	**25**	14
	1 cup prepared	290	**90**	28
Roasted Garlic & Olive Oil with Tomatoes	¾ cup (60 g)	220	**25**	5
	1 cup prepared	270	**80**	18

GRAINS AND PASTA

FOOD	AMOUNT	CALORIES TOTAL	CALORIES FAT	CALORIES SAT-FAT
Lipton (cont.)				
Roasted Garlic Chicken Flavor	¾ cup (56 g)	210	**20**	9
	1 cup prepared	290	**90**	23
Rice-A-Roni				
Pasta Roni				
Angel Hair Pasta with Herbs	56 g mix	200	**25**	5
	1 cup prepared	320	**120**	30
Angel Hair Pasta with Lemon & Butter	70 g mix	250	**30**	9
	1 cup prepared	360	**140**	30
Angel Hair Pasta with Parmesan Cheese	56 g mix	210	**40**	9
	1 cup prepared	320	**130**	31
Broccoli	56 g mix	200	**30**	5
	1 cup prepared	340	**140**	36
Fettuccine Alfredo	70 g mix	270	**60**	18
	1 cup prepared	470	**230**	54
Garlic & Olive Oil with Vermicelli	70 g mix	250	**30**	5
	1 cup prepared	360	**140**	27
Herb & Butter	56 g mix	200	**25**	5
	1 cup prepared	380	**170**	40
Homestyle Chicken	56 g mix	190	**15**	0
	1 cup prepared	230	**50**	45
Parmesano	70 g mix	260	**35**	9
	1 cup prepared	390	**150**	36
Shells & White Cheddar	70 g mix	270	**50**	23
	1 cup prepared	380	**210**	45
White Cheddar & Broccoli	70 g mix	270	**50**	18
	1 cup prepared	400	**170**	43
Tuna Helper (Betty Crocker)				
Cheesy Pasta	¾ cup mix (44 g)	170	**25**	9
	1 cup prepared	280	**100**	23
Creamy Broccoli	⅔ cup mix (50 g)	190	**40**	14
	1 cup prepared	310	**110**	27

GRAINS AND PASTA

FOOD	AMOUNT	CALORIES TOTAL	FAT	SAT-FAT
Tuna Helper (Betty Crocker) (cont.)				
Creamy Pasta	¾ cup mix (46 g)	190	**50**	14
	1 cup prepared	300	**120**	30
Pancakes and Waffles: *see also* **FAST FOODS; FROZEN, MICROWAVE, AND REFRIGERATED FOODS; RESTAURANT FOODS**				
Pancake Mix				
Buttermilk (Flap-Stax)	½ cup mix (63 g)	240	**35**	9

MEATS (BEEF, GAME, LAMB, PORK, AND VEAL)

The following cuts of meat are braised, roasted, or broiled, unless otherwise noted. *See also* **SAUSAGES AND LUNCHEON MEATS** for cold cuts made from beef and pork products and **FROZEN, MICROWAVE, AND REFRIGERATED FOODS.**

FOOD	AMOUNT	CALORIES TOTAL	FAT	SAT-FAT
BEEF				
Remember that most of the following calorie figures are for ***only 1 ounce*** *of beef!*				
Arm pot roast, braised				
lean and fat	1 oz	99	**66**	27
lean only	1 oz	68	**25**	11
Backribs	2 ribs (140 g)	360	**220**	90
Bottom round steak, braised				
lean and fat	1 oz	74	**38**	14
lean only	1 oz	67	**25**	9
Brisket flat half, braised				
lean and fat	1 oz	116	**89**	37
lean only	1 oz	74	**40**	17
Chuck steak, braised				
lean and fat	1 oz	108	**78**	32
lean only	1 oz	77	**39**	18
Club steak, broiled				
lean and fat	1 oz	129	**104**	50
lean only	1 oz	69	**33**	16
Flank steak, braised				
lean and fat	1 oz	73	**39**	18
lean only	1 oz	69	**35**	16

MEATS (BEEF, GAME, LAMB, PORK, AND VEAL)

FOOD	AMOUNT	CALORIES TOTAL	CALORIES FAT	CALORIES SAT-FAT
Ground beef, raw				
extra lean	1 oz	66	**44**	17
lean	1 oz	75	**53**	21
regular	1 oz	88	**68**	28
Ground beef, broiled, medium				
extra lean	1 oz	72	**42**	16
lean	1 oz	77	**47**	18
regular	1 oz	82	**53**	21
Ground beef, pan-fried, medium				
extra lean	1 oz	72	**42**	16
lean	1 oz	78	**49**	19
regular	1 oz	87	**58**	23
Porterhouse steak, broiled				
lean and fat	1 oz	85	**54**	22
lean only	1 oz	62	**28**	13
Rib roast				
lean and fat	1 oz	108	**81**	34
lean only	1 oz	68	**35**	17
Round, broiled				
lean and fat	1 oz	78	**47**	19
lean only	1 oz	55	**20**	7
Rump roast				
lean and fat	1 oz	98	**70**	33
lean only	1 oz	59	**24**	7
Shortribs, braised				
lean and fat	1 oz	133	**107**	45
lean only	1 oz	84	**46**	20
Sirloin steak, broiled				
lean and fat	1 oz	79	**46**	19
lean only	1 oz	59	**22**	12
T-bone steak, broiled				
lean and fat	1 oz	92	**63**	26
lean only	1 oz	61	**26**	11
Tenderloin steak, broiled				
lean and fat	1 oz	75	**44**	18
lean only	1 oz	58	**24**	9
Top round, broiled				
lean and fat	1 oz	60	**22**	8

MEATS (BEEF, GAME, LAMB, PORK, AND VEAL)

FOOD	AMOUNT	CALORIES TOTAL	CALORIES FAT	CALORIES SAT-FAT
Top round, broiled (cont.)				
lean only	1 oz	54	**16**	6
MIXED BEEF DISHES				
Beef Barbecue (Brookwood Farms)	¼ cup (63 g)	350	**220**	90
Beef and vegetable stew	1 cup	220	**99**	40
Chili con carne, canned	1 cup	340	**144**	52
Lasagne	1/12 casserole	570	**210**	142
Spaghetti w/meatballs and tomato sauce, canned	1 cup	260	**90**	22
VARIETY MEATS AND BY-PRODUCTS				
Brain, simmered	1 oz	45	**32**	7
Liver, braised	1 oz	46	**12**	5
Liver Paté	½ cup	289	**204**	104
	1 tbsp	36	**26**	13
Tripe	1 oz	28	**10**	5
Tongue, simmered	1 oz	81	**53**	23
GAME MEATS				
*Remember that the following calorie figures are for **only 1 ounce** of game!*				
Antelope, roasted	1 oz	42	**7**	2
Bear, simmered	1 oz	73	**34**	NA
Beefalo, composite of cuts, roasted	1 oz	53	**16**	7
Bison, roasted	1 oz	41	**6**	2
Boar, wild, roasted	1 oz	45	**11**	3
Buffalo, water, roasted	1 oz	37	**5**	2
Caribou, roasted	1 oz	47	**11**	4
Deer, roasted	1 oz	45	**8**	3
Elk, roasted	1 oz	41	**5**	2
Goat, roasted	1 oz	41	**8**	2
Horse, roasted	1 oz	50	**15**	5
Moose, roasted	1 oz	38	**2**	1
Muskrat, roasted	1 oz	52	**23**	NA
Opossum, roasted	1 oz	63	**26**	NA
Rabbit				
domesticated, composite of cuts roasted	1 oz	44	**16**	5

MEATS (BEEF, GAME, LAMB, PORK, AND VEAL)

FOOD	AMOUNT	CALORIES TOTAL	CALORIES FAT	CALORIES SAT-FAT
Rabbit, domesticated (cont.)				
stewed	1 oz	58	**21**	6
wild, stewed	1 oz	49	**9**	3
Raccoon, roasted	1 oz	72	**37**	NA
Squirrel, roasted	1 oz	39	**9**	1
LAMB				
Remember that most of the following calorie figures are for ***only 1 ounce*** *of lamb!*				
Foreshank, braised				
lean and fat	1 oz	69	**34**	14
lean only	1 oz	53	**15**	6
Leg, whole (shank and sirloin), roasted				
lean and fat	1 oz	73	**42**	18
lean only	1 oz	54	**20**	7
Leg, shank half, roasted				
lean and fat	1 oz	64	**32**	13
lean only	1 oz	51	**17**	6
Leg, sirloin half, roasted				
lean and fat	1 oz	83	**53**	23
lean only	1 oz	58	**23**	8
Loin, roasted				
lean and fat	1 oz	88	**60**	25
lean only	1 oz	57	**25**	9
Chop, with bone, broiled				
lean and fat	1 chop (3.5 oz)	357	**265**	145
lean only	1 chop (3.5 oz)	189	**74**	37
Rib, broiled or roasted				
lean and fat	1 oz	102	**76**	33
lean only	1 oz	67	**34**	12
Chop, with bone, broiled				
lean and fat	1 chop (3.5 oz)	398	**315**	148
lean only	1 chop (3.5 oz)	212	**105**	40
Shoulder, whole (arm and blade)				
braised				
lean and fat	1 oz	97	**63**	27
lean only	1 oz	80	**40**	16
Cubed lamb for stew or kabob (leg and shoulder), lean only				
braised	1 oz	63	**22**	8
broiled	1 oz	53	**19**	7

MEATS (BEEF, GAME, LAMB, PORK, AND VEAL)

FOOD	AMOUNT	CALORIES TOTAL	FAT	SAT-FAT
Ground lamb				
broiled	1 oz	81	**50**	26
PORK				
Remember that many of the following calorie figures are for ***only 1 ounce*** *of pork!*				
Loin				
whole, broiled				
lean and fat, without bone	1 oz	98	**69**	25
	1 chop (104 g)	284	**202**	73
lean only	1 oz	73	**39**	13
	1 chop (104 g)	169	**91**	31
blade, pan-fried				
lean and fat	1 oz	117	**94**	34
	1 chop (104 g)	368	**296**	107
lean only	1 oz	85	**51**	19
	1 chop (89 g)	177	**111**	39
center loin, pan fried				
lean and fat	1 oz	106	**78**	28
	1 chop (112 g)	333	**244**	88
lean only	1 oz	75	**41**	14
	1 chop (112 g)	178	**96**	33
center rib, broiled				
lean and fat	1 oz	97	**67**	24
	1 chop (104 g)	264	**183**	66
lean only	1 oz	73	**38**	13
	1 chop (104 g)	162	**85**	29
sirloin, broiled				
lean and fat	1 oz	94	**64**	23
	1 chop (106 g)	278	**191**	69
lean only	1 oz	69	**35**	12
	1 chop (106 g)	165	**83**	29
tenderloin, roasted				
lean only	1 oz	47	**12**	4
top loin, broiled				
lean and fat	1 oz	102	**85**	26
	1 chop (104 g)	295	**211**	76
lean only	1 oz	73	**39**	13
	1 chop (104 g)	165	**86**	30

MEATS (BEEF, GAME, LAMB, PORK, AND VEAL)

FOOD	AMOUNT	CALORIES TOTAL	CALORIES FAT	CALORIES SAT-FAT
Shoulder cut				
whole, roasted				
lean and fat	1 oz	92	**65**	24
lean only	1 oz	69	**38**	13
arm picnic, roasted				
lean and fat	1 oz	94	**67**	24
lean only	1 oz	65	**32**	11
blade, Boston, roasted				
lean and fat	1 oz	91	**64**	23
lean only	1 oz	73	**43**	15
Spareribs, cooked				
lean and fat	1 oz	113	**77**	30
Pork Ribs (Spare the Ribs)	5 oz (140 g)	380	**260**	72
Pork Baby Back Ribs (Lloyd's)	3 ribs (140 g)	330	**190**	72
PORK PRODUCTS, CURED (*see also* **SAUSAGES AND LUNCHEON MEATS**)				
Bacon, cooked	1 strip	36	**28**	10
Breakfast strips, cooked	1 strip	52	**37**	13
Canadian bacon, grilled	1 slice	43	**18**	6
Ham, boneless				
Extra lean (5% fat)	1 slice (28 g)	41	**14**	5
Regular (11% fat)	1 slice (28 g)	52	**27**	9
Ham, canned				
Extra lean (4% fat)	1 slice (28 g)	39	**12**	4
Regular (13% fat)	1 slice (28 g)	64	**39**	13
Ham, center slice				
Country-style				
Lean and fat	1 oz	57	**33**	12
Lean	1 oz	55	**21**	1
Salt pork, raw	1 oz	212	**205**	75
PORK DISHES, MIXED				
Barbecue Pork	½ cup	320	**220**	72
Hash	1 cup	410	**250**	117

MEATS (BEEF, GAME, LAMB, PORK, AND VEAL)

FOOD	AMOUNT	CALORIES TOTAL	CALORIES FAT	CALORIES SAT-FAT
VARIETY MEATS & BY-PRODUCTS				
Backfat, raw	1 oz	230	**226**	226
Chitterlings, simmered	1 oz	86	**73**	37
Feet				
pickled	1 oz	58	**41**	14
simmered, without bone	1 oz	55	**32**	11
Liver paté	2 oz (56 g)	200	**160**	54
For cold cuts made from pork products, *see* **SAUSAGES AND LUNCHEON MEATS.**				
VEAL				
Remember that the following calorie figures are for ***only 1 ounce*** *of veal!*				
Breast, lean and fat, braised	1 oz	86	**54**	26
Cutlet, lean and fat, braised	1 oz	62	**27**	12
Ground, broiled	1 oz	49	**19**	8
Leg				
top round, lean and fat, braised	1 oz	60	**16**	6
top round, lean, braised	1 oz	57	**13**	5
Loin				
lean and fat, braised	1 oz	81	**44**	16
lean only, braised	1 oz	64	**23**	7
lean and fat, roasted	1 oz	61	**31**	16
lean only, roasted	1 oz	50	**18**	7
Rib				
lean and fat, braised	1 oz	71	**32**	13
lean only, braised	1 oz	62	**20**	7
lean and fat, roasted	1 oz	65	**36**	14
lean only, roasted	1 oz	50	**19**	6
Shoulder				
arm, lean and fat, braised	1 oz	67	**26**	10
arm, lean only, braised	1 oz	57	**14**	4
blade, lean and fat, braised	1 oz	64	**26**	9
blade, lean, braised	1 oz	56	**17**	5

MEATS (BEEF, GAME, LAMB, PORK, AND VEAL)

FOOD	AMOUNT	CALORIES TOTAL	CALORIES FAT	CALORIES SAT-FAT
Sirloin				
lean and fat, braised	1 oz	72	**34**	13
lean only, braised	1 oz	58	**17**	5
Veal cubed for stew (leg and shoulder)				
lean only braised	1 oz	53	**11**	3

NUTS AND SEEDS

FOOD	AMOUNT	CALORIES TOTAL	CALORIES FAT	CALORIES SAT-FAT
NUTS				
Almonds				
slivered or sliced	½ cup	400	**315**	30
whole, dry-roasted	1 oz (24 nuts)	167	**132**	13
	½ cup	405	**320**	30
Almond butter				
plain	1 tbsp	101	**85**	8
honey-cinnamon	1 tbsp	96	**75**	7
Almond paste	1 oz	127	**69**	7
	1 cup	1012	**556**	53
Beechnuts, dried	1 oz	164	**128**	15
Brazil nuts, shelled, dried	1 oz (8 med. nuts)	186	**169**	41
Butternuts, dried	1 oz	174	**146**	3
Cashew nuts				
dry-roasted	1 oz (18 med. nuts)	163	**118**	23
	1 cup	787	**572**	113
oil-roasted	1 oz (18 med. nuts)	163	**123**	24
	1 cup	748	**564**	112
jumbo	12 nuts (30 g)	190	**130**	27
Cashew butter, plain	1 oz	167	**126**	25
	1 tbsp	94	**71**	14
Chestnuts, Chinese				
raw	1 oz	64	**3**	0
dried	1 oz	103	**5**	1
boiled and steamed	1 oz	44	**2**	0
roasted	1 oz	68	**3**	0

NUTS AND SEEDS

FOOD	AMOUNT	CALORIES TOTAL	CALORIES FAT	CALORIES SAT-FAT
Chestnuts, European				
raw, unpeeled	1 oz	60	**6**	1
raw, peeled	1 oz	56	**3**	1
dried, unpeeled	1 oz	106	**11**	2
dried, peeled	1 oz	105	**10**	2
boiled and steamed	1 oz	37	**4**	1
roasted	1 oz (3 nuts)	70	**6**	1
	1 cup	350	**28**	5
Chestnuts, Japanese				
raw	1 oz	44	**1**	0
dried	1 oz	102	**3**	0
boiled and steamed	1 oz	16	**1**	0
roasted	1 oz	57	**2**	0
Coconut meat				
dried, creamed	1 oz	194	**177**	157
dried, sweetened, flaked	1 oz	135	**82**	73
	1 cup	351	**214**	190
dried, toasted	1 oz	168	**120**	107
fresh frozen with sugar	2 tbsp (13 g)	45	**25**	18
fresh, shredded or grated	1 oz	101	**86**	76
	1 cup	283	**241**	214
Coconut cream				
fresh	1 tbsp	49	**47**	42
	1 cup	792	**749**	664
canned	1 tbsp	36	**30**	27
	1 cup	568	**472**	419
Coconut milk				
fresh	1 tbsp	35	**32**	29
	1 cup	552	**515**	457
canned	1 tbsp	30	**29**	26
	1 cup	445	**434**	385
frozen	1 tbsp	30	**28**	25
	1 cup	486	**449**	398
Filberts (hazelnuts)				
dried	1 oz	179	**160**	12
	1 cup, chopped	727	**648**	48

NUTS AND SEEDS

FOOD	AMOUNT	CALORIES TOTAL	FAT	SAT-FAT
Filberts (cont.)				
dry-roasted	1 oz	188	**169**	12
oil-roasted	1 oz	187	**163**	12
Hickory nuts, dried	1 oz	187	**165**	18
Macadamia nuts				
dried	1 oz	199	**188**	28
oil-roasted	1 oz (24 halves)	204	**196**	29
	1 cup	962	**923**	139
Mixed nuts				
dry-roasted, with peanuts	1 oz	169	**131**	18
	1 cup	814	**634**	85
oil-roasted, with peanuts	1 oz	175	**144**	22
	1 cup	876	**720**	112
oil-roasted, without peanuts	1 oz	175	**144**	23
	1 cup	886	**728**	118
Peanuts, shelled				
dry-roasted	1 oz (35 kernels)	161	**126**	17
	1 cup	827	**646**	90
	1 tbsp	50	**35**	9
oil-roasted	1 oz (35 kernels)	165	**126**	18
	1 cup	841	**642**	89
Planters Peanuts				
Hot Spicy Peanuts (to heat)	37 pieces (28 g)	160	**120**	18
Snack Mix (to heat)	¼ cup (28 g)	140	**70**	9
Sweet 'N Crunchy	18 pieces (28 g)	140	**60**	9
Peanut butter	2 tbsp	190	**148**	24
Jif				
Extra Crunchy	2 tbsp (32 g)	190	**130**	27
Simply Creamy	2 tbsp (31 g)	190	**130**	27
reduced fat	2 tbsp (36 g)	190	**110**	23
Peter Pan				
Creamy	2 tbsp (32 g)	190	**140**	31
reduced fat	2 tbsp (36 g)	180	**90**	23
Crunchy	2 tbsp (32 g)	190	**130**	27

NUTS AND SEEDS

FOOD	AMOUNT	CALORIES TOTAL	CALORIES FAT	CALORIES SAT-FAT
Peanut butter (cont.)				
Reese's Creamy	2 tbsp (32 g)	200	**140**	27
Skippy				
Creamy	2 tbsp (32 g)	190	**140**	31
reduced fat	2 tbsp (36 g)	190	**100**	23
Super Chunk	2 tbsp (32 g)	190	**140**	31
reduced fat	2 tbsp (35 g)	190	**100**	23
Pecans				
dried	1 oz	190	**173**	14
dry-roasted	1 oz (14 halves)	187	**165**	13
	⅓ cup	230	**210**	18
oil-roasted	1 oz	195	**182**	15
Pine nuts	¼ cup	190	**140**	32
Pistachio nuts				
dried	1 oz	164	**124**	16
dry-roasted	1 oz (47 kernels)	172	**135**	17
Walnuts	¼ cup (32 g)	210	**180**	14
walnut halves	⅓ cup (33 g)	210	**190**	18
walnut pieces	¼ cup (30 g)	190	**170**	18
black walnuts	¼ cup (30 g)	200	**150**	9
English, dried	1 oz (14 halves)	182	**158**	14
	1 cup	770	**668**	60
SEEDS				
Poppy	1 tsp	15	**11**	NA
	1 tbsp	66	**51**	NA
Pumpkin and squash				
whole, roasted	1 oz	127	**50**	9
kernels, dried	1 oz	154	**117**	22
kernels, roasted	1 oz	148	**108**	20
Sesame seed kernels, dried	1 tbsp	47	**39**	6
	1 cup	882	**73** **9**	104
Sunflower				
whole	1 oz	80	**59**	8
kernels				
dried	1 oz	162	**127**	13

NUTS AND SEEDS

FOOD	AMOUNT	CALORIES TOTAL	FAT	SAT-FAT
Sunflower (cont.)				
dry-roasted	1 oz	165	**127**	13
	1 tbsp	48	**34**	7
	1 cup	745	**574**	60
oil-roasted	1 oz	175	**147**	15
	1 tbsp	47	**44**	5
toasted	1 oz	176	**145**	15
	1 tbsp	51	**43**	5
Tahini	1 tbsp	90	**72**	10
Watermelon, dried	1 oz	158	**121**	25

POULTRY

See also **SAUSAGES AND LUNCHEON MEATS** for cold cuts made from poultry products and **FROZEN, MICROWAVE, AND REFRIGERATED FOODS.**

Remember that many of the following calorie figures are for ***only 1 ounce*** *of poultry!*

FOOD	AMOUNT	CALORIES TOTAL	FAT	SAT-FAT
CHICKEN				
Back				
meat and skin				
raw	½ back (99 g)	316	**256**	74
	1 oz	90	**73**	21
fried, batter-dipped	½ back (120 g)	397	**237**	63
	1 oz	94	**56**	15
fried, flour-coated	½ back (72 g)	238	**134**	36
	1 oz	94	**53**	14
roasted	½ back (53 g)	159	**100**	28
	1 oz	85	**54**	15
meat only				
raw	½ back (51 g)	70	**27**	7
	1 oz	39	**15**	4
fried	½ back (58 g)	167	**80**	22
	1 oz	82	**39**	10

POULTRY

FOOD	AMOUNT	CALORIES TOTAL	FAT	SAT-FAT
Back, meat only (cont.)				
roasted	½ back (51 g)	70	**27**	7
	1 oz	39	**15**	4
Breast				
meat and skin				
raw	1 breast (145 g)	250	**121**	35
	1 oz	49	**24**	7
fried, batter-dipped	1 breast (140 g)	364	**166**	44
	1 oz	74	**34**	9
fried, flour-coated	1 breast (98 g)	218	**78**	22
	1 oz	63	**23**	6
roasted	1 breast (98 g)	193	**69**	19
	1 oz	56	**20**	6
meat only				
raw	1 breast (118 g)	129	**13**	3.5
	1 oz	31	**3**	1
fried	1 breast (86 g)	161	**36**	10
	1 oz	53	**12**	3
roasted	1 breast (86 g)	129	**13**	3.5
	1 oz	31	**3**	1
Drumstick				
meat and skin				
raw	1 drumstick (73 g)	117	**57**	16
	1 oz	46	**22**	6
fried, batter-dipped	1 drumstick (72 g)	193	**102**	27
	1 oz	76	**40**	10
fried, flour-coated	1 drumstick (49 g)	120	**60**	16
	1 oz	69	**35**	9
roasted	1 drumstick (52 g)	112	**52**	14
	1 oz	61	**28**	8
meat only				
raw	1 drumstick (62 g)	74	**19**	5
	1 oz	34	**9**	2
fried	1 drumstick (42 g)	82	**31**	8
	1 oz	55	**21**	5
roasted	1 drumstick (44 g)	74	**19**	5
	1 oz	34	**9**	2

POULTRY

FOOD	AMOUNT	CALORIES TOTAL	CALORIES FAT	CALORIES SAT-FAT
Gizzard				
raw	1 gizzard (37 g)	44	**14**	4
	1 oz	33	**11**	3
simmered	1 cup	222	**48**	14
	1 oz	43	**9**	3
Ground Chicken				
Fresh (Perdue)	1 oz	48	**30**	9
Frozen (Longacre)	1 oz	55	**30**	9
Leg				
meat and skin				
raw	1 leg (167 g)	312	**182**	51
	1 oz	53	**31**	9
fried, batter-dipped	1 leg (158 g)	431	**230**	61
	1 oz	77	**41**	11
fried, flour-coated	1 leg (112 g)	285	**145**	39
	1 oz	72	**37**	10
roasted	1 leg (114 g)	265	**138**	38
	1 oz	66	**34**	9
meat only				
raw	1 leg (130 g)	156	**45**	11
	1 oz	34	**10**	2
fried	1 leg (94 g)	195	**79**	21
	1 oz	59	**24**	6
roasted	1 leg (95 g)	156	**45**	11
	1 oz	34	**10**	2
Liver				
raw	1 liver (32 g)	40	**11**	4
	1 oz	35	**10**	3
simmered	1 cup	219	**69**	23
	1 oz	44	**14**	5
Neck				
meat and skin				
raw	1 neck (50 g)	148	**118**	33
	1 oz	84	**67**	18
fried, batter dipped	1 neck (52 g)	172	**110**	29
	1 oz	94	**60**	16
fried, flour coated	1 neck (36 g)	119	**76**	21
	1 oz	94	**60**	16

POULTRY

FOOD	AMOUNT	CALORIES TOTAL	FAT	SAT-FAT
Neck (cont.)				
simmered	1 neck (38 g)	94	**62**	17
	1 oz	70	**46**	13
meat only				
raw	1 neck (20 g)	31	**16**	4
	1 oz	44	**22**	6
fried	1 neck (22 g)	50	**23**	6
	1 oz	65	**30**	8
simmered	1 neck (18 g)	32	**13**	3
	1 oz	44	**21**	5
Thigh				
meat and skin				
raw	1 thigh (94 g)	199	**129**	36
	1 oz	60	**39**	11
fried, batter-dipped	1 thigh (86 g)	238	**128**	34
	1 oz	78	**42**	11
fried, flour-coated	1 thigh (62 g)	162	**84**	23
	1 oz	74	**38**	11
roasted	1 thigh (62 g)	153	**86**	24
	1 oz	70	**40**	11
meat only				
raw	1 thigh (69 g)	82	**24**	6
	1 oz	34	**10**	3
fried	1 thigh (52 g)	113	**48**	13
	1 oz	62	**26**	7
roasted	1 thigh (52 g)	82	**24**	6
	1 oz	34	**10**	3
Wing				
meat and skin				
raw	1 wing (49 g)	109	**70**	20
	1 oz	63	**41**	11
fried, batter-dipped	1 wing (49 g)	159	**96**	26
	1 oz	92	**56**	15
fried, flour-coated	1 wing (32 g)	103	**64**	17
	1 oz	91	**57**	15
roasted	1 wing (34 g)	99	**60**	17
	1 oz	82	**50**	14

POULTRY

FOOD	AMOUNT	CALORIES TOTAL	CALORIES FAT	CALORIES SAT-FAT
Wing (cont.)				
meat only				
raw	1 wing (29 g)	36	**9**	2
	1 oz	36	**9**	2
fried	1 wing (20 g)	42	**16**	5
	1 oz	60	**23**	6
roasted	1 wing (21 g)	36	**9**	2
	1 oz	36	**9**	2
MIXED CHICKEN DISHES (*see also* FROZEN, MICROWAVE, AND REFRIGERATED FOODS)				
Chicken à la king	1 cup	470	**306**	116
Chicken and noodles	1 cup	365	**162**	46
Chicken potpie	⅓ pie	545	**279**	93
DUCK, DOMESTICATED				
meat and skin				
raw	½ duck (634 g)	2561	**2245**	754
	1 oz	115	**100**	33
roasted	½ duck (382 g)	1287	**975**	332
	1 oz	96	**72**	25
meat only				
raw	½ duck (303 g)	399	**162**	63
	1 oz	37	**15**	6
roasted	½ duck (221 g)	445	**223**	83
	1 oz	57	**29**	11
liver, raw	1 liver (44 g)	60	**18**	6
	1 oz	39	**12**	4
DUCK, WILD				
meat and skin, raw	½ duck (270 g)	571	**369**	125
	1 oz	60	**39**	13
breast meat only, raw	1 breast (83 g)	102	**32**	10
	1 oz	35	**11**	3
GOOSE, DOMESTICATED				
meat and skin				
raw	½ goose (1319 g)	4893	**3991**	1161
	1 oz	105	**86**	25

POULTRY

FOOD	AMOUNT	CALORIES TOTAL	CALORIES FAT	CALORIES SAT-FAT
meat and skin (cont.)				
roasted	½ goose (774 g)	2362	**1527**	479
	1 oz	86	**56**	18
meat only				
raw	½ goose (766 g)	1237	**492**	192
	1 oz	46	**18**	7
roasted	½ goose (591 g)	1406	**674**	242
	1 oz	67	**32**	12
liver, raw	1 liver (94 g)	125	**36**	14
	1 oz	38	**11**	4
PHEASANT				
meat and skin, raw	½ pheasant (400 g)	723	**335**	97
	1 oz	51	**24**	7
meat only, raw	½ pheasant (352 g)	470	**115**	39
	1 oz	38	**9**	3
breast meat only, raw	1 breast (182 g)	243	**53**	18
	1 oz	38	**8**	3
leg meat only, raw	1 leg (107 g)	143	**41**	14
	1 oz	38	**11**	4
QUAIL				
meat and skin, raw	1 quail (109 g)	210	**118**	33
	1 oz	54	**31**	9
meat only, raw	1 quail (92 g)	123	**38**	11
	1 oz	38	**12**	3
breast meat only, raw	1 breast (56 g)	69	**15**	4
	1 oz	35	**8**	2
SQUAB (PIGEON)				
meat and skin, raw	1 squab (199 g)	584	**426**	151
	1 oz	83	**61**	21
meat only, raw	1 squab (168 g)	239	**113**	30
	1 oz	40	**19**	5
breast meat only, raw	1 breast (101 g)	135	**41**	11
	1 oz	38	**12**	3

POULTRY

FOOD	AMOUNT	CALORIES TOTAL	FAT	SAT-FAT
TURKEY				
Dark meat, roasted				
meat and skin	1 oz	52	**18**	5
meat only	1 oz	46	**11**	4
Light meat, roasted				
meat and skin	1 oz	46	**12**	3
meat only	1 oz	40	**3**	1
Back				
meat and skin				
raw	½ back (183 g)	275	**120**	35
	1 oz	43	**18**	5
roasted	½ back (130 g)	265	**120**	35
	1 oz	58	**26**	8
meat only				
raw	½ back (150 g)	180	**47**	16
	1 oz	34	**9**	3
roasted	½ back (96 g)	180	**47**	16
	1 oz	34	**9**	3
Breast				
meat and skin				
raw	1 oz	35	**7**	2
roasted	1 oz	43	**8**	2
meat only				
raw	1 oz	31	**2**	1
roasted	1 oz	31	**2**	1
Cutlet, braised	1 oz	31	**2**	1
Ground Turkey				
Fresh				
Breast, 99% fat free (Shady Brook Farm)	112 g (4 oz)	120	**5**	0
Perdue	1 oz	40	**18**	6
lean (Shady Brook Farm)	112 g (4 oz)	160	**70**	18
Frozen (Longacre)	1 oz	53	**33**	7
Turkey Meatballs (Shady Brook Farms)	3 meatballs (85 g)	130	**60**	23

POULTRY

FOOD	AMOUNT	CALORIES TOTAL	CALORIES FAT	CALORIES SAT-FAT
Leg				
meat and skin				
raw	1 leg (349 g)	412	**112**	34
	1 oz	33	**9**	3
roasted	1 leg (245 g)	418	**119**	37
	1 oz	48	**14**	4
meat only				
raw	1 leg (329 g)	356	**70**	24
	1 oz	31	**6**	2
roasted	1 leg (224 g)	355	**76**	24
	1 oz	45	**10**	3
Wing				
meat and skin				
raw	1 wing (128 g)	203	**89**	24
	1 oz	45	**20**	5
roasted	1 wing (90 g)	186	**80**	22
	1 oz	59	**25**	7
meat only				
raw	1 wing (90 g)	96	**9**	3
	1 oz	30	**3**	1
roasted	1 wing (60 g)	96	**9**	3
	1 oz	30	**3**	1

RESTAURANT FOODS

FOOD	AMOUNT	CALORIES TOTAL	CALORIES FAT	CALORIES SAT-FAT
AU BON PAIN				
Breads				
Bagels				
Cinnamon Raisin	1	280	**9**	5
Plain, Onion, or Sesame	1	270	**9**	5
Loaf				
Baguette	1 loaf	810	**18**	5
Cheese	1 loaf	1670	**261**	81
Four Grain	1 loaf	1420	**99**	5

RESTAURANT FOODS

FOOD	AMOUNT	CALORIES TOTAL	CALORIES FAT	CALORIES SAT-FAT
Breads, loaf (cont.)				
Onion Herb	1 loaf	1430	**117**	5
Parisienne	1 loaf	1490	**36**	5
Muffins				
Blueberry	1	390	**99**	36
Bran	1	390	**99**	27
Carrot	1	450	**198**	45
Corn	1	460	**153**	27
Cranberry Walnut	1	350	**117**	18
Oat Bran Apple	1	400	**90**	18
Pumpkin	1	410	**144**	18
Whole Grain	1	440	**144**	18
Rolls				
Alpine	1	220	**27**	5
Country Seed	1	220	**36**	5
Hearth	1	250	**18**	5
Petit Pain	1	220	**9**	5
Pumpernickel	1	210	**18**	5
Rye	1	230	**18**	5
3 Seed Raisin	1	250	**36**	5
Vegetable	1	230	**45**	5
Sandwich				
Braided Roll	1 roll	387	**99**	27
Croissant	1 roll	300	**126**	72
French Sandwich	1 roll	320	**9**	5
Hearth Sandwich	1 roll	370	**27**	5
Multigrain Slice	2 slices	391	**27**	9
Pita Pocket	1 pocket	80	**9**	5
Rye Slice	2 slices	374	**36**	9
Cookies				
Chocolate Chip	1 serving	280	**135**	81
Chocolate Chunk Pecan	1 serving	290	**153**	54
Oatmeal Raisin	1 serving	250	**81**	27
Peanut Butter	1 serving	290	**135**	54
White Chocolate Chunk Pecan	1 serving	300	**153**	54
Croissants				
Dessert				
Almond	1	420	**225**	108

RESTAURANT FOODS

FOOD	AMOUNT	CALORIES TOTAL	FAT	SAT-FAT
Croissants (cont.)				
Apple	1	250	**90**	54
Blueberry Cheese	1	380	**180**	108
Chocolate	1	400	**216**	126
Cinnamon Raisin	1	390	**117**	72
Coconut Pecan	1	440	**207**	108
Hazelnut Chocolate	1	480	**252**	126
Plain	1	220	**90**	54
Strawberry or Raspberry Cheese	1	400	**180**	108
Sweet Cheese	1	420	**207**	126
Hot Filled				
Ham & Cheese	1	370	**180**	108
Spinach & Cheese	1	290	**144**	90
Turkey & Cheddar	1	410	**198**	117
Turkey & Havarti	1	410	**189**	117
Salads				
Chicken Tarragon Garden	1	310	**135**	18
Cracked Pepper Chicken Garden	1	100	**18**	5
Grilled Chicken Garden	1	110	**18**	5
Large Garden	1	40	**9**	5
Shrimp Garden	1	102	**18**	5
Small Garden	1	20	**9**	5
Tuna Garden	1	350	**225**	36
Salad Dressings				
Balsamic Vinaigrette	2.25 oz	311	**297**	45
Champagne Vinaigrette	2.25 oz	251	**234**	36
County Blue Cheese	2.25 oz	325	**279**	54
Honey with Poppy Seed	2.25 oz	351	**315**	54
Low Cal Italian	2.25 oz	68	**54**	5
Olive Oil Caesar	2.25 oz	255	**144**	NA
Parmesan & Pepper	2.25 oz	235	**189**	45
Sesame French	2.25 oz	339	**243**	36
Tomato Basil	2.25 oz	66	**9**	0
Sandwich Fillings				
Cheese				
Brie Cheese	1 serving	300	**216**	135

RESTAURANT FOODS

FOOD	AMOUNT	CALORIES TOTAL	CALORIES FAT	CALORIES SAT-FAT
Sandwich Fillings (cont.)				
Cheddar Cheese	1 serving	110	**81**	45
Herb Cheese	1 serving	290	**261**	162
Provolone Cheese	1 serving	155	**113**	66
Swiss Cheese	1 serving	330	**216**	135
Meats				
Albacore Tuna Salad	1 serving	310	**216**	36
Bacon	1 serving	140	**108**	36
Chicken Tarragon	1 serving	270	**135**	18
Country Ham	1 serving	150	**63**	27
Cracked Pepper Chicken	1 serving	120	**18**	5
Grilled Chicken	1 serving	130	**36**	5
Roast Beef	1 serving	180	**72**	36
Smoked Turkey	1 serving	100	**9**	5
Soups				
Beef Barley	1 cup	75	**18**	5
	1 bowl	112	**23**	9
Chicken Noodle	1 cup	79	**9**	5
	1 bowl	119	**15**	5
Clam Chowder	1 cup	289	**162**	81
	1 bowl	433	**243**	126
Cream of Broccoli	1 cup	201	**153**	72
	1 bowl	302	**234**	108
Garden Vegetarian	1 cup	29	**9**	5
	1 bowl	44	**9**	5
Minestrone	1 cup	105	**9**	5
	1 bowl	158	**15**	5
Split Pea	1 cup	176	**9**	5
	1 bowl	264	**15**	5
Tomato Florentine	1 cup	61	**9**	5
	1 bowl	92	**15**	5
Vegetarian Chili	1 cup	139	**27**	5
	1 bowl	208	**36**	5
BOSTON MARKET				
¼ white meat chicken				
without skin & wing	140 g	170	**35**	9
with skin & wing	152 g	280	**110**	32

RESTAURANT FOODS

FOOD	AMOUNT	CALORIES TOTAL	CALORIES FAT	CALORIES SAT-FAT
¼ dark meat chicken				
without skin	95 g	190	**90**	27
with skin	125 g	320	**190**	54
½ chicken with skin	277 g	590	**300**	90
Baked Italian pasta	¾ cup (170 g)	190	**80**	32
BBQ baked beans	¾ cup (201 g)	270	**45**	18
Brownie	1 piece (95 g)	450	**240**	63
Butternut squash	¾ cup (193 g)	160	**60**	36
Caesar salad				
entree	1 (283 g)	510	**380**	99
without dressing	1 (225 g)	230	**110**	54
side	1 (113 g)	200	**159**	41
Chicken Caesar salad	1 (369 g)	650	**410**	108
Chicken gravy	28 g	15	**10**	0
Chicken noodle soup	1 cup (257 g)	130	**40**	9
Chicken potpie	1 pie (425 g)	780	**410**	117
Chicken salad sandwich	1 (327 g)	680	**270**	45
Chicken sandwich				
with cheese & sauce	1 (352 g)	750	**300**	108
without cheese & sauce	1 (281 g)	430	**40**	9
Chicken tortilla soup	1 cup (238 g)	220	**100**	36
Chocolate chip cookie	1 (79 g)	340	**150**	54
Chunky chicken salad	1¾ cup (58 g)	370	**240**	41
Cinnamon apple pie	⅕ pie (136 g)	390	**200**	36
Corn bread	1 loaf (68 g)	200	**50**	14
Creamed spinach	¾ cup (181 g)	260	**180**	117
Fruit salad	¾ cup (156 g)	70	**5**	0
Green bean casserole	¾ cup (170 g)	130	**80**	41
Ham & turkey club				
with cheese & sauce	1 (379 g)	890	**390**	180
without cheese & sauce	1 (266 g)	420	**50**	14
Ham sandwich				
with cheese & sauce	1 (337 g)	750	**310**	108
without cheese & sauce	1 (266 g)	440	**70**	23
Hearth honey ham	142 g	210	**80**	32
Homestyle mashed potatoes	⅔ cup (161 g)	190	**80**	54
with gravy	¾ cup (189 g)	210	**90**	54

RESTAURANT FOODS

FOOD	AMOUNT	CALORIES TOTAL	CALORIES FAT	CALORIES SAT-FAT
Hot cinnamon apples	¾ cup (181 g)	250	**40**	5
Macaroni & cheese	¾ cup (192 g)	280	**100**	54
Meat loaf & chunky tomato sauce	227 g	370	**160**	72
Meat loaf & brown gravy	198 g	390	**200**	72
Meat loaf sandwich				
with cheese	1 (383 g)	860	**290**	144
without cheese	1 (351 g)	690	**190**	63
New potatoes	¾ cup (131 g)	130	**20**	0
Old fashioned potato salad	¾ cup (176 g)	340	**210**	36
Rice Pilaf	⅔ cup (145 g)	180	**45**	9
Savory stuffing	¾ cup (174 g)	310	**110**	18
Skinless rotisserie turkey breast	142 g	170	**10**	5
Steamed vegetables	⅔ cup (105 g)	35	**5**	0
Turkey sandwich				
with cheese & sauce	1 (337 g)	710	**260**	90
without cheese & sauce	1 (266 g)	400	**30**	9
Whole kernel corn	¾ cup (146 g)	180	**40**	5

CHAIN FAMILY-STYLE RESTAURANTS, such as Applebee's, Bennigans, Chili's, TGI Friday's, Grady's American Grill, Hard Rock Cafe, Houlihan's, Houston's, and Ruby Tuesday*

FOOD	AMOUNT	TOTAL	FAT	SAT-FAT
Appetizers				
Buffalo Wings	12 wings	700	**432**	144
Chili	1½ cups	350	**144**	72
Fried Mozzarella Sticks	9 sticks	830	**459**	252
Stuffed Potato Skins	8 skins	1120	**711**	360
Entrees				
Beef				
BBQ Baby Back Ribs	14 ribs	770	**486**	189
Hamburger	1	660	**324**	153
Mushroom Cheeseburger	1	900	**513**	252
Sirloin Steak	7 oz	410	**180**	90
Steak Fajitas with Tortillas	1 order	860	**279**	108
with guacamole, sour cream, & cheese		1190	**567**	252

*Adapted from *Nutrition Action Health Letter,* October 1996.

RESTAURANT FOODS

FOOD	AMOUNT	CALORIES TOTAL	CALORIES FAT	CALORIES SAT-FAT
Entrees (cont.)				
Chicken				
Bacon & Cheese Grilled Chicken Sandwich	1	650	**270**	108
Chicken Caesar Salad with dressing	4 cups	660	**414**	99
Chicken Fajitas with Tortillas	1 order	840	**216**	54
with guacamole, sour cream, & cheese		1170	**504**	198
Chicken Fingers	5 pieces	620	**306**	117
Grilled Chicken	6 oz	270	**72**	27
with loaded baked potato & vegetables		950	**378**	207
Oriental Chicken Salad with dressing	4 cups	750	**441**	108
Side Orders				
Cole Slaw	1 cup	170	**126**	18
French Fries	2 cups	590	**279**	108
Loaded Baked Potato	1	620	**279**	171
Onion Rings	11	900	**576**	207
Vegetable of the Day	1 cup	60	**27**	9
Dessert				
Fudge Brownie Sundae	10 oz	1130	**513**	270
CHILI'S				
Chicken fajitas	1 order	870	**306**	NA
Chicken sandwich	1	1082	**450**	NA
Diet by Chocolate cake with hot fudge sauce	1 slice	370	**18**	5
Grilled chicken platter	1	757	**189**	NA
Guiltless Grill				
chicken fajitas	1 order	690	**54**	18
chicken platter	1	450	**32**	9
chicken salad	1	254	**30**	10
chicken sandwich	1	485	**47**	14
Tuna sandwich	1	950	**342**	NA
CHINESE RESTAURANT				
Barbecued pork (not fried)	1 whole dish	1374	**986**	355

RESTAURANT FOODS

FOOD	AMOUNT	CALORIES TOTAL	FAT	SAT-FAT
Barbecued spareribs	1 whole dish	1863	**1232**	480
Beef with vegetables	1 whole dish	1572	**1068**	263
Chicken with cashews	1 whole dish	1765	**1075**	186
Chicken with vegetables	1 whole dish	1224	**652**	102
Chinese Noodle Soup	1 serving	265	**81**	NA
Egg rolls	1	152	**103**	30
Hot and sour soup	1 serving	165	**72**	NA
Hunan shrimp (not fried)	1 whole dish	1068	**755**	100
Kung Pao beef	1 whole dish	2458	**1706**	444
Kung Pao chicken	1 whole dish	1806	**1134**	158
Kung Pao shrimp	1 whole dish	1068	**755**	115
Moo shu pork	1 whole dish	1383	**1053**	258
Orange beef	1 whole dish	1710	**1216**	342
Pork with vegetables	1 whole dish	1574	**1219**	338
Sweet and sour pork	1 whole dish	1845	**1509**	389
Sweet and sour shrimp	1 whole dish	1069	**805**	130
Szechuan pork	1 whole dish	1694	**1339**	356
Velvet corn soup	1 serving	115	**27**	NA
Wonton soup	1 serving	283	**108**	NA
COFFEE BAR COFFEES				
If whipped cream is added to your coffee, add 60 total calories and 45 fat calories.				
COFFEE BEANERY				
Cafe Mocha				
with whole milk	8 fl oz	94	**45**	NA
with 2% milk	8 fl oz	76	**27**	NA
with nonfat milk	8 fl oz	54	**0**	NA
Cappuccino				
with whole milk	12 fl oz	296	**81**	NA
with 2% milk	12 fl oz	267	**54**	NA
with nonfat milk	12 fl oz	232	**9**	NA
Espresso	2.4 fl oz	0	**0**	NA
Latte				
with whipped cream and grated chocolate	16 fl oz	350	**180**	NA
with whole milk	16 fl oz	263	**126**	NA
with 2% milk	16 fl oz	211	**72**	NA
with nonfat milk	16 fl oz	151	**9**	NA

RESTAURANT FOODS

FOOD	AMOUNT	CALORIES TOTAL	CALORIES FAT	CALORIES SAT-FAT
GLORIA JEAN'S				
(only made with 2% milk)				
Cafe Mocha	8 fl oz	222	**36**	NA
Grande	16 fl oz	312	**63**	NA
Iced	12 fl oz	282	**54**	NA
Espresso	2.7 fl oz	0	**0**	NA
Latte	8 fl oz	76	**27**	NA
Grande	16 fl oz	166	**54**	NA
STARBUCKS				
Coffee Drinks				
Caffe Latte				
Short				
with whole milk	8 fl oz	140	**60**	41
with lowfat milk	8 fl oz	110	**35**	23
with nonfat milk	8 fl oz	80	**0**	0
with soy milk	8 fl oz	70	**35**	0
Tall				
with whole milk	12 fl oz	210	**100**	63
with lowfat milk	12 fl oz	170	**50**	36
with nonfat milk	12 fl oz	120	**5**	0
with soy milk	12 fl oz	110	**60**	5
Grande				
with whole milk	16 fl oz	270	**130**	81
with lowfat milk	16 fl oz	220	**70**	45
with nonfat milk	16 fl oz	160	**10**	5
with soy milk	16 fl oz	150	**70**	0
Venti				
with whole milk	20 fl oz	350	**160**	108
with lowfat milk	20 fl oz	270	**90**	54
with nonfat milk	20 fl oz	200	**10**	5
with soy milk	20 fl oz	190	**90**	9
Caffe Mocha with whipping cream				
Short				
with whole milk	8 fl oz	250	**150**	90
with lowfat milk	8 fl oz	220	**120**	72
with nonfat milk	8 fl oz	200	**100**	63

RESTAURANT FOODS

FOOD	AMOUNT	CALORIES TOTAL	CALORIES FAT	CALORIES SAT-FAT
Caffe Mocha (cont.)				
Tall				
with whole milk	12 fl oz	340	**190**	117
with lowfat milk	12 fl oz	300	**150**	90
with nonfat milk	12 fl oz	260	**100**	63
Grande				
with whole milk	16 fl oz	410	**220**	135
with lowfat milk	16 fl oz	370	**160**	99
with nonfat milk	16 fl oz	320	**110**	72
Venti				
with whole milk	20 fl oz	500	**250**	162
with lowfat milk	20 fl oz	440	**190**	117
with nonfat milk	20 fl oz	380	**120**	72
Caffe Mocha (iced) with whipping cream				
Short				
with whole milk	8 fl oz	180	**110**	72
with lowfat milk	8 fl oz	170	**90**	54
with nonfat milk	8 fl oz	150	**80**	45
Tall				
with whole milk	12 fl oz	290	**160**	99
with lowfat milk	12 fl oz	260	**130**	81
with nonfat milk	12 fl oz	230	**100**	63
Grande				
with whole milk	16 fl oz	390	**220**	135
with lowfat milk	16 fl oz	350	**180**	108
with nonfat milk	16 fl oz	310	**130**	81
Caffe Rhumba				
Tall				
with lowfat milk	12 fl oz	250	**45**	36
Grande				
with lowfat milk	16 fl oz	330	**60**	45
Venti				
with lowfat milk	20 fl oz	410	**70**	63
Cappuccino				
Short				
with whole milk	8 fl oz	100	**45**	31
with lowfat milk	8 fl oz	80	**25**	14

RESTAURANT FOODS

FOOD	AMOUNT	CALORIES TOTAL	FAT	SAT-FAT
Cappuccino (cont.)				
with nonfat milk	8 fl oz	60	**0**	0
with soy milk	8 fl oz	50	**25**	0
Tall				
with whole milk	12 fl oz	140	**60**	41
with lowfat milk	12 fl oz	110	**35**	23
with nonfat milk	12 fl oz	80	**0**	0
with soy milk	12 fl oz	70	**35**	0
Grande				
with whole milk	16 fl oz	180	**80**	54
with lowfat milk	16 fl oz	140	**45**	27
with nonfat milk	16 fl oz	110	**5**	0
with soy milk	16 fl oz	100	**50**	5
Venti				
with whole milk	20 fl oz	200	**90**	63
with lowfat milk	20 fl oz	160	**50**	32
with nonfat milk	20 fl oz	120	**5**	0
with soy milk	20 fl oz	110	**50**	5
Espresso				
Solo	1 fl oz	5	**0**	0
Doppio	2 fl oz	10	**0**	0
Con Panna				
Solo	1 fl oz	30	**25**	18
Doppio	2 fl oz	35	**25**	18
Macchiato				
Solo				
with whole milk	1 fl oz	15	**5**	0
with lowfat milk	1 fl oz	10	**0**	0
with nonfat milk	1 fl oz	10	**0**	0
Doppio				
with whole milk	2 fl oz	20	**5**	0
with lowfat milk	2 fl oz	15	**0**	0
with nonfat milk	2 fl oz	15	**0**	0
Frappuccino				
Tall				
with lowfat milk	12 fl oz	200	**25**	14
Grande				
with lowfat milk	16 fl oz	270	**35**	23

RESTAURANT FOODS

FOOD	AMOUNT	CALORIES TOTAL	CALORIES FAT	CALORIES SAT-FAT
Frappuccino (cont.)				
Venti				
with lowfat milk	20 fl oz	340	**45**	27
Power Frappuccino				
Tall	12 fl oz	290	**20**	9
Grande	16 fl oz	350	**25**	14
Venti	20 fl oz	410	**25**	18
Power Mocha Frappuccino				
Tall	12 fl oz	320	**25**	14
Grande	16 fl oz	390	**25**	18
Venti	20 fl oz	460	**30**	23
Cocoa Drinks				
Cocoa with whipping cream				
Short				
with whole milk	8 fl oz	260	**160**	99
with lowfat milk	8 fl oz	230	**130**	81
with nonfat milk	8 fl oz	210	**100**	63
Tall				
with whole milk	12 fl oz	350	**190**	126
with lowfat milk	12 fl oz	310	**150**	90
with nonfat milk	12 fl oz	270	**100**	63
Grande				
with whole milk	16 fl oz	440	**230**	144
with lowfat milk	16 fl oz	390	**170**	108
with nonfat milk	16 fl oz	330	**110**	72
Venti				
with whole milk	20 fl oz	530	**270**	171
with lowfat milk	20 fl oz	460	**200**	126
with nonfat milk	20 fl oz	390	**120**	72
Tea Drinks				
Chai Tea Latte hot				
Short				
with whole milk	8 fl oz	160	**60**	36
with lowfat milk	8 fl oz	130	**30**	18
with nonfat milk	8 fl oz	110	**0**	0
Tall				
with whole milk	12 fl oz	220	**80**	54
with lowfat milk	12 fl oz	190	**45**	27
with nonfat milk	12 fl oz	150	**0**	0

RESTAURANT FOODS

FOOD	AMOUNT	CALORIES TOTAL	FAT	SAT-FAT
Chai Tea Latte hot (cont.)				
Grande				
with whole milk	16 fl oz	310	**120**	72
with lowfat milk	16 fl oz	260	**60**	41
with nonfat milk	16 fl oz	210	**5**	0
Venti				
with whole milk	20 fl oz	400	**150**	99
with lowfat milk	20 fl oz	330	**80**	54
with nonfat milk	20 fl oz	270	**10**	5
Cakes, Cookies, and Pastries				
Biscotti				
Chocolate Hazelnut				
standard size	1 (29 g)	110	**45**	18
mini size	1 (12 g)	50	**20**	9
Vanilla Almond				
standard size	1 (28 g)	110	**40**	14
mini size	1 (14 g)	50	**20**	9
Brownies & Bars				
Blondies	1 (76 g)	350	**170**	81
Fantasy Bar	⅓ bar (35 g)	180	**80**	41
Raspberry Crumb Bar	⅓ bar (35 g)	160	**70**	27
Ultra Chocolate Brownie	1 bar (98 g)	418	**146**	75
Ultra Hazelnut Brownie	1 bar (35 g)	433	**174**	72
Walnut Brownie	1 brownie (98 g)	450	**240**	99
Cookie				
Oatmeal Raisin	1 (85 g)	400	**130**	45
Croissants				
Almond	1 (74 g)	320	**150**	81
Butter	1 (99 g)	410	**210**	144
Chocolate	1 (99 g)	400	**200**	136
Cupcakes				
Carrot Cake	1 (114 g)	440	**210**	81
Chocolate Chocolate	1 (113 g)	380	**140**	63
Focaccia				
Tomato & Cheese	1 (170 g)	640	**385**	90

RESTAURANT FOODS

FOOD	AMOUNT	CALORIES TOTAL	CALORIES FAT	CALORIES SAT-FAT
Cakes, Cookies, and Pastries (cont.)				
Muffins				
Chocolate Chunk	1 muffin (145 g)	580	**280**	81
Cranberry Orange	1 muffin (145 g)	520	**300**	36
Lemon Poppyseed	1 muffin (145 g)	540	**210**	45
Blueberry	1 muffin (145 g)	490	**230**	63
Blueberry, lowfat	1 muffin (113 g)	250	**10**	0
Pecan Rolls	1 (145 g)	450	**150**	54
Rugelach				
Cinnamon Raisin (traditional)	3 dolcini (64 g)	380	**270**	108
Scones				
Apricot reduced fat	1 (128 g)	320	**50**	41
Blueberry (no cholesterol)	1 (128 g)	380	**100**	54
Chocolate	1 (1298 g)	420	**140**	81
Cinnamon	1 (128 g)	420	**130**	63
Maple Oat Nut	1 (124 g)	540	**250**	126
Multigrain (50% reduced fat)	1 (128 g)	320	**50**	41
Very Blueberry	1 (124 g)	530	**210**	126
DELI SANDWICH SHOPS				
Bacon, lettuce, and tomato	8 oz	599	**333**	108
Chicken salad (plain bread)	10 oz	537	**288**	54
with mayo	10 oz	655	**414**	72
Corned beef with mustard	9 oz	497	**180**	72
Egg salad (plain bread)	10 oz	546	**279**	90
with mayo	10 oz	664	**396**	108
Grilled cheese (plain bread)	5 oz	511	**297**	153
Ham with mustard	9 oz	563	**243**	90
with mayo	9 oz	666	**360**	108
Reuben	14 oz	916	**450**	180
Roast beef with mustard	9 oz	462	**108**	36
with mayo	9 oz	565	**216**	54
Tuna salad (plain bread)	11 oz	716	**387**	72
with mayo	11 oz	833	**504**	90
Turkey Club	13 oz	737	**306**	90
Turkey with mustard	9 oz	370	**54**	18
with mayo	9 oz	473	**171**	36
Vegetarian	12 oz	753	**360**	126

RESTAURANT FOODS

FOOD	AMOUNT	CALORIES TOTAL	CALORIES FAT	CALORIES SAT-FAT
DENNY'S				
Baked potato	1	180	**0**	0
Banana/Strawberry Medley	½ cup	170	**9**	0
Biscuit	1	217	**63**	NA
BLT Sandwich	1	492	**306**	NA
Blueberry Muffin	1	309	**126**	NA
Catfish	1 entree	576	**432**	NA
Buttermilk Pancakes, plain	3	410	**54**	18
Chicken Strips	4 oz	240	**90**	NA
Chili	8 oz	238	**135**	NA
Cinnamon Roll	1	450	**126**	NA
Club Sandwich	1	590	**180**	NA
Coleslaw	1 cup	119	**86**	NA
Country Gravy	1 oz	140	**72**	NA
Eggs Benedict	1	658	**320**	NA
French Fries	1 order	303	**142**	NA
French Toast	2 slices	729	**504**	NA
Fried Chicken, entree only	4 pieces	463	**270**	NA
Fried Shrimp, entree only	1	230	**135**	NA
Grilled Cheese Sandwich	1	454	**261**	NA
Grilled Chicken, entree only	1	130	**36**	9
Grilled Chicken Sandwich	1	439	**108**	NA
Guacamole	1 oz	60	**55**	NA
Hamburger				
Bacon Swiss burger	1	819	**468**	NA
Denny burger	1	629	**340**	NA
San Fran burger	1	872	**432**	NA
Works burger	1	944	**549**	NA
Hashed Browns	4 oz	164	**18**	NA
Liver with Bacon and Onions, entree only	2 slices	334	**130**	NA
Mozzarella Sticks	1 piece	88	**60**	NA
Omelet				
Denver	1	567	**243**	NA
Ultimate	1	577	**369**	NA
Veggie Cheese	1	350	**180**	NA
Onion Rings	3 rings	258	**135**	NA
Patty Melt	1	761	**423**	NA

RESTAURANT FOODS

FOOD	AMOUNT	CALORIES TOTAL	CALORIES FAT	CALORIES SAT-FAT
Rice Pilaf	⅓ cup	89	**21**	NA
Salads				
California Grilled Chicken, no dressing	1	280	**90**	9
Chef	1	492	**180**	NA
Chicken salad, no shell	1	207	**36**	NA
Garden, no dressing	1	115	**36**	9
Taco, no shell	1	514	**180**	NA
Tuna salad	1	340	**162**	NA
Sausage	1 link	113	**90**	NA
Senior Grilled Chicken, entree only	1	130	**36**	9
Soups				
Cheese	1 bowl	309	**198**	NA
Chicken Noodle	1 bowl	45	**9**	0
Clam Chowder	1 bowl	235	**126**	NA
Cream of Potato	1 bowl	175	**81**	63
Split Pea	1 bowl	231	**45**	NA
Spaghetti with Tomato Sauce	1 order	600	**72**	0
Stir-fry, entree only	1	328	**99**	NA
Stuffing	½ cup	180	**81**	NA
Super Bird Sandwich	1	625	**216**	NA
Steak				
Chicken Fried Steak, entree only, no gravy	2 pieces	252	**131**	NA
Hamburger Steak	1 entree	669	**484**	NA
New York Steak, entree only	1	582	**324**	NA
Top Sirloin Steak, entree only	1	223	**57**	NA
Tortilla Shell, fried	1	439	**270**	NA
Turkey, no gravy, entree only	1	505	**130**	NA
Waffle	1	261	**94**	NA
INTERNATIONAL HOUSE OF PANCAKES				
Pancakes				
Buttermilk	1 pancake (56 g)	108	**28**	6
Buckwheat	1 pancake (63 g)	134	**45**	11
Country Griddle	1 pancake (63 g)	134	**34**	9
Egg	1 pancake (56 g)	102	**45**	11
Harvest Grain 'N Nut	1 pancake (63 g)	160	**74**	12

RESTAURANT FOODS

FOOD	AMOUNT	CALORIES TOTAL	CALORIES FAT	CALORIES SAT-FAT
Foods prepared with Eggstro'dnaire				
Broccoli & Mushroom Omelette	1 omelette	310	**62**	NA
Breakfast Burrito	1 burrito	456	**109**	NA
Chicken Fajita Burrito	1 burrito	523	**89**	NA
French Toast	1 piece	99	**18**	NA
Waffles				
Regular	1 waffle (112 g)	305	**133**	30
Belgian				
Regular	1 waffle (168 g)	408	**177**	100
Harvest Grain 'N Nut	1 waffle (168 g)	445	**251**	107
ITALIAN RESTAURANT				
Appetizers				
Antipasto	1.5 lbs	629	**423**	132
Entrees				
Eggplant Parmigiana with spaghetti	2.5 cups	1208	**558**	145
Fettuccine Alfredo	2.5 cups	1498	**873**	434
Lasagna	2 cups	958	**477**	192
Linguine with red clam sauce	3 cups	892	**207**	36
Linguine with white clam sauce	3 cups	907	**261**	45
Spaghetti with meat sauce	3 cups	918	**225**	92
Spaghetti with meatballs	3.5 cups	1155	**351**	94
Spaghetti with sausage	2.5 cups	1043	**351**	92
Spaghetti with tomato sauce	3.5 cups	849	**153**	34
Veal Parmigiana with spaghetti	1.5 cups	1064	**396**	128
Side Dishes				
Fried calamari	3 cups	1037	**630**	83
Garlic bread	8 oz	822	**360**	90
Spaghetti with tomato sauce	1.5 cups	409	**72**	16
MEXICAN RESTAURANT				
Appetizers				
Beef and cheese nachos with sour cream and guacamole	1 serving	1362	**801**	250
Cheese quesadilla with sour cream and guacamole	1 serving	900	**531**	220
Cheese Nachos	1 serving	807	**500**	225

RESTAURANT FOODS

FOOD	AMOUNT	CALORIES TOTAL	CALORIES FAT	CALORIES SAT-FAT
Entrees				
Beef Burrito	1 serving	833	**360**	121
with beans, rice, sour cream, and guacamole	1 serving	1639	**711**	248
Beef Chimichanga	1 serving	802	**423**	113
with beans, rice, sour cream, and guacamole	1 serving	1607	**774**	241
Beef Enchilada	1 serving	324	**171**	67
two enchiladas with beans and rice	1 serving	1253	**522**	140
Chicken Fajitas and Flour Tortillas	1 serving	839	**216**	54
with beans, rice, sour cream, and guacamole	1 serving	1661	**567**	173
Chile Rellenos	1 serving	487	**342**	45
two chile rellenos with beans and rice	1 serving	1578	**864**	173
Chicken Enchilada	1 serving	329	**162**	103
two enchiladas with beans and rice	1 serving	1264	**513**	270
Crispy Chicken Taco	1 serving	219	**99**	27
two tacos with beans and rice	1 serving	1042	**378**	119
Taco Salad				
with sour cream and guacamole	1 serving	1099	**639**	177
Side Dishes				
Rice	¾ cup	229	**34**	5
Refried beans	¾ cup	375	**146**	60
Tortilla chips	50 chips	645	**432**	81
OLIVE GARDEN				
Breadsticks				
garlic	1 stick	160	**32**	14
plain	1 stick	140	**14**	0
Capellini pomodoro	1 dinner	520	**144**	32
	1 lunch	340	**72**	18
Capellini primavera	1 dinner	380	**63**	27
	1 lunch	270	**45**	18
Garden salad, no dressing	1 order	70	**9**	0

RESTAURANT FOODS

FOOD	AMOUNT	CALORIES TOTAL	CALORIES FAT	CALORIES SAT-FAT
Grilled chicken with peppers	1 dinner	470	**81**	32
Julius				
banana	16 oz	210	**9**	5
banana Julius Smoothy	16 oz	270	**54**	45
Cool Cappuccino Julius Java	16 oz	390	**63**	63
Cool Mocha Julius Java	16 oz	460	**90**	72
lemon, orange, peach, pineapple,		190		
or strawberry Julius	16 oz	220	**5**	0
pinata colada Julius Smoothy	16 oz	330	**54**	45
strawberry Julius Smoothy	16 oz	330	**54**	45
tropical Julius Smoothy	16 oz	330	**54**	45
Minestrone soup	6 oz	80	**9**	0
Pasta e fagioli soup	6 oz	140	**45**	14
Raspberry sorbetto	170 g	110	**0**	0
Shrimp primavera	1 dinner	420	**108**	27
	1 lunch	320	**90**	18
Spaghetti				
with marinara sauce	1 dinner	500	**81**	14
	1 lunch	340	**54**	9
with Sicilian sauce	1 dinner	530	**108**	14
	1 lunch	370	**72**	9
with tomato sauce	1 dinner	550	**90**	14
	1 lunch	390	**63**	9
Venetian grilled chicken	1 dinner	240	**45**	14
RAX				
Baked Potato				
Barbecue with 2 oz cheese	1 serving	730	**216**	NA
Chili with 2 oz cheese	1 serving	700	**207**	NA
Plain	1 serving	270	**0**	0
with margarine	1 serving	370	**99**	NA
with sour cream topping	1 serving	400	**99**	NA
with 3 oz cheese & bacon	1 serving	780	**252**	NA
with 3 oz cheese & broccoli	1 serving	760	**234**	NA
Barbecue Sandwich	1	420	**126**	NA
Beans				
Garbanzo	½ cup	360	**45**	NA
Kidney	1 cup	220	**9**	NA
Breadstick, sesame	28 g	150	**90**	NA

RESTAURANT FOODS

FOOD	AMOUNT	CALORIES TOTAL	CALORIES FAT	CALORIES SAT-FAT
Chili Topping	84 g	80	**18**	NA
Coleslaw	98 g	70	**36**	NA
Chocolate Chip Cookie	1	130	**54**	NA
French Fries				
large	1 order	390	**180**	NA
regular	1 order	260	**117**	NA
Hot Chocolate	1 serving	110	**99**	NA
Milkshakes, without whipped topping				
Chocolate	1 serving	560	**117**	NA
Strawberry	1 serving	560	**117**	NA
Vanilla	1 serving	500	**126**	NA
Pasta & Noodles				
Pasta Shells	98 g	170	**36**	NA
Pasta/Vegetable Blend	98 g	100	**36**	NA
Rainbow Rotini	98 g	180	**36**	NA
Potato Salad	1 cup	260	**153**	NA
Pudding, all flavors	98 g	140	**54**	NA
Refried Beans	84 g	120	**36**	NA
Roast Beef	1 serving	140	**54**	NA
Salads				
Chef, no dressing	1 serving	230	**126**	NA
Garden, no dressing	1 serving	160	**99**	NA
Garden, Lighterside	1 serving	134	**54**	NA
Macaroni	98 g	160	**63**	NA
Pasta	98 g	80	**9**	NA
Potato	1 cup	260	**63**	NA
Three Bean	½ cup	100	**9**	NA
Salad dressing (*see* **FATS AND OILS**)				
Sandwiches				
Beef				
BBQ Beef, Bacon, Chicken	1 serving	720	**441**	NA
Philly Beef & Cheese	1 serving	480	**198**	NA
Fish	1 serving	460	**153**	NA
Ham & Cheese	1	430	**207**	NA
Roast Beef				
large	1 serving	570	**315**	NA
regular	1 serving	320	**99**	NA
small	1 serving	260	**126**	NA

RESTAURANT FOODS

FOOD	AMOUNT	CALORIES TOTAL	CALORIES FAT	CALORIES SAT-FAT
Sandwiches (cont.)				
Turkey Bacon Club	1 serving	670	**387**	NA
Sauces				
Cheese				
Nacho	98 g	470	**198**	NA
Regular	98 g	420	**153**	NA
Spaghetti				
Regular	98 g	80	**9**	NA
With Meat	98 g	150	**72**	NA
Spicy Meat	98 g	80	**36**	NA
Taco	98 g	30	**9**	NA
Soups				
Chicken Noodle	98 g	40	**9**	NA
Cream of Broccoli	98 g	50	**18**	NA
Soy Nuts	28 g	120	**63**	NA
Spaghetti	98 g	140	**36**	NA
Taco Shell	1 shell	40	**18**	NA
Tortilla	1 tortilla	110	**18**	NA
Turkey Bits	56 g	70	**27**	NA
Whipped Topping	1 dollop	50	**36**	NA
RED LOBSTER				
Alaskan snow crab legs	1 order (454 g)	200	**99**	54
Bay platter	1	680	**243**	81
Bayou-style seafood gumbo	170 g	180	**45**	9
Broiled fish fillet sandwich	1	230	**86**	5
Broiled flounder fillets	1 order (142 g)	150	**54**	27
Broiled rock lobster	1 order (369 g)	250	**45**	18
Fish fillet sandwich	1	230	**85**	,9
Grilled chicken breast				
dinner menu	228 g	340	**108**	36
lunch menu	114 g	170	**54**	18
Grilled chicken (114 g) and 10 shrimp	1 order	490	**180**	54
Grilled Chicken sandwich	1	340	**90**	36
Grilled shrimp salad, lite dressing	1 order	170	**72**	9
Grilled shrimp skewers	120 shrimp	290	**81**	36
Ice cream	1 order (128 g)	260	**126**	81
Live Maine lobster	1 order (511 g)	200	**45**	18

RESTAURANT FOODS

FOOD	AMOUNT	CALORIES TOTAL	CALORIES FAT	CALORIES SAT-FAT
Rice pilaf	114 g	140	**27**	4
Seafood Lover's platter	1	650	**243**	108
Sherbet	1 order (128 g)	180	**27**	18
Shrimp				
cocktail	6 shrimp	90	**18**	4
in the shell	170 g	130	**18**	5
scampi	11 shrimp	310	**207**	126
Today's fresh catch *(for lunch portions, halve the calories and fat calories)*				
Atlantic cod	1 dinner (10 oz)	300	**108**	54
Atlantic salmon	1 dinner (10 oz)	460	**306**	108
Catfish	1 dinner (10 oz)	440	**270**	108
Coho salmon	1 dinner (10 oz)	480	**252**	90
Grouper	1 dinner (10 oz)	300	**108**	54
Haddock	1 dinner (10 oz)	320	**108**	54
King salmon	1 dinner (10 oz)	580	**360**	72
Mahi mahi	1 dinner (10 oz)	320	**108**	54
Ocean perch	1 dinner (10 oz)	360	**162**	90
Orange roughy	1 dinner (10 oz)	440	**270**	54
Rainbow trout	1 dinner (10 oz)	440	**252**	72
Red rockfish	1 dinner (10 oz)	280	**108**	54
Sea bass	1 dinner (10 oz)	360	**144**	72
Snapper	1 dinner (10 oz)	320	**108**	54
Sole	1 dinner (10 oz)	320	**108**	54
Swordfish	1 dinner (10 oz)	300	**162**	108
Walleye pike	1 dinner (10 oz)	340	**108**	54
Yellow lake perch	1 dinner (10 oz)	340	**108**	54
SWISS CHALET				
Apple Pie	1 serving	413	**171**	36
Back Rib	½ rib	405	**234**	81
Back Rib	full rib	810	**468**	162
Caesar Salad Entrée	1 serving	454	**342**	36
Caesar Salad Appetizer	1 serving	345	**171**	36
Chicken				
White (with skin)	¼ chicken	381	**198**	36
White (skinless)	¼ chicken	225	**72**	18
Dark (with skin)	¼ chicken	313	**153**	45
Dark (skinless)	¼ chicken	232	**90**	27
Chicken (with skin)	½ chicken	694	**351**	81

RESTAURANT FOODS

FOOD	AMOUNT	CALORIES TOTAL	FAT	SAT-FAT
Chicken Pot Pie	1 pie	494	**216**	45
Chicken Salad & Roll	1 serving	466	**198**	36
Roll	1 roll	116	**9**	0
MISCELLANEOUS RESTAURANT FOODS				
Beef Gyro	1 sandwich (122 g)	340	**190**	99
Caesar salad*	1 salad	660	**414**	NA
Fettuccine with creamed spinach*	1 serving	1050	**738**	NA
Focaccia club sandwich*	1 sandwich	1222	**585**	NA
Gnocchi*	1 serving	700	**423**	NA
Lasagna*	1 serving	960	**477**	NA
Omelet	for 1	337	**250**	**110**
Cheese	for 1	377	**276**	**126**
Denver	for 1	425	**290**	**125**
Porterhouse steak dinner*	1 dinner	1860	**1125**	NA
Risotto	1 serving	1280	**990**	NA
Tuna salad sandwich*	1 sandwich	720	**387**	NA

SALAD BAR FOODS

FOOD	AMOUNT	CALORIES TOTAL	FAT	SAT-FAT
Bacon bits	1 tbsp (7 g)	30	**10**	0
Baked beans	½ cup	160	**36**	5
Breadsticks, mini, sesame	2 (7 g)	35	**10**	0
Sauces				
nacho cheese	¼ cup	120	**90**	54
Cheese, shredded				
cheddar	⅓ cup	110	**80**	45
mozzarella	¼ cup	80	**45**	27
Chow mein noodles	½ cup	140	**60**	NA
Cottage cheese (4% fat)	½ cup	120	**45**	27
Crackers				
Oyster	1 pkg (14 g)	60	**25**	5
Saltine	2 crackers (14 g)	25	**5**	0
Croutons	2 tbsp (7 g)	35	**20**	0

*Source: Marian Burros, *New York Times*.

SALAD BAR FOODS

FOOD	AMOUNT	CALORIES TOTAL	CALORIES FAT	CALORIES SAT-FAT
Eggs, chopped, hard boiled	28 g	45	**25**	9
Olives, green or ripe	2 tbsp (16 g)	30	**25**	0
Puddings & Desserts				
Bread pudding	½ cup	170	**36**	NA
Chocolate mousse	½ cup	160	**45**	NA
Chocolate pudding	½ cup	110	**18**	9
Lemon mousse	½ cup	160	**45**	NA
Rice pudding	½ cup	120	**15**	9
Strawberry Creme dessert	1 container (99 g)	100	**10**	9
Strawberry Fruit dessert	1 container (113 g)	90	**0**	0
Tapioca Pudding	1 container (113 g)	120	**15**	9
Salads				
Carrot Waldorf	½ cup	190	**108**	18
Creamy coleslaw	½ cup	220	**60**	18
Pasta				
Chicken	¾ cup	320	**225**	27
Fiesta	⅔ cup	240	**80**	18
Macaroni	½ cup	370	**220**	27
Penne mozzarella	¾ cup	190	**70**	18
Rigati Garden	½ cup	170	**80**	9
Seafood	¾ cup	290	**190**	23
Sicilian Tortellini	1 cup	280	**50**	18
Spaghetti	¾ cup	310	**160**	36
Tuna supreme	½ cup	240	**160**	18
Vegetable Pasta				
(cholesterol free)	⅔ cup	170	**15**	5
with broccoli	⅔ cup	190	**50**	5
Polynesian	½ cup	160	**45**	14
Potato	½ cup	210	**80**	9
Red skin potato	⅔ cup	260	**155**	45
Three bean	⅓ cup	90	**5**	0
Waldorf	½ cup	250	**225**	30
Salad dressings (*see* **FATS AND OILS**)				
Sauces				
Salsa	2 tbsp	10	**0**	0
Soup bar				
Bean and ham	1 cup	180	**25**	9
Beef barley	1 cup	127	**24**	NA

SALAD BAR FOODS

FOOD	AMOUNT	CALORIES TOTAL	CALORIES FAT	CALORIES SAT-FAT
Soup bar (cont.)				
Beef stew	1 cup	270	**130**	45
Chicken corn noodle	1 cup	150	**20**	5
Chicken noodle	1 cup	170	**30**	14
Chicken rice	1 cup	50	**10**	0
Chili con carne with beans	1 cup	320	**100**	45
Clam chowder New England style	1 cup	130	**35**	18
Crab	1 cup	73	**24**	NA
Cream of broccoli	1 cup	140	**60**	36
Cream of potato with bacon	1 cup	130	**30**	14
Vegetable beef	1 cup	190	**25**	9
Vegetable crab	1 cup	80	**20**	14
Sunflower seeds	¼ cup (30 g)	160	**110**	14
Suremi (imitation crabmeat)	½ cup (85 g)	80	**0**	0
Tofu	½ cup (85 g)	90	**45**	5
Tortilla chips, regular & blue corn	8 chips (28 g)	140	**60**	9
Turkey, diced	56 g	115	**80**	23
SALAD TOPPINGS				
Bac-o's, bits or chips (Betty Crocker)	1½ tbsp (7 g)	30	**10**	0
Bac'n Pieces (McCormick)	1½ tbsp (7 g)	30	**15**	0
Croutons				
Pepperidge Farm				
Cheese & Garlic	9 croutons (7 g)	35	**15**	0
Cracked Pepper & Parmesan	6 croutons (7 g)	35	**10**	0
Seasoned	9 croutons (7 g)	35	**15**	0
Zesty Italian	6 croutons (7 g)	35	**15**	0
Real Bacon Bits (Hormel)	1 tbsp (7 g)	30	**15**	9
Salad Crispins, Mini Croutons (Hidden Valley)				
Italian Parmesan	1 tbsp (7 g)	35	**10**	0
Original Ranch	1 tbsp (7 g)	35	**10**	0

SAUCES, GRAVIES, AND DIPS

FOOD	AMOUNT	CALORIES TOTAL	CALORIES FAT	CALORIES SAT-FAT
SAUCES				
Barbecue				
Chinese	2 tbsp	45	**2**	0
Masterpiece	2 tbsp	40–60	**0**	0
Kraft	2 tbsp	40	**0**	0
Open Pit	2 tbsp	50	**5**	0
Bean Sauce	1 tbsp	23	**8**	0
Black Bean Garlic Sauce	1 tbsp	25	**0**	0
Bearnaise	1 tbsp	53	**48**	29
	½ cup	423	**383**	232
Browning & Seasoning (Kitchen Bouquet)	1 tsp	15	**0**	0
Cheese Sauce	1 tbsp	31	**21**	14
	½ cup	250	**170**	110
Chili Paste with Garlic	1 Tbsp	10	**9**	0
Clam Sauce, White	½ cup	120	**80**	14
Cream	1 tbsp	28	**22**	14
	½ cup	225	**175**	110
Curry Cream	1 tbsp	40	**31**	20
	½ cup	317	**250**	160
Fish Sauce	1 Tbsp	0	**0**	0
Hoisin Sauce	2 Tbsp	60	**10**	0
Hollandaise	1 tbsp	82	**80**	47
	½ cup	660	**627**	377
Horseradish (Kraft)	1 tbsp	20	**15**	0
Hunan Sauce	4 tsp	25	**10**	0
Korean Bulkogi Marinade Sauce	30 g	54	**1**	0
Kung Pao Sauce	4 tsp	35	**15**	0
Louis	1 tbsp	63	**60**	14
	½ cup	504	**480**	112
Nacho Cheese Sauce (Kaukauna)	2 tbsp	90	**60**	18
Nacho Topping (Tostitos)				
Beef Fiesta	¼ cup	120	**70**	27
Chicken Quesadilla	¼ cup	90	**50**	18
Pasta & Spaghetti Sauces				
Contadina				
Alfredo	¼ cup	180	**140**	90
Light	¼ cup	80	**45**	27

SAUCES, GRAVIES, AND DIPS

FOOD	AMOUNT	CALORIES TOTAL	CALORIES FAT	CALORIES SAT-FAT
Pasta & Spaghetti Sauces (cont.)				
Garden Vegetable	½ cup	40	**0**	0
Marinara	½ cup	80	**35**	9
Mushroom Alfredo	¼ cup	100	**60**	45
Pesto with Basil	¼ cup	290	**220**	63
reduced fat	¼ cup	230	**170**	36
Roasted Garlic Marinara	½ cup	60	**15**	5
DiGiorno				
Alfredo	¼ cup	180	**160**	63
Marinara	¼ cup	180	**160**	63
Healthy Choice				
all flavors	½ cup	50	**0**	0
Prego				
Diced Onion & Garlic	½ cup	110	**45**	5
Flavored with meat	½ cup	140	**50**	14
Fresh Mushrooms	½ cup	150	**45**	14
Garden combination	½ cup	90	**15**	5
Mushroom Parmesan	½ cup	120	**30**	9
Mushroom Supreme	½ cup	130	**40**	5
Roasted Garlic & Herb	½ cup	110	**30**	5
Three cheese	½ cup	100	**20**	9
Tomato & Basil	½ cup	110	**30**	5
Tomato Parmesan	½ cup	120	**25**	9
Tomato, onion, & garlic	½ cup	110	**30**	9
Traditional	½ cup	140	**40**	14
Ragú				
Cheese Creations				
Double Cheese	¼ cup	110	**90**	36
Roasted Garlic Parmesan	¼ cup	120	**100**	36
Spicy Cheddar & Tomato	¼ cup	50	**20**	14
Chunky Garden Style				
Mushroom & Green Pepper	½ cup	110	**30**	5
Roasted Red Pepper & Onion	½ cup	110	**30**	5
Super Chunky Mushroom	½ cup	120	**30**	5
Super Garlic	½ cup	100	**20**	0
Super Vegetable Primavera	½ cup	110	**30**	5

SAUCES, GRAVIES, AND DIPS

FOOD	AMOUNT	CALORIES TOTAL	CALORIES FAT	CALORIES SAT-FAT
Pasta & Spaghetti Sauces (cont.)				
Tomato, Basil & Italian Cheese	½ cup	110	**25**	9
Tomato, Garlic & Onion	½ cup	120	**30**	5
Robust Blend, Hearty				
Parmesan & Romano	½ cup	120	**30**	9
Red Wine & Herbs	½ cup	100	**25**	0
Sauteed Onion & Garlic	½ cup	120	**35**	5
Spicy Red Pepper	½ cup	110	**15**	0
Old World Style				
flavored with meat	½ cup	80	**30**	9
Mushroom	½ cup	80	**25**	5
Traditional	½ cup	80	**25**	5
Light				
Chunky Mushroom & Garlic	½ cup	70	**0**	0
Tomato & Basil	½ cup	50	**0**	0
Safeway Select Verdi				
Alfredo	½ cup	120	**100**	63
Light	½ cup	160	**100**	63
Creamy Sundried Tomato Pesto	½ cup	210	**150**	90
Pizza Sauce				
Boboli	½ pouch	40	**0**	0
Plum Sauce	2 Tbsp	90	**0**	0
Soy	1 tbsp	11	**0**	0
Double Black Soy	1 tbsp	15	**0**	0
Stir Fry Sauce	4 tsp	40	**0**	0
Tartar				
Fat-free (Kraft)	2 tbsp	25	**0**	0
Regular (Hellman's)	1 tbsp	70	**70**	9
Thai Peanut Stir-Fry & Dipping Sauce	2 tbsp	70	**25**	5
Tomato (*see under* Pasta & Spaghetti Sauces above)				
White	1 tbsp	24	**17**	11
	½ cup	195	**138**	88
Worcestershire	1 tbsp	0	**0**	0

SAUCES, GRAVIES, AND DIPS

FOOD	AMOUNT	CALORIES TOTAL	CALORIES FAT	CALORIES SAT-FAT
GRAVIES				
Gravies by type				
Beef, canned	½ cup	62	**25**	13
Gravies by brand name				
Franco-American				
Beef	¼ cup	30	**15**	5
Chicken	¼ cup	40	**25**	9
Fat free				
all flavors	¼ cup	15	**0**	0
Mushroom	¼ cup	20	**10**	0
Turkey	¼ cup	25	**10**	0
Heinz				
Classic chicken	¼ cup	25	**10**	0
Fat-free, all	¼ cup	15	**0**	0
Rich mushroom	¼ cup	20	**5**	0
Roasted turkey	¼ cup	30	**15**	0
Savory beef	¼ cup	25	**10**	0
Zesty onion	¼ cup	25	**10**	0
DIPS				
Bacon Horseradish				
Heluva Good	2 tbsp	60	**45**	27
Bean Dip (Frito-Lay)	2 tbsp	40	**10**	0
Cheddar Cheese, mild (Utz)	2 tbsp	45	**25**	14
Jalapeño (Frito-Lay)	2 tbsp	50	**30**	9
Jalapeño & Cheddar (Utz)	2 tbsp	30	**25**	9
Chesapeake Clam Dip				
Breakstone's	2 tbsp	50	**40**	27
Chili Cheese Flavor (Fritos)	2 tbsp	45	**30**	9
French Onion				
Frito-Lay's	2 tbsp	60	**45**	27
Heluva Good	2 tbsp	60	**45**	27
Fat Free	2 tbsp	25	**0**	0
French Onion Dip				
Kraft	2 tbsp	45	**35**	23
Lucerne	2 tbsp	70	**60**	27
Green Onion (Lucerne)	2 tbsp	50	**45**	27

SAUCES, GRAVIES, AND DIPS

FOOD	AMOUNT	CALORIES TOTAL	CALORIES FAT	CALORIES SAT-FAT
Guacamole				
Calavo (all flavors)	2 tbsp	60	**45**	9
Lucerne	2 tbsp	90	**80**	23
Hummus	2 tbsp	57	**30**	27
Jalapeño				
Frito-Lay	2 tbsp	50	**30**	9
Jalapeño & Cheddar				
Utz	2 tbsp	30	**25**	9
Jalapeño Cheese Sauce				
Pablo's	2 tbsp	150	**90**	32
Nacho Cheese Dip (Snyder's of Hanover)	2 tbsp	30	**25**	9
New England Clam				
Heluva Good	2 tbsp	50	**40**	27
Ranch				
Heluva Good	2 tbsp	60	**45**	27
Hidden Valley	28 g prepared	70	**50**	36
Lucerne	2 tbsp	110	**100**	18
Salsa				
Chunky (Herr's)	2 tbsp	12	**0**	0
Chunky (Utz)	2 tbsp	60	**0**	0
Dip (Pace)	2 tbsp	10	**0**	0
Dip (Tostitos)	2 tbsp	15	**0**	0
Mexican (Kaukauna)	2 tbsp	15	**0**	0
Mild (Rojo's)	2 tbsp	10	**0**	0
Salsa and Cream Cheese (Kaukauna)	2 tbsp	70	**50**	NA
Salsa Con Queso (Kaukauna)	2 tbsp	70	**40**	27
Salsa con Queso (Tostitos)	2 tbsp	40	**20**	5
Southwest, mild (Safeway)	2 tbsp	10	**0**	0
Sour Cream & Onion Dip				
Herr's	2 tbsp	60	**45**	27
Utz	2 tbsp	60	**45**	27
Vegetable Dip				
Heluva Good	2 tbsp	60	**45**	27

SAUSAGES AND LUNCHEON MEATS

FOOD	AMOUNT	CALORIES TOTAL	FAT	SAT-FAT
Bacon	2 slices (11 g)	60	**45**	18
Hickory Smoked (Smithfield)	2 slices (15 g)	90	**70**	27
Thick Sliced (Gwaltney)	1 slice (8 g)	45	**35**	9
Turkey (Louis Rich)	1 slice (14 g)	30	**20**	5
Barbecue loaf, pork, beef	28 g	49	**23**	8
	1 slice (22 g)	40	**18**	7
Beer 'n Bratwurst (Johnsonville)	1 grilled link (85 g)	290	**230**	81
Beerwurst, beer salami				
beef	28 g	92	**75**	31
	1 slice (22 g)	75	**61**	25
pork	28 g	67	**48**	16
	1 slice (22 g)	55	**39**	13
Berliner, pork, beef	28 g	65	**44**	15
	1 slice (22 g)	53	**36**	13
Bockwurst, raw	28 g	87	**70**	26
	1 link (64 g)	200	**161**	59
Bologna				
Beef (Hebrew National)	56 g	180	**150**	54
Beef (Oscar Mayer)	1 slice (28 g)	90	**70**	36
Chicken (Gwaltney)	1 slice (32 g)	80	**50**	18
Chicken, Pork, & Beef (Thorn Apple)	1 slice (37 g)	120	**90**	18
Pork	1 slice (28 g)	70	**51**	18
Pork & Beef (Oscar Mayer)	1 slice (28 g)	90	**70**	27
Pork, Chicken, & Beef (Oscar Mayer)	1 slice (28 g)	90	**70**	27
Light	1 slice (28 g)	60	**35**	14
Pork & Turkey (Gwaltney)	1 slice (38 g)	120	**100**	36
Turkey (Louis Rich)	1 slice (28 g)	50	**35**	9
Bratwurst, pork, beef	28 g	92	**71**	25
	1 link (70 g)	226	**175**	63
Bratwurst, pork, cooked	28 g	85	**66**	24
	1 link (84 g)	256	**198**	71
Braunschweiger, pork				
Jones	56 g	150	**110**	36
Jones Sandwich slices	1 slice (34 g)	110	**90**	27
Kahn's	56 g	180	**140**	81
Oscar Mayer	1 slice (56 g)	190	**150**	54

SAUSAGES AND LUNCHEON MEATS

FOOD	AMOUNT	CALORIES TOTAL	CALORIES FAT	CALORIES SAT-FAT
Breakfast strips, beef, cured cooked	1 slice (11 g)	51	**35**	15
Canadian bacon, grilled	1 slice	43	**18**	6
Cheese Dog (Oscar Mayer)	1 frank (45 g)	140	**120**	45
Chicken breast				
Deli Thin				
Fat Free (Oscar Mayer)	4 slices (52 g)	40	**0**	0
Oven roasted (Louis Rich)	5 slices (55 g)	60	**15**	5
Chicken roll, white meat (Tysons)	3 slices (55 g)	90	**50**	18
Chipped Beef	28 g	50	**20**	9
Corned Beef				
Hormel	56 g	130	**60**	27
thin sliced (Hebrew National)	4 slices (56 g)	90	**40**	18
Corned beef brisket				
Cooked	28 g	71	**48**	16
Loaf, jellied	1 slice (28 g)	43	**16**	7
Thorn Apple Valley	84 g	190	**150**	63
Corned Beef Hash				
Libby's	1 cup (252 g)	420	**220**	99
Cured Beef				
Oven Roasted (Hillshire Farms)	6 slices (57 g)	50	**5**	0
Dried beef, cured (beef jerky)	28 g	47	**10**	4
Dutch brand loaf, pork, beef	1 slice (28 g)	68	**45**	16
Frankfurter				
Beef	1 frank (57g)	190	**150**	63
Beef Franks (Hebrew National)	1 frank (48 g)	150	**120**	45
Big 8's Jumbo Beef Hot Dogs (Gwaltney)	1 frank (56 g)	190	**150**	63
Esskay Beef Franks	1 frank (56 g)	170	**130**	36
Oscar Mayer Beef Franks	1 frank (45 g)	150	**120**	54
Quarter pound (Hebrew National)	1 frank (114 g)	350	**300**	108

SAUSAGES AND LUNCHEON MEATS

FOOD	AMOUNT	CALORIES TOTAL	FAT	SAT-FAT
Frankfurter (cont.)				
Safeway Jumbo Beef Franks	1 frank (57 g)	170	**140**	63
Smithfield Jumbo Beef Hot Dogs	1 frank (56 g)	190	**150**	63
Chicken				
Gwaltney Great Dogs	1 frank (56 g)	140	**90**	27
Wampler-LongAcre	1 frank (56 g)	120	**100**	27
Chicken, Pork & Beef				
Safeway Jumbo Franks	1 frank (57 g)	180	**150**	54
Corn Dogs (Ball Park)	1 corn dog (75 g)	220	**110**	27
Turkey				
Louis Rich	1 frank (57 g)	110	**70**	23
Safeway Jumbo Turkey Franks	1 frank (57 g)	120	**80**	23
Turkey & Chicken				
Louis Rich Bun-Length	1 frank (57 g)	110	**70**	23
Turkey & Pork Wiener (Oscar Mayer)	1 frank (45 g)	150	**120**	41
Jumbo Wiener	1 frank (57 g)	180	**150**	54
Low-fat and Fat-free Franks				
Ball Park				
Fat Free Beef Franks	1 frank (50 g)	45	**0**	0
Lite Franks	1 frank (50 g)	100	**70**	18
Healthy Choice				
Beef Franks	1 frank (50 g)	60	**15**	5
Turkey, Pork, Beef Franks	1 frank (40 g)	50	**10**	5
Oscar Mayer				
Fat Free Hot Dogs	1 frank (50 g)	40	**0**	0
Light Beef Franks	1 frank (57 g)	110	**70**	31
Ham				
Baked (Oscar Mayer)	3 slices (63 g)	60	**10**	5
Boiled (Oscar Mayer)	3 slices (63 g)	70	**15**	5
Chopped	1 slice (28 g)	50	**30**	14
	1 slice (21 g)	50	**36**	12
Cured (Hormel)	85 g	100	**45**	14
Danish (Plumrose)	2 slices (56 g)	65	**25**	9

SAUSAGES AND LUNCHEON MEATS

FOOD	AMOUNT	CALORIES TOTAL	CALORIES FAT	CALORIES SAT-FAT
Ham (cont.)				
Deli Thins				
Baked Ham	4 slices (52 g)	50	**10**	0
Honey Ham	4 slices (52 g)	50	**15**	5
Minced	28 g	75	**53**	18
	1 slice (21 g)	55	**39**	14
Salad spread	28 g	61	**40**	13
	1 tbsp	32	**21**	7
Smoked				
Esskay	85 g	120	**50**	18
Hickory Smoked	3 slices (89 g)	160	**100**	36
Oscar Mayer	3 slices (63 g)	60	**20**	9
Smok-a-Roma	57 g	120	**70**	27
Turkey Ham (*see* Turkey, below)				
Ham and cheese loaf or roll				
Oscar Mayer	1 slice (28 g)	60	**40**	23
Ham and cheese spread	28 g	69	**47**	22
	1 tbsp	37	**25**	12
Headcheese, pork	1 slice (28 g)	60	**40**	13
Honey loaf, pork, beef	1 slice (28 g)	36	**11**	4
Honey roll sausage, beef	28 g	52	**27**	10
	1 slice (22 g)	42	**22**	8
Kielbasa				
pork, beef (Eckrich)	56 g	180	**140**	63
	1 slice (25 g)	81	**64**	23
beef polska (Hillshire Farm)	2 oz (56 g)	190	**150**	72
Healthy Choice	2 oz (56 g)	70	**15**	5
turkey (Mr. Turkey & Hillshire Farm)	56 g	90	**45**	23
Knockwurst, beef (Hebrew National)	1 link (85 g)	260	**210**	81
Lebanon bologna, beef	28 g	64	**38**	16
	1 slice (22 g)	52	**31**	13
Liver cheese, pork	28 g	86	**65**	23
Liver pudding, pork	45 g	170	**110**	9
Liverwurst (*see* Braunschweiger)				

SAUSAGES AND LUNCHEON MEATS

FOOD	AMOUNT	CALORIES TOTAL	CALORIES FAT	CALORIES SAT-FAT
Luncheon meat				
beef, loaved	1 slice (28 g)	87	**67**	29
beef, thin sliced	28 g	35	**8**	3
	5 slices (21 g)	26	**6**	2
pork, beef	1 slice (28 g)	100	**82**	30
pork, canned	28 g	95	**77**	28
	1 slice (21 g)	70	**57**	20
Luxury loaf, pork	1 slice (28 g)	40	**12**	4
Mortadella, beef, pork	28 g	88	**65**	24
	1 slice (14 g)	47	**34**	13
Mother's loaf, pork	28 g	80	**57**	20
	1 slice (21 g)	59	**42**	15
Safeway	1 slice (35 g)	100	**70**	23
Olive loaf	1 slice (28 g)	70	**45**	15
Pastrami				
beef	1 slice (28 g)	99	**74**	27
turkey	1 slice (28 g)	40	**16**	9
Paté				
Chicken liver	1 tbsp	26	**15**	4
Goose liver	28 g	131	**112**	NA
	1 tbsp	60	**51**	NA
Pork	28 g	100	**80**	27
Peppered Beef (Carl Buddig)	71 g	100	**45**	18
Peppered loaf, pork, beef	1 slice (28 g)	42	**16**	6
Pepperoni				
Hormel	15 slices (28 g)	140	**120**	54
	1 sausage (252 g)	1248	**993**	364
Bridgford	1 oz (28 g)	130	**110**	36
Pickle and pimento loaf (Oscar Mayer)	1 slice (28 g)	70	**50**	18
Picnic loaf, pork, beef	1 slice (28 g)	66	**42**	15
Pork Cracklins (fried pork fat with skin)	½ oz (14 g)	80	**50**	9
Potted Meat Food Product	¼ cup (58 g)	110	**80**	27
Prosciutto				
Citterio	2 slices (30 g)	70	**40**	14

SAUSAGES AND LUNCHEON MEATS

FOOD	AMOUNT	CALORIES TOTAL	CALORIES FAT	CALORIES SAT-FAT
Salami				
Beef (Hebrew National)	56 g	170	**130**	54
Cotto (Oscar Mayer)	1 slice (28 g)	60	**40**	18
Hard, pork & beef (Oscar Mayer)	3 slices (27 g)	100	**70**	27
Genoa salami	3 slices (30 g)	100	**70**	27
Sandwich spread, pork, beef	28 g	67	**44**	15
	1 tbsp	35	**23**	8
Sausage				
Beef sausage (Jones Dairy Farm)	2 links (45 g)	170	**140**	NA
Beerwurst Sausage	2 oz (56 g)	160	**130**	45
Biscuits (Jimmy Dean)	2 (96 g)	330	**190**	63
Blood sausage	28 g	107	**88**	34
	1 slice (25 g)	95	**78**	30
Cajun Brand Andouille Sausage (Aidells Sausage Company)	1 link (90 g)	200	**140**	63
Farmer Summer Sausage	2 oz (56 g)	200	**160**	63
Ham sausage (Smithfield)	48 g	180	**140**	54
Italian sausage, cooked, pork	28 g	92	**66**	23
	1 link (5/lb)	216	**155**	55
	1 link (4/lb)	268	**192**	68
Johnsonville	1 grilled link (85 g)	290	**230**	81
Usinger's	1 (84 g)	270	**230**	90
Italian Turkey Sausage (Shady Brook Farms)	1 link (64 g)	100	**45**	14
Liver sausage, liverwurst, pork	28 g	93	**73**	27
	1 slice (17 g)	59	**46**	17
Luncheon sausage, pork and beef	28 g	74	**53**	19
	1 slice (22 g)	60	**43**	16
New England brand sausage, pork, beef	28 g	46	**19**	6
	1 slice (22 g)	37	**16**	5

SAUSAGES AND LUNCHEON MEATS

FOOD	AMOUNT	CALORIES TOTAL	CALORIES FAT	CALORIES SAT-FAT
Sausage (cont.)				
New Mexico Brand Smoked Turkey & Chicken Sausage (Aidells Sausage Company)	1 link (90 g)	190	**130**	41
Polish sausage, beef (Hebrew National)	1 link (85 g)	240	**190**	90
Polish sausage, pork	28 g	92	**73**	26
	1 sausage (8 oz)	739	**587**	211
Pork Sausage				
Bob Evans	2 patties pan-fried (53 g)	230	**180**	63
Country	28 g	120	**100**	36
Hot				
Gwaltney	39 g	150	**130**	45
Jamestown	36 g	170	**140**	45
Links (Parks)	2 links (42 g)	170	**150**	54
Smoked	1 link (85 g)	290	**230**	81
Pork and beef sausage, cooked	1 patty (28 g)	112	**92**	33
	1 link (13 g)	52	**33**	15
Smoked link sausage				
Hot Links	1 (76 g)	250	**200**	99
Lite (Hillshire Farms)	2 oz (56 g)	110	**70**	32
pork	28 g	110	**81**	29
	1 link (67 g)	265	**194**	69
	1 link (16 g)	62	**46**	16
pork and beef	28 g	95	**77**	27
	1 link (67 g)	229	**186**	65
	1 link (16 g)	54	**44**	15
Summer sausage, beef	3 slices (57 g)	180	**140**	63
Vienna sausage (Hormel)	28 g	70	**60**	36
Scrapple				
Parks	2 oz (56 g)	90	**45**	14
Rapa	2 oz (56 g)	120	**70**	27
Spam	56 g	170	**140**	54
Lite	56 g	110	**70**	27

SAUSAGES AND LUNCHEON MEATS

FOOD	AMOUNT	CALORIES TOTAL	CALORIES FAT	CALORIES SAT-FAT
Tongue, beef				
raw	28 g	63	**41**	18
cook, simmered	28 g	80	**53**	23
Tripe, beef, raw	28 g	28	**10**	5
Turkey breast, processed				
Oven Roasted (Oscar Mayer)	3 slices (63 g)	70	**15**	0
Fat Free Deli Thin (Oscar Mayer)	4 slices (52 g)	40	**0**	0
Smoked white (Louis Rich)	1 slice (28 g)	30	**10**	0
Turkey Cold Cuts				
bacon (Louis Rich)	1 slice (14 g)	30	**20**	5
bologna (Louis Rich)	1 slice (28 g)	50	**35**	9
ham				
Chopped	1 slice (28 g)	40	**20**	9
Jennie-O	56 g	80	**40**	7
Louis Rich	1 slice (28 g)	35	**10**	0
pastrami	1 slice (28 g)	40	**16**	9
salami (Louis Rich)	1 slice (28 g)	45	**25**	9
Cotted Salami (Louis Rich)	1 slice (28 g)	90	**50**	18
Turkey Roll				
light and dark meat	1 slice (28 g)	42	**18**	5
light meat	1 slice (28 g)	42	**18**	5

SNACK FOODS

FOOD	AMOUNT	CALORIES TOTAL	CALORIES FAT	CALORIES SAT-FAT
BREADSTICKS				
Grissini-style, garlic (Stella D'oro)	3 sticks (15 g)	60	**0**	0
Thin Bread Sticks (Pepperidge Farm)				
Cheddar cheese	7 sticks (16 g)	70	**25**	9
CHIPS, CRISPS, ETC.				
Bagel Chips				
Burns & Ricker				
Cinnamon raisin	7 pieces (30 g)	130	**30**	5
fat-free	7 pieces (30 g)	110	**0**	0

SNACK FOODS

FOOD	AMOUNT	CALORIES TOTAL	CALORIES FAT	CALORIES SAT-FAT
Bagel Chips (cont.)				
Garlic				
fat-free	7 pieces (30 g)	100	**0**	0
Roasted garlic				
bite size	⅔ cup (30 g)	140	**40**	5
New York Style				
bite size				
Garlic	28 chips (31 g)	140	**30**	5
Ranch	⅔ cup (28 g)	120	**35**	5
Rondele				
Cinnamon with honey & raisins	7 pieces (28 g)	130	**30**	5
Roasted garlic	7 pieces (28 g)	130	**30**	5
fat-free	9 pieces (30 g)	100	**0**	0
Banana Chips	⅓ cup (27 g)	140	**60**	60
Bugles				
Original Baked	1½ cups (30 g)	130	**30**	5
Original	1⅓ cups (30 g)	160	**80**	72
Caramel Corn Clusters (Utz)	1⅛ cups (28 g)	142	**18**	0
Cheese Balls (Utz)	27 balls (28 g)	170	**90**	18
Cheese Curls				
Utz	30 curls (28 g)	150	**90**	18
Weight Watchers	1 pkg (14 g)	70	**25**	9
Cheese Doodles (Wise)				
Crunchy	½ cup (28 g)	150	**80**	23
Puffed	19 pieces (28 g)	150	**70**	23
Cheetos	12 pieces (28 g)	160	**90**	23
Corn Nuts	⅓ cup (28 g)	130	**35**	9
French Fried Onions (French's)	2 tbsp (7 g)	45	**30**	9
Fritos	32 chips (28 g)	160	**90**	14
Bar-B-Q	29 chips (28 g)	150	**80**	14
Scoops	11 chips (28 g)	160	**90**	9
Jax (Bachman)	25 pieces (30 g)	150	**70**	9
Onion Rings (Wise)	39 rings (28 g)	140	**50**	14
Pita Chips (Pechter's)				
Original	2 chips (14 g)	80	**45**	9
Sesame Seed	2 chips (14 g)	80	**45**	9
Toasted Garlic	2 chips (14 g)	70	**40**	9

SNACK FOODS

FOOD	AMOUNT	CALORIES TOTAL	CALORIES FAT	CALORIES SAT-FAT
Plantain Chips (Goya)	38 pieces (30 g)	170	**100**	18
Potato Chips				
Lay's				
Baked	32 g	130	**15**	0
Bar-B-Q	15 chips (28 g)	150	**90**	27
Classic	20 chips (28 g)	150	**90**	27
Sour Cream & Onion	17 chips (28 g)	160	**100**	27
Wavy Lay's				
Original	11 chips (28 g)	150	**90**	23
Wow	20 chips (28 g)	75	**0**	0
Pringles				
Original	14 crisps (28 g)	160	**90**	27
Ridges	12 crisps (28 g)	150	**90**	23
Right Crisps	16 crisps (28 g)	140	**60**	18
Sour Cream & Onion	14 crisps (28 g)	160	**90**	23
fat-free	16 crisps (28 g)	70	**0**	0
Ruffles (Frito-Lay)				
Cheddar & sour cream	11 chips (28 g)	160	**90**	27
Ranch	13 chips (28 g)	150	**80**	NA
The Works!	14 chips (28 g)	160	**100**	23
Wow	17 chips (28 g)	75	**0**	0
Utz	20 chips (28 g)	150	**80**	18
Baked Potato Crisps	12 chips (28 g)	110	**15**	0
Bar-B-Q Ripple Cut	20 chips (28 g)	150	**90**	23
reduced fat	22 chips (28 g)	140	**60**	14
Grandma Utz's Handcooked	20 chips (28 g)	140	**70**	14
Kettle Classics	20 chips (28 g)	150	**80**	14
Ripple Cut	20 chips (28 g)	150	**90**	23
reduced fat	24 chips (28 g)	140	**60**	14
Sour Cream & Onion, Ripple Cut	20 chips (28 g)	160	**90**	27
Potato Sticks (French's)	¾ cup (30 g)	180	**110**	18
Terra Chips				
mixed vegetables	28 g	140	**70**	9
Tortilla Chips				
Doritos	11 chips (28 g)	140	**70**	9
3Ds (all flavors)	27 pieces (28 g)	140	**60**	14
Doritos Wow	11 chips (28 g)	90	**10**	0
Guiltless Gourmet	18 chips (28 g)	110	**15**	0

SNACK FOODS

FOOD	AMOUNT	CALORIES TOTAL	CALORIES FAT	CALORIES SAT-FAT
Tortilla Chips (cont.)				
Utz White Corn Tortillas	12 chips (28 g)	140	**60**	9
Baked Tortilla	8 chips (28 g)	120	**15**	0
Nacho Tortilla	12 chips (28 g)	140	**60**	9
Tostitos	6 chips (28 g)	130	**50**	9
Baked Tostitos	9 chips (28 g)	110	**5**	0
Baked, Salsa & Cream Cheese flavor, bite size	16 chips (28 g)	120	**25**	5
Veggie Rings (Good Health)				
Potato Onion	28 g	140	**60**	5
Veggie Stix (Good Health)				
Mixed Vegetables	28 g	140	**60**	5
CRACKERS				
Austin				
Cheese Crackers on Cheese	6 (39 g)	200	**100**	24
Cheese Crackers & Creamy Peanut Butter	6 (39 g)	200	**100**	18
Cheese Peanut Butter Cracker Sandwiches	4 (26 g)	140	**70**	14
Cream Cheese & Chives	6 (39 g)	190	**90**	23
Toasty Crackers & Peanut Butter	6 (39 g)	190	**90**	18
Toasty Peanut Butter Cracker Sandwiches	4 (26 g)	140	**60**	14
Wheat 'n Cheddar	6 (39 g)	200	**100**	27
Delicious				
Snack Crackers	8 (30 g)	140	**60**	9
Cheddar Cheese	28 (30 g)	150	**60**	18
Garden Vegetable	13 (31 g)	150	**60**	14
Devonsheer				
Melba Rounds				
Garlic	5 pieces (15 g)	60	**10**	0
Onion	5 pieces (15 g)	50	**0**	0
Plain	5 pieces (15 g)	50	**0**	0
Sesame	3 pieces (14 g)	50	**10**	0
12-Grain	5 pieces (15 g)	50	**0**	0
Vegetable	5 pieces (15 g)	50	**0**	0

SNACK FOODS

FOOD	AMOUNT	CALORIES TOTAL	FAT	SAT-FAT
Devonsheer (cont.)				
Melba Toast				
Plain	3 pieces (14 g)	50	**0**	0
Sesame	3 pieces (14 g)	50	**10**	0
Herr's				
Sandwiches				
Cheese in Cheese	6 (39 g)	200	**100**	23
Cheese Peanut Butter	6 (39 g)	200	**100**	18
Toast Peanut Butter	1 pkg (39 g)	190	**90**	18
Keebler				
Club				
Original	4 (14 g)	70	**25**	9
reduced fat	5 (16 g)	70	**20**	0
Sandwich Crackers				
Cheese & Peanut Butter	1 package (38 g)	190	**80**	18
Club & Cheddar	1 package (36 g)	190	**100**	23
Toast & Peanut Butter	1 package (38 g)	190	**80**	18
Munch 'ems				
Cheddar	41 (30 g)	130	**40**	9
Mesquite BBQ	41 (30 g)	140	**40**	9
Original	41 (30 g)	130	**40**	9
Ranch	41 (30 g)	130	**40**	9
Sour Cream & Onion	41 (30 g)	140	**40**	9
Toasteds				
Buttercrisp	5 (16 g)	80	**30**	9
Onion	5 (16 g)	80	**30**	5
Rye	9 (29 g)	140	**60**	NA
Sesame	5 (16 g)	80	**30**	5
Wheat	5 (16 g)	80	**25**	5
Town House				
Original	5 (16 g)	80	**40**	9
reduced fat	6 (15 g)	70	**20**	5
Wheatables				
Ranch	29 (30 g)	130	**35**	9
Savory Original	26 (30 g)	150	**60**	18
reduced fat	29 (30 g)	130	**30**	9
White Cheddar	27 (30 g)	130	**35**	9

SNACK FOODS

FOOD	AMOUNT	CALORIES TOTAL	CALORIES FAT	CALORIES SAT-FAT
Keebler (cont.)				
Zesta Saltines				
Fat-free	5 (14 g)	50	**0**	0
Original	5 (15 g)	60	**20**	5
Kraft Handi-Snacks				
Cheez 'n Breadsticks	1 unit (31 g)	120	**60**	27
Cheez 'n Crackers	1 unit (27 g)	110	**60**	27
Cheez 'n Pretzels	1 unit (29 g)	100	**45**	27
Nacho Stix 'n Cheez	1 unit (31 g)	110	**60**	27
Lu				
Le Petite Beurre Butter Biscuits	4 crackers (33 g)	150	**35**	27
Manischewitz				
Matzo				
American	1 matzo (28 g)	110	**15**	5
Egg 'n Onion	1 matzo (28 g)	100	**10**	0
Whole Wheat	1 matzo (28 g)	110	**5**	0
Milk Lunch				
New England Biscuits	4 crackers (32 g)	140	**35**	9
Safeway Snack Crackers				
Bacon-Flavored	11 crackers (30 g)	150	**60**	14
Cheddar Cheese	28 crackers (30 g)	150	**60**	18
Chicken-Flavored	10 crackers (31 g)	160	**80**	18
Garden Vegetable	13 crackers (31 g)	150	**60**	14
Onion	11 crackers (30 g)	140	**50**	14
Ranch	10 crackers (30 g)	160	**80**	14
Sesame Cheddar	13 crackers (31 g)	150	**70**	14
Sesame Wheat Snack	13 crackers (31 g)	150	**60**	NA
Snack Crackers	10 crackers (30 g)	140	**45**	9
Sour Cream 'n Chives	9 crackers (31 g)	160	**80**	18
Wheat	15 crackers (30 g)	140	**50**	14
reduced fat	16 crackers (30 g)	120	**30**	5
Woven Wheats	7 crackers (31 g)	140	**45**	9
Nabisco				
Air Crisps				
Cheese Nips	32 crackers (30 g)	130	**35**	9
Potato Barbecue	22 crisps (28 g)	120	**35**	5
Pretzel	23 crisps (28 g)	110	**0**	0
Ritz	24 crackers (30 g)	140	**45**	9

SNACK FOODS

FOOD	AMOUNT	CALORIES TOTAL	CALORIES FAT	CALORIES SAT-FAT
Nabisco (cont.)				
Ritz Sour Cream & Onion	23 crackers (30 g)	140	**40**	9
Wheat Thins Ranch	23 pieces (30 g)	140	**40**	9
Better Cheddars	22 crackers (30 g)	150	**70**	2
reduced fat	24 crackers (30 g)	140	**50**	14
Cheese Nips	29 crackers (30 g)	150	**60**	14
reduced fat	31 crackers (30 g)	130	**35**	9
Chicken in a Biskit	12 crackers (30 g)	160	**80**	14
Harvest Crisps				
5-grain	13 crackers (31 g)	130	**30**	5
Garden Vegetable	15 crackers (30 g)	130	**30**	5
Italian Herb	13 crackers (31 g)	130	**30**	5
Premium Saltines				
Fat-Free	5 crackers (15 g)	60	**0**	0
Low Sodium	5 crackers (14 g)	60	**15**	0
Original	5 crackers (14 g)	60	**15**	0
Unsalted Tops	5 crackers (14 g)	60	**15**	0
With Multigrain	5 crackers (14 g)	60	**15**	0
Rice Crackers	½ cup (30 g)	110	**0**	0
Ritz Bits				
Cheese Sandwiches	14 sandwiches (31 g)	170	**90**	23
Peanut Butter Sandwiches	14 sandwiches (31 g)	150	**70**	14
Ritz Crackers	5 crackers (16 g)	80	**35**	5
reduced fat	5 crackers (15 g)	70	**15**	0
Whole Wheat	5 crackers (15 g)	70	**20**	0
SnackWell's Snack Crackers				
French Onion	38 crackers (30 g)	130	**25**	5
Ranch	38 crackers (30 g)	130	**25**	5
Salsa Cheddar	32 crackers (30 g)	120	**15**	0
Wheat Crackers	5 crackers (15 g)	70	**15**	0
Sociables	7 crackers (15 g)	80	**35**	5
Sweet Crispers				
Caramel	18 crisps (31 g)	140	**25**	5
Chocolate	18 crisps (31 g)	130	**25**	5
Cinnamon	18 crisps (31 g)	130	**25**	0
Honey	18 crisps (31 g)	130	**20**	0
Triscuit Wafers	7 wafers (31 g)	140	**45**	9
Deli-style Rye	7 wafers (32 g)	140	**45**	9

SNACK FOODS

FOOD	AMOUNT	CALORIES TOTAL	FAT	SAT-FAT
Nabisco (cont.)				
Garden Herb	6 wafers (28 g)	130	**40**	9
Reduced Fat	8 wafers (32 g)	130	**25**	5
Thin Crisps	15 crackers (30 g)	130	**45**	9
Uneeda Biscuit	2 crackers (15 g)	60	**15**	0
Vegetable Thins	14 crackers (31 g)	160	**80**	14
Waverly Crackers	5 crackers (15 g)	70	**30**	5
Wheat Thins				
Multigrain	17 crackers (30 g)	130	**35**	5
Original	16 crackers (29 g)	140	**50**	9
Reduced Fat	18 crackers (29 g)	120	**35**	5
Wheatsworth Stone-Ground Wheat crackers	5 crackers (16 g)	80	**30**	5
Pepperidge Farm				
Distinctive				
Butter Thins	4 crackers (15 g)	70	**25**	9
Hearty Wheat	3 crackers (16 g)	80	**30**	0
Quartet	4 crackers (15 g)	70	**20**	5
Sesame	3 crackers (14 g)	70	**25**	0
Three-Cracker	4 crackers (15 g)	70	**20**	0
Goldfish				
Cheddar Cheese	55 pieces (30 g)	140	**50**	14
Extra Cheddar Cheese	51 pieces (30 g)	140	**50**	14
Nacho	51 pieces (30 g)	140	**60**	9
Original	55 pieces (30 g)	140	**60**	18
Parmesan Cheese	60 pieces (30 g)	140	**50**	14
Pizza-Flavored	55 pieces (30 g)	140	**60**	14
Pretzel	43 pieces (30 g)	120	**25**	5
Sour Cream & Onion	51 pieces (30 g)	150	**60**	9
Snack Sticks				
Pumpernickel	15 sticks (31 g)	120	**15**	0
Sesame	8 sticks (29 g)	130	**50**	5
Three-Cheese	8 sticks (29 g)	140	**50**	18
Ry Krisp				
Natural	2 crackers (15 g)	60	**0**	0
Seasoned	2 crackers (14 g)	60	**10**	0
SnackWell's Crackers: *see under* Nabisco				

SNACK FOODS

FOOD	AMOUNT	CALORIES TOTAL	CALORIES FAT	CALORIES SAT-FAT
Sunshine				
Cheez-It				
Big Cheez-It				
reduced fat	15 crackers (30 g)	140	**40**	9
Heads and Tails	37 crackers (30 g)	140	**50**	14
Hot & Spicy	27 crackers (30 g)	160	**80**	18
Nacho	28 crackers (30 g)	150	**60**	14
reduced fat	29 crackers (30 g)	140	**40**	9
Sandwich crackers				
Cheese	1 pkg (36 g)	200	**110**	23
Peanut Butter	1 pkg (38 g)	190	**90**	18
White Cheddar	27 crackers (30 g)	160	**80**	18
Hi-Ho, all flavors	4 crackers (14 g)	70	**35**	9
Krispy Saltines				
fat-free	5 crackers (14 g)	50	**0**	0
Mild Cheddar	5 crackers (15 g)	60	**20**	5
Original	5 crackers (14 g)	60	**10**	0
FROZEN SNACKS				
Anchor				
Stuffed Jalapeños Cream Cheese Poppers	6 pieces (136 g)	360	**200**	90
Farm Rich				
Double Cheese Pizza Dippers	3 sticks (90 g)	210	**80**	36
Italian Four Cheese Sticks	2 sticks (45 g)	150	**80**	27
Giorgio Pierogies (lowfat)				
Potato & Cheddar Cheese–Filled Pierogies	3 (132 g)	220	**25**	14
Potato & Onion–Filled Pierogies	3 (132 g)	230	**25**	9
Hanover				
Baked Soft Pretzels	1 pretzel (61 g)	160	**0**	0
Inland Valley Munchskin Meals				
Cheese & Bacon Potato Skins	2 topped potato skins (113 g)	250	**140**	63
J & J Snack Foods				
Cinnamon Raisin Minis	2 pretzels (57 g) + 1 icing pkt (7 g)	190	**15**	0
SuperPretzel	1 pretzel (64 g)	170	**9**	0

SNACK FOODS

FOOD	AMOUNT	CALORIES TOTAL	CALORIES FAT	CALORIES SAT-FAT
Mrs. T's Pierogies (lowfat)				
Potato & Cheddar Cheese–Filled Pasta Pockets	3 (120 g)	180	**20**	9
Potato & Onion–Filled Pasta Pockets	3 (120 g)	180	**15**	0
FRUIT SNACKS				
Fruit by the Foot all flavors	1 roll (21 g)	80	**10**	5
Fruit Roll-ups all flavors	2 rolls (28 g)	110	**10**	0
GRANOLA BARS, BREAKFAST BARS, AND POWER BARS				
Balance				
Almond Brownie	1 bar (50 g)	190	**50**	14
Chocolate	1 bar (50 g)	190	**50**	27
Honey Peanut	1 bar (50 g)	200	**50**	23
Boulder Bar				
Apple Cinnamon	1 bar (71 g)	190	**10**	0
Original Chocolate	1 bar (71 g)	190	**10**	0
Clif				
Chocolate Chip	1 bar (68 g)	250	**27**	5
Crunchy Peanut Bar	1 bar (68 g)	250	**36**	9
Real Berry	1 bar (68 g)	250	**18**	5
Kellogg's				
Nutrigrain Lowfat Cereal Bar				
Apple Cinnamon	1 bar (37 g)	140	**25**	5
Blueberry	1 bar (37 g)	140	**25**	5
Rice Crispies Treats				
Chocolate Chip	1 bar (22 g)	90	**25**	9
Crispy Marshmallow Squares	1 bar (22 g)	90	**20**	5
Kudos				
Chocolate Chip	1 bar (28 g)	120	**40**	23
with Snickers Chunks	1 bar (23 g)	100	**30**	9
with M&M's	1 bar (23 g)	90	**25**	9
Mountain Lift	1 bar (60 g)	220	**40**	36
Nature Valley Oats 'n Honey	1 bar (21 g)	90	**30**	5

SNACK FOODS

FOOD	AMOUNT	CALORIES TOTAL	CALORIES FAT	CALORIES SAT-FAT
Power Bar				
Banana	1 bar (65 g)	230	**20**	5
Chocolate	1 bar (65 g)	230	**20**	5
Malt Nut	1 bar (65 g)	230	**25**	5
Mocha	1 bar (65 g)	230	**25**	10
Oatmeal Raisin	1 bar (65 g)	230	**25**	5
Wild Berry	1 bar (65 g)	230	**25**	5
Quaker				
Cap'n Crunch's				
Crunch Berries Treats	1 bar (22 g)	90	**20**	5
Peanut Butter Crunch Treats	1 bar (22 g)	90	**25**	5
Chewy				
Chocolate Chip	1 bar (28 g)	120	**35**	14
Cookies 'n Cream	1 bar (28 g)	110	**25**	5
Peanut Butter & Chocolate Chunk	1 bar (28 g)	120	**25**	9
Fruit & Oatmeal, all flavors	1 bar (37 g)	140	**25**	0
Lowfat, all flavors	1 bar (28 g)	110	**20**	5
Safeway Healthy Advantage Cereal Bars				
all flavors	1 bar (37 g)	140	**25**	5
SnackWell's (Nabisco)				
Fat-Free Cereal Bars, all flavors	1 bar (37 g)	120	**0**	0
Granola Bars, all flavors	1 bar (28 g)	120	**25**	5
Sunfelt				
Almond	1 bar (28 g)	130	**60**	18
Chocolate Chip	1 bar (35 g)	160	**60**	27
Oatmeal Raisin or Raisin	1 bar (35 g)	150	**50**	18
Oats and Honey	1 bar (28 g)	120	**45**	18
JERKY AND PORK RINDS				
Beef Jerky				
Giant Jerk (Slim Jim)	1 (16 g)	70	**35**	18
Hickory Smoked (Pemmican)	1 bag (35 g)	90	**10**	5
Pork Rinds				
Utz Pork Cracklins	½ oz (14 g)	90	**70**	27

SNACK FOODS

FOOD	AMOUNT	CALORIES TOTAL	CALORIES FAT	CALORIES SAT-FAT
LUNCH PACKS				
Lunchables (Oscar Mayer)				
with dessert & drink				
96% Fat-Free Ham & Cheddar	1 pkg (164 g)	390	**100**	45
Beef Tacos	1 pkg (161 g)	470	**120**	54
Bologna & American Cheese	1 pkg (107 g)	530	**240**	117
Cheese & Salsa Nachos	1 pkg (135 g)	550	**220**	54
Lean Ham & American Cheese	1 pkg (107 g)	460	**180**	81
Lean Turkey Breast & American Cheese	1 pkg (107 g)	440	**170**	81
Lean Turkey Breast & Cheddar	1 pkg (107 g)	440	**140**	63
Pizza (3 extra-cheesy pizzas)	1 pkg (136 g)	460	**140**	72
Pizza (3 pepperoni-flavored sausage pizzas)	1 pkg (136 g)	460	**140**	72
Pizza Dunks (4 soft breadsticks)	1 pkg (143 g)	510	**130**	72
Pizza Swirls	1 pkg (130 g)	480	**160**	90
without drink				
Beef Tacos	1 pkg (152 g)	310	**100**	45
Bologna & American Cheese + Chocolate Chip Cookie	1 pkg (118 g)	480	**300**	117
Cheese & Salsa Nachos	1 pkg (125 g)	380	**190**	41
Lean Ham & Cheddar	1 pkg (128 g)	360	**200**	99
Lean Ham & Cheddar + Vanilla Sandwich Cookie	1 pkg (129 g)	420	**200**	81
Lean Turkey Breast & American Cheese + Fudge Cookie	1 pkg (119 g)	360	**170**	81
Lean Turkey Breast & Cheddar	1 pkg (128 g)	350	**180**	99
Pizza (3 extra-cheesy pizzas)	1 pkg (128 g)	300	**120**	63
Pizza (3 pepperoni-flavored sausage pizzas)	1 pkg (128 g)	310	**130**	63
Safeway Snacks				
Bologna & American Cheese + Peanut Butter Cookies	1 pkg (112 g)	370	**230**	90
Fat-Free Ham & Swiss + Chocolate Chip Cookies	1 pkg (112 g)	350	**170**	72
Fat-Free Turkey Breast & Cheddar + Chocolate Chip Cookies	1 pkg (112 g)	350	**170**	72

SNACK FOODS

FOOD	AMOUNT	CALORIES TOTAL	CALORIES FAT	CALORIES SAT-FAT
NUTS AND SEEDS (*see also* **NUTS AND SEEDS,** page 439)				
Colossal Pistachios	¼ cup without shells	190	**130**	18
Honey Nut Crunch	3 tbsp 27 g)	150	**70**	14
Roasted Salted Cashews	¼ cup (33 g)	210	**140**	27
Roasted Salted Sunflower Kernels	2 tbsp (28 g)	170	**140**	18
POPCORN				
Air-popped, no added fat	1 cup popped	30	**0**	0
Commercially popped				
Bachman All Natural				
Air-popped Lite	5 cups (30 g)	120	**15**	0
Cheese Popcorn	3 cups (30 g)	160	**80**	NA
with white cheddar cheese	2½ cups (30 g)	160	**80**	9
Boston's				
Caramel Popcorn, fat-free	⅔ cup (30 g)	100	**0**	0
Lite Popcorn	4 cups (30 g)	140	**50**	5
White Cheddar (40% less fat)	2¾ cups (29 g)	140	**50**	14
Crunch 'n Munch				
Almond Supreme	½ cup (30 g)	140	**40**	0
Toffee fat-free	¾ cup (28 g)	110	**0**	0
Toffee Popcorn with Peanuts	⅔ cup (31 g)	140	**35**	9
Fiddle Faddle				
Caramel Popcorn w/peanuts	¾ cup (30 g)	140	**50**	27
fat-free	1 cup (30 g)	110	**0**	0
Olde Tyme air-popped buttery popcorn	3⅛ cups (28 g)	140	**70**	5
Smartfood				
White Cheddar	1¾ cups (28 g)	160	**90**	18
reduced fat	3 cups (30 g)	140	**50**	14
Weight Watchers (Smart Snackers)				
Butter flavor	1 pkg (19 g)	90	**20**	0
Butter Toffee	1 pkg (26 g)	110	**25**	9
Caramel	1 pkg (26 g)	100	**10**	0
White Cheddar	1 pkg (19 g)	90	**35**	9

SNACK FOODS

FOOD	AMOUNT	CALORIES		
		TOTAL	FAT	SAT-FAT
Commercially popped (cont.)				
Wise				
Original Butter Popcorn	3 cups (28 g)	150	**90**	18
White Cheddar Popcorn	2 cups (28 g)	160	**100**	23
Microwave				
Jolly Time				
Blast O Butter	3½ cups popped	150	**100**	23
Healthy Pop	5 cups popped	90	**0**	0
Light	5 cups popped	120	**50**	9
Newman's Own Oldstyle Picture Show	3½ cups popped	170	**100**	18
Orville Redenbacher's				
Butter	4½ cups popped	170	**110**	23
Natural flavor	1 cup popped	30	**18**	0
Reden Budders	4 cups popped	170	**110**	23
Smart Pop, butter	5 cups popped	90	**20**	0
Pop Secret				
Butter	4 cups popped	180	**110**	27
Video Club Generic				
Butter flavor Light	5 cups (39 g unpopped)	170	**60**	9
Natural	4½ cups (39 g unpopped)	200	**110**	23
Movie Theater Popcorn				
Popped in coconut oil	kid's (5 cups)	300	**180**	126
	small (7 cups)	398	**243**	171
	med. (11 cups)	647	**387**	279
	med. (16 cups)	901	**540**	387
	large (20 cups)	1161	**693**	495
Popped in coconut oil with butter topping	kid's (5 cups)	472	**333**	198
	small (7 cups)	632	**450**	261
	med. (11 cups)	910	**639**	369
	med. (16 cups)	1221	**873**	504
	large (20 cups)	1642	**1134**	657
Popped in canola shortening	small (7 cups)	361	**198**	63
	medium (11 cups)	627	**342**	108
	large (16 cups)	850	**468**	144

SNACK FOODS

FOOD	AMOUNT	CALORIES TOTAL	CALORIES FAT	CALORIES SAT-FAT
TOASTER PASTRIES				
Kellogg's Pop Tarts				
Apple cinnamon	1 pastry (52 g)	210	**50**	9
Blueberry	1 pastry (52 g)	200	**45**	9
Brown sugar cinnamon	1 pastry (50 g)	210	**60**	9
Frosted blueberry or cherry	1 pastry (52 g)	200	**45**	9
Frosted brown sugar cinnamon	1 pastry (50 g)	210	**60**	14
Frosted chocolate fudge	1 pastry (52 g)	200	**45**	9
Frosted raspberry	1 pastry (52 g)	210	**45**	9
Frosted strawberry	1 pastry (52 g)	200	**45**	9
S'Mores	1 pastry (52 g)	200	**50**	9
Strawberry	1 pastry (52 g)	200	**50**	9
Wild berry	1 pastry (54 g)	210	**50**	14
Wild Tropical Blast	1 pastry (54 g)	210	**45**	18
Wild Watermelon	1 pastry (54 g)	210	**50**	14
Kellogg's Low-fat Pop Tarts				
Blueberry	1 pastry (52 g)	190	**25**	5
Frosted brown sugar cinnamon	1 pastry (50 g)	190	**25**	5
Frosted chocoate fudge	1 pastry (52 g)	180	**25**	5
Frosted strawberry	1 pastry (52 g)	190	**25**	5
Safeway				
Frosted strawberry	1 pastry (52 g)	200	**60**	18
Toastem Pop-Ups				
Frosted chocolate fudge	1 pastry (52 g)	200	**45**	9
Frosted wild berry	1 pastry (52 g)	200	**45**	9
PRETZELS				
Combos				
Cheddar Cheese Pretzel	⅓ cup (28 g)	130	**45**	9
Pizzeria Pretzel	⅓ cup (28 g)	130	**40**	5
Snyder's of Hanover				
Honey Mustard & Onion	⅓ cup (28 g)	140	**60**	9
Oat Bran	3 pretzels (28 g)	100	**5**	0
Old Tyme	3 pretzels (30 g)	120	**10**	0
Old Tyme Stix	28 stix (30 g)	110	**10**	0
Sourdough Hard Pretzel, Fat-free	1 pretzel (28 g)	100	**0**	0

SNACK FOODS

FOOD	AMOUNT	CALORIES TOTAL	FAT	SAT-FAT
Snyders of Hanover (cont.)				
Sourdough Nibblers, Fat-Free	16 nibblers (30 g)	120	**0**	0
Ultra-Thin, Fat-Free	11 pretzels (30 g)	110	**0**	0
Unsalted Mini, Fat-Free	20 minis (30 g)	110	**0**	0
Snyder's of Hanover Nibblers				
Garlic Bread	13 pieces (30 g)	130	**20**	5
Honey Mustard & Onion	13 pieces (30 g)	130	**25**	0
Utz				
Sourdough Hard Pretzels, Fat-Free	1 pretzel (23 g)	90	**0**	0
Sourdough Nuggets	10 pretzels (28 g)	100	**0**	0
Sourdough Specials	4 pretzels (27 g)	110	**15**	0
Stix Pretzels	12 stix (28 g)	110	**5**	0
Nibs	½ cup (30 g)	110	**15**	0
Thin (Rold Gold)	10 pretzels (28 g)	110	**0**	0
RICE, CORN, AND POPCORN CAKES				
Corn Cakes				
Quaker				
Butter	1 cake (9 g)	35	**0**	0
Caramel Corn	1 cake (13 g)	50	**0**	0
Popcorn Cakes				
Orville Redenbacher's Mini Cakes				
Caramel	6 cakes (15 g)	60	**0**	0
Chocolate Peanut Crunch	6 cakes (15 g)	60	**10**	0
Nacho	8 cakes (15 g)	60	**5**	0
Peanut Caramel Crunch	6 cakes (16 g)	60	**5**	0
Orville Redenbacher's Cakes				
Butter	2 cakes (17 g)	60	**10**	0
Caramel	1 cake (11 g)	40	**0**	0
Milk Chocolate	1 cake (10 g)	40	**5**	0
Rice Cakes				
Quaker				
all flavors	1 cake (13 g)	35–50	**0**	0
SNACK MIXES				
California Mix	¼ cup (33 g)	130	**45**	14
Chex Mix (Ralston)				
Bold 'n Zesty	½ cup (30 g)	140	**50**	9
Cheddar	½ cup (30 g)	130	**45**	9

SNACK FOODS

FOOD	AMOUNT	CALORIES TOTAL	FAT	SAT-FAT
Chex Mix (Ralston) (cont.)				
Traditional	⅔ cup (30 g)	130	**35**	5
Cranberry Nut & Fruit Mix	3 tbsp (26 g)	110	**45**	9
Doo Dads Snack Mix (Nabisco)	½ cup (32 g)	150	**60**	9
Fiesta Fun Mix	½ cup	300	**180**	27
Goldfish				
Honey Mustard	½ cup	180	**90**	14
Nutty Deluxe	½ cup	180	**80**	14
Original	½ cup (35 g)	170	**70**	14
Roasted Peanuts	½ cup	170	**70**	14
Savory	½ cup (32 g)	150	**60**	9
Seasoned	½ cup	170	**70**	14
Nut & Fruit Mix	3 tbsp (28 g)	120	**45**	9
Oriental Party Mix	⅓ cup (31 g)	170	**100**	14
Party Mix				
Oriental	½ cup	300	**180**	27
Pastamore	½ cup	260	**100**	18
Sesame Walnut	½ cup	300	**180**	27
Smokehouse	½ cup	260	**160**	18
Cheez-It (Sunshine)				
Nacho	½ cup (30 g)	130	**40**	9
Party Mix	½ cup (30 g)	140	**45**	9
Reduced Fat	½ cup (30 g)	130	**30**	5
Snak-ens Snack Mix				
Chicago Style Pizza	½ cup (30 g)	140	**45**	9
Mustard	½ cup (30 g)	120	**20**	0
Original	½ cup (34 g)	170	**80**	14
Swiss Mix	3 tbsp (29 g)	130	**50**	23
Trail Mix	½ cup	300	**160**	27
Deluxe Super	½ cup	300	**120**	45
Tropical	½ cup	300	**120**	72

VENDING MACHINE FOOD (LANCE)

FOOD	AMOUNT	CALORIES TOTAL	FAT	SAT-FAT
Cakes				
Brownies	1¾ oz/pkg	200	**81**	9
Dunking Sticks	5½ oz/pkg	380	**180**	54
Fig Cake	2⅛ oz/pkg	210	**27**	9

SNACK FOODS

FOOD	AMOUNT	CALORIES TOTAL	FAT	SAT-FAT
Cakes (cont.)				
Oatmeal Cake	2 oz/pkg	240	**99**	27
Raisin Cake	2 oz/pkg	230	**90**	27
Candy				
Chocolaty Peanut Bar	2 oz/pkg	320	**162**	54
Peanut Bar	1¾ oz/pkg	260	**126**	27
Chips, etc.				
Cheese Balls	1⅛ oz/pkg	190	**117**	27
Corn Chips				
BBQ	1¾ oz/pkg	260	**144**	36
Plain	1¾ oz/pkg	270	**153**	27
Crunchy Cheese Twists	1½ oz/pkg	260	**144**	36
Gold-N-Chee	1⅜ oz/pkg	180	**81**	18
6-pack tray	6 oz/pkg	780	**432**	54
Jalapeño Cheese Tortilla Chips	1⅛ oz/pkg	160	**72**	18
Nacho Tortilla Chips	1⅛ oz/pkg	160	**72**	18
Potato Chips				
Baked	1⅛ oz	130	**15**	0
BBQ	1⅛ oz/pkg	190	**108**	27
Cajun style	2 oz/pkg	320	**180**	36
Handcooked	35 chips (50 g)	260	**140**	45
Plain	1⅛ oz/pkg	190	**135**	36
Sour cream & onion	1⅛ oz/pkg	190	**108**	27
Cookies				
Apple-Cinnamon Cookies	2 oz/pkg	240	**72**	18
Apple-Oatmeal Cookies	1.65 oz/pkg	190	**63**	18
Blueberry Cookies	2 oz/pkg	240	**72**	18
Bonnie Sandwich	1³⁄₁₆ oz/pkg	160	**63**	18
Choc-O-Lunch	1⁵⁄₁₆ oz/pkg	180	**63**	18
	4½ oz/pkg	585	**203**	41
Choc-O-Mint	1¼ oz/pkg	180	**90**	27
Coated Graham	1⁵⁄₁₆ oz/pkg	200	**90**	36
Fig Bar	1½ oz/pkg	150	**18**	9
Fudge/Chocolate Chip Cookies	2 oz/pkg	260	**90**	36
Malt	1¼ oz/pkg	190	**99**	18
Nekot	1½ oz/pkg	210	**90**	18
Nut-O-Lunch	4½ oz/pkg	630	**243**	81
Oatmeal Cookies	2 oz/pkg	260	**90**	18

SNACK FOODS

FOOD	AMOUNT	CALORIES TOTAL	FAT	SAT-FAT
Cookies (cont.)				
Peanut Butter Creme-filled Wafer	1¾ oz/pkg	240	**90**	27
Soft Chocolate Chip Cookies	2 oz/pkg	260	**90**	36
Strawberry Cookies	2 oz/pkg	240	**72**	18
Van-O-Lunch	1⁵⁄₁₆ oz/pkg	180	**63**	18
	4½ oz/pkg	630	**162**	41
Crackers				
Captain's Wafers with Cream Cheese & Chives	1⁵⁄₁₆ oz/pkg	170	**81**	18
Cheese-on-Wheat	1⁵⁄₁₆ oz/pkg	180	**81**	18
Golden Toast Cheese (Cheetos)	6 crackers (45.3 g)	240	**130**	36
Lanchee	1¼ oz/pkg	180	**99**	18
Nacho Cheesier (Doritos)	6 crackers (45.3 g)	240	**120**	36
Nip-Chee	1⁵⁄₁₆ oz/pkg	130	**81**	18
Peanut Butter Wheat	1⁵⁄₁₆ oz/pkg	190	**99**	18
Rye-Chee	1⁷⁄₁₆ oz/pkg	190	**81**	18
Spicy Gold-N-Chee	10 oz/pkg	1400	**540**	180
Thin Wheat Snacks	10 oz/pkg	1600	**720**	180
Toast Peanut Butter (Peter Pan)	6 crackers (41 g)	210	**100**	23
Toastchee	1⅜ oz/pkg	190	**99**	18
Toasty	1¼ oz/pkg	180	**90**	18
Nut products				
Cashews	1⅛ oz/pkg	190	**135**	27
long tube	2½ oz/pkg	400	**288**	54
Peanuts				
Honey toasted	1⅜ oz/pkg	230	**153**	27
Roasted (shell)	1¾ oz/pkg	190	**135**	27
Salted	1⅛ oz/pkg	190	**135**	27
tube	3 oz/pkg	480	**360**	72
Pistachios	1⅛ oz/pkg	180	**126**	18
Pie				
Pecan Pie	3 oz/pkg	350	**135**	27
Popcorn				
Cheese	⅞ oz/pkg	130	**72**	9
Plain	1 oz/pkg	160	**90**	18
Pork Skins				
BBQ	½ oz/pkg	80	**45**	18

SNACK FOODS

FOOD	AMOUNT	CALORIES TOTAL	CALORIES FAT	CALORIES SAT-FAT
Pork Skins (cont.)				
Plain	½ oz/pkg	80	**45**	18
Pretzel Twist	1½ oz/pkg	150	**9**	0
MISCELLANEOUS SNACKS				
Cajun Hot Snacks	⅓ cup (34 g)	180	**110**	14
Yogurt Pretzels	6 pieces (40 g)	150	**80**	72
Yogurt Raisins	3 tbsp (35 g)	150	**50**	45

SOUPS

If you are keeping track of total fat calories and your soup is made with whole milk, add 100 fat calories and 206 total calories per can. With 2% milk, add 58 fat calories and 166 total calories per can. If you are keeping track of saturated fat and your soup is made with whole milk, add 62 sat-fat calories and 206 total calories per can. With 2% milk, add 37 sat-fat calories and 166 total calories per can.

FOOD	AMOUNT	CALORIES TOTAL	CALORIES FAT	CALORIES SAT-FAT
CONDENSED				
Prepared with water				
Campbell's				
Bean with Bacon	1 cup prepared	180	**45**	18
Beef Broth	1 cup prepared	15	**0**	0
Beef Noodle	1 cup prepared	70	**25**	9
Broccoli Cheese 98% fat-free	1 cup prepared	80	**25**	14
Cheddar Cheese	1 cup prepared	130	**70**	32
Chicken Alphabet	1 cup prepared	80	**20**	9
Chicken and Stars	1 cup prepared	70	**20**	5
Chicken Broth	1 cup prepared	30	**20**	5
Chicken Gumbo	1 cup prepared	60	**15**	5
Chicken Noodle	1 cup prepared	70	**20**	9
Chicken Noodle (Healthy Request)	1 cup prepared	70	**20**	5
Chicken NoodleO's	1 cup prepared	80	**25**	9

SOUPS

FOOD	AMOUNT	CALORIES TOTAL	FAT	SAT-FAT
Campbell's (cont.)				
Chicken Rice (Healthy Request)	1 cup prepared	60	**25**	9
Chicken Vegetable	1 cup prepared	80	**20**	5
Chicken Vegetable (Healthy Request)	1 cup prepared	80	**20**	5
Chicken with Rice	1 cup prepared	70	**25**	9
Chicken with White & Wild Rice	1 cup prepared	70	**20**	5
Chicken Won Ton	1 cup prepared	45	**10**	0
Consomme Beef	1 cup prepared	25	**0**	0
Cream of Asparagus	1 cup prepared	110	**60**	18
Cream of Broccoli	1 cup prepared	100	**50**	23
Cream of Broccoli 98% fat-free	1 cup prepared	80	**25**	9
Cream of Celery	1 cup prepared	110	**60**	23
Cream of Celery (Healthy Request)	1 cup prepared	70	**20**	5
Cream of Celery 98% fat-free	1 cup prepared	70	**25**	9
Cream of Chicken	1 cup prepared	130	**70**	27
Cream of Chicken 98% fat-free	1 cup prepared	80	**25**	14
Cream of Chicken (Healthy Request)	1 cup prepared	70	**20**	9
Cream of Chicken & Mushroom	1 cup prepared	130	**80**	23
Cream of Chicken with Herbs	1 cup prepared	80	**35**	14
Cream of Mushroom	1 cup prepared	110	**60**	23
Cream of Mushroom 98% fat-free	1 cup prepared	70	**25**	9
Cream of Mushroom (Healthy Request)	1 cup prepared	70	**25**	9
Cream of Mushroom with Roasted Garlic	1 cup prepared	70	**25**	9
Cream of Potato	1 cup prepared	90	**25**	14

SOUPS

FOOD	AMOUNT	CALORIES TOTAL	CALORIES FAT	CALORIES SAT-FAT
Campbell's (cont.)				
Cream of Roasted Chicken (Healthy Request)	1 cup prepared	80	**25**	9
Cream of Shrimp	1 cup prepared	100	**60**	18
Double Noodle	1 cup prepared	100	**25**	9
French Onion	1 cup prepared	70	**25**	0
Golden Mushroom	1 cup prepared	80	**25**	9
Green Pea	1 cup prepared	180	**25**	9
Hearty Pasta & Vegetable (Healthy Request)	1 cup prepared	90	**10**	5
Herbed Potato (Healthy Request)	1 cup prepared	80	**20**	9
Homestyle Chicken Noodle	1 cup prepared	70	**25**	14
Italian Tomato	1 cup prepared	100	**5**	0
Manhattan Clam Chowder	1 cup prepared	100	**25**	9
Minestrone	1 cup prepared	100	**20**	5
New England Clam Chowder	1 cup prepared	100	**25**	9
New England Clam Chowder 98% fat-free	1 cup prepared	90	**20**	5
Old-Fashioned Tomato Rice	1 cup prepared	120	**20**	5
Pepper Pot	1 cup prepared	100	**45**	18
Split Pea with Ham & Bacon	1 cup prepared	180	**30**	18
Tomato	1 cup prepared	80	**0**	0
Tomato (Healthy Request)	1 cup prepared	90	**15**	5
Tomato Bisque	1 cup prepared	130	**25**	14
Turkey Noodle	1 cup prepared	80	**25**	9
Vegetable	1 cup prepared	80	**10**	5
Vegetable (Healthy Request)	1 cup prepared	90	**10**	5
Vegetable Beef	1 cup prepared	80	**20**	9
Vegetable Beef (Healthy Request)	1 cup prepared	80	**20**	9
Vegetarian Vegetable	1 cup prepared	90	**10**	0

SOUPS

FOOD	AMOUNT	CALORIES TOTAL	CALORIES FAT	CALORIES SAT-FAT
DEHYDRATED				
Knorr's				
Black Bean	1 package (53 g)	200	**10**	0
Chicken Flavor Vegetable	1 package (30 g)	100	**0**	0
Hearty Lentil	1 package (57 g)	220	**0**	0
Navy Bean	1 package (38 g)	140	**0**	0
Potato Leek	1 package (34 g)	120	**0**	0
Marachan Instant Lunch				
Beef flavor	1 container (64 g)	290	**110**	54
with Shrimp	1 container (64 g)	290	**110**	54
Nissin Top Ramen				
Baked Ramen Noodle Soup (98% fat-free)				
all flavors	½ block (37 g)	140	**10**	0
	1 cup (58 g)	210	**10**	0
Cup Noodles Ramen Noodle Soup				
Chicken flavor	1 container (64 g)	300	**130**	63
Chicken Vegetable flavor	1 container (64 g)	300	**130**	54
Creamy Chicken flavor	1 container (64 g)	300	**120**	54
French Onion flavor	1 container (64 g)	300	**110**	54
Teriyaki Chicken flavor	1 container (64 g)	300	**130**	54
Twin Pack, with Shrimp	1 cup (34 g)	150	**60**	27
Oodles of Noodles, all flavors	½ pkg (43 g)	180–190	**60**	32
Soup Starter (Wyler's)				
Beef Vegetable	⅛ pkg (28 g dry)	90	**5**	0
Chicken Noodle	⅛ pkg (24 g dry)	80	**5**	0
Hearty Beef Stew	⅐ pkg (23 g dry)	80	**0**	0
Hearty Chicken Vegetable	⅐ pkg (23 g dry)	70	**0**	0
HOMEMADE OR RESTAURANT				
Cream of Mushroom Soup	1 cup	170	**155**	98
French Onion Soup	1 cup	350	**125**	65
Gazpacho	1 cup	111	**80**	11
New England Clam Chowder	1 cup	230	**145**	65
Vichyssoise	1 cup	315	**210**	140

SOUPS

FOOD	AMOUNT	CALORIES TOTAL	FAT	SAT-FAT
READY TO SERVE				
Campbell's				
Chunky Soup				
Beef	10¾ oz (305 g)	200	**45**	14
Beef Pasta	1 cup	150	**25**	9
Cheese Tortellini	1 cup	110	**20**	9
Chicken and Pasta	10¾ oz	150	**40**	14
Chicken Broccoli Cheese	1 cup	200	**110**	45
Chicken Corn Chowder	1 cup	250	**140**	63
Chicken Mushroom Chowder	1 cup	210	**110**	36
Chicken with Rice	1 cup	140	**30**	9
Clam Chowder, Manhattan Style	1 cup	130	**35**	9
Classic Chicken Noodle	1 cup	130	**25**	9
Hearty Bean 'n Ham	1 cup	190	**20**	9
Hearty Chicken with Vegetables	1 cup	90	**20**	5
New England Clam Chowder	10¾ oz	300	**160**	63
Pepper Steak	1 cup	140	**25**	9
Potato Ham Chowder	1 cup	220	**130**	72
Sirloin Burger	1 cup	180	**60**	32
Split Pea 'n Ham	1 cup	190	**25**	9
Tomato Ravioli	1 cup	150	**25**	14
Vegetable	10¾ oz	160	**35**	9
Healthy Request				
Chicken Broth	1 cup	20	**0**	0
Hearty Chicken Vegetable	1 cup	120	**20**	5
Home Cookin'				
Bean and Ham	1 cup	180	**15**	5
Chicken Rice	1 cup	140	**15**	5
Chicken Vegetable	1 cup	130	**35**	9
Chicken with Egg Noodles	10¾ oz	110	**25**	14
Country Mushroom Rice	1 cup	80	**5**	0

SOUPS

FOOD	AMOUNT	CALORIES TOTAL	FAT	SAT-FAT
Home Cookin' (cont.)				
Country Vegetable	1 cup	110	**10**	0
Cream of Mushroom	1 cup	80	**20**	9
Creamy Potato	1 cup	180	**80**	23
Fiesta	1 cup	130	**25**	5
New England Clam Chowder	1 cup	120	**30**	9
Tomato Garden	1 cup	130	**30**	9
Vegetable Beef	1 cup	120	**20**	9
Microwavable Ready-to-Serve Soups (Campbell's)				
Chicken Noodle	10.5 oz	130	**35**	9
Vegetable Beef	10.5 oz	140	**5**	0
Manischewitz				
Matzo Ball Soup	1 cup	110	**45**	18

SWEETS

FOOD	AMOUNT	CALORIES TOTAL	FAT	SAT-FAT
BARS				
Tasty Kake Snack Bars				
Chocolate Chip	1 bar (57 g)	250	**110**	27
Iced Fudge	1 bar (57 g)	250	**90**	14
Iced Lemon	1 bar (57 g)	260	**90**	9
Iced Strawberry	1 bar (57 g)	260	**90**	9
Oatmeal Raisin	1 bar (57 g)	250	**80**	23
BROWNIES				
Brownie				
with nuts	1 brownie (40 g)	180	**80**	18
without nuts	1 brownie (40 g)	160	**60**	18
Fudge Brownies (Little Debbie)	1 brownie (61 g)	270	**120**	23
Brownie Bites (Hostess)	3 pieces (37 g)	170	**80**	18
Nonfat (Entenmann's)	1 brownie (40 g)	110	**0**	0

SWEETS

FOOD	AMOUNT	CALORIES TOTAL	CALORIES FAT	CALORIES SAT-FAT
BROWNIE MIXES BY BRAND NAME				
Betty Crocker				
Dark Chocolate Supreme	30 g mix	129	**15**	5
	prep w. eggs & oil	170	**70**	9
Fudge Brownies	28 g mix	110	**15**	5
	prep w. eggs & oil	170	**70**	14
Original Supreme	32 g mix	130	**15**	5
	prep w. eggs & oil	160	**50**	9
Sweet Rewards Low-fat Fudge Brownie mix	32 g mix	130	**25**	9
Turtle Supreme	29 g mix	120	**20**	5
	prep w. eggs & oil	170	**70**	14
Walnut Supreme	28 g mix	120	**30**	5
	prep w. eggs & oil	180	**80**	9
Duncan Hines				
Chewy Fudge Brownie	¼ cup mix (31 g)	130	**25**	5
	baked	160	**60**	14
Dark 'n Chunky	¼ cup mix (30 g)	140	**30**	9
	baked	160	**70**	14
Double Fudge	¼ cup mix (34 g)	140	**25**	5
	baked	170	**60**	9
Milk Chocolate Chunk	¼ cup mix (33 g)	140	**35**	14
	baked	170	**60**	14
Pillsbury				
Rich & Moist Fudge Brownie	31 g mix	130	**25**	5
	baked w. oil & eggs	190	**80**	15
Thick 'n Fudgy Cheesecake Swirl	27 g mix	130	**40**	14
	baked w. oil & eggs	170	**80**	25
Chocolate Chunk	27g mix	120	**30**	14
	baked w. oil & eggs	160	**60**	18
Double Chocolate	28 g mix	120	**25**	5
	baked w. oil & eggs	150	**50**	10
Walnut	32 g mix	150	**45**	9
	baked w. oil & eggs	190	**90**	14
Washington				
Fudge Brownie	¼ cup mix (31 g)	130	**35**	14
	baked w. oil & eggs	150	**45**	18

SWEETS

FOOD	AMOUNT	CALORIES TOTAL	CALORIES FAT	CALORIES SAT-FAT
BUNS				
Breakfast Buns	1 bun (55 g)	170	**25**	9
Butterfly Buns	1 bun (55 g)	190	**35**	9
Honey Bun (Little Debbie)	1 bun (50 g)	220	**100**	23
Hot Cross Buns	1 bun	168	**56**	16
Iced Honey Bun (Hostess)	1 bun (99 g)	420	**220**	54
Pecan Twirls	1 (28 g)	110	**35**	5
Rum Buns (Hadley Farm)	1 bun (85 g)	290	**80**	18
	1 bun (55 g)	200	**60**	14
Sticky Bun	1 (55 g)	210	**70**	14
Cinnamon Pecan Sticky	1 roll (71 g)	220	**60**	14
Entenmann's Light	1 bun (61 g)	160	**25**	5
CAKES, BAKERY, HOMEMADE, AND RESTAURANT				
Almond Danish Coffee Cake	⅛ cake (57 g)	230	**100**	18
Almond Poppy	1/10 cake (80 g)	320	**160**	27
Angel Food	1/12 cake	125	**0**	0
Angel Food Loaf	2" slice (55 g)	150	**5**	0
Angel Food Ring	⅙ cake (47 g)	130	**5**	0
Apple Danish Coffee Cake	⅛ cake (57 g)	160	**60**	18
Apple Strudel Cake	1 slice (64 g)	190	**90**	27
Baked Alaska	1/12 of cake	263	**112**	60
Banana Nut Loaf	2" slice (76 g)	270	**120**	27
Bavarian Chocolate Cake	2" square (80 g)	340	**180**	63
Black Forest (7" diam)	1 slice (80 g)	330	**170**	41
Blueberry Cheese Coffee Cake	⅛ cake (57 g)	170	**70**	18
Boston Cream Cake	⅙ cake (113 g)	320	**120**	27
Carrot Cake (7" diam)	1 slice (80 g)	300	**130**	36
Cheesecake	1/12 cake	280	**162**	89
	2" wedge (250 g)	820	**540**	342
Blueberry-Topped Cheesecake	⅙ cake (113 g)	350	**180**	81
French Cheesecake	4" wedge (125 g)	370	**200**	54
Cherry Puddin' Cake	⅛ cake (74 g)	290	**130**	23
Chocolate Crunch Ring	⅛ cake (71 g)	310	**140**	36
Chocolate Fudge (7" diam)	1 slice (80 g)	300	**140**	32
Chocolate Marble Loaf	⅕ cake (80 g)	340	**150**	32

SWEETS

FOOD	AMOUNT	CALORIES TOTAL	FAT	SAT-FAT
Chocolate Sheet Cake with Vanilla Icing	1 slice (80 g)	320	**160**	36
Cinnamon Stix	1 stick (57 g)	200	**60**	14
Creme Cakes				
Blueberry	1 slice (80 g)	280	**120**	23
Chocolate	1 slice (80 g)	300	**140**	27
Maple Nut	1 slice (80 g)	310	**140**	23
Strawberry	1 slice (80 g)	290	**120**	23
Vanilla	1 slice (80 g)	300	**140**	23
Crumb Cakes				
Apple Cinnamon Walnut	1 slice (80 g)	320	**140**	31
Blueberry	1 slice (80 g)	320	**140**	31
Cupcake, low fat	1 (80 g)	340	**140**	36
Creme-filled Chocolate	2 cakes (64 g)	200	**25**	9
Creme-filled Vanilla	2 cakes (64 g)	200	**25**	9
Devil's Food Fudge Cake	⅙ cake (85 g)	340	**140**	36
Devil's Food Layer Cake w/ white icing	2½" wedge (80 g)	360	**170**	45
Fruitcake	1 slice (125 g)	470	**170**	32
German Apple Strudel	⅛ cake (53 g)	140	**60**	18
German Chocolate Cake				
(10" diam)	1⁄16 cake	521	**277**	27
(7" diam)	1 slice (80 g)	300	**140**	36
Golden Coconut (7" diam)	1 slice (80 g)	320	**160**	36
Hazelnut Torte	1⁄16 torte	315	**187**	56
Ladyfingers	12 (85 g)	280	**35**	14
Lemon Crunch Ring	⅛ cake (71 g)	270	**110**	23
Lemon Supreme (7" diam)	1 slice (80 g)	320	**160**	36
Marble Cake (7" diam)	1 slice (80 g)	330	**160**	36
Pecan Cinnamon Ring	⅛ cake (71 g)	300	**130**	23
Pecan Danish Ring	⅛ cake (57 g)	230	**120**	27
Pecan Twirls Sweet Rolls	2 twirls (56 g)	220	**80**	9
Pineapple Upside-Down Cake (7" diam)	1 slice (80 g)	210	**80**	14
Pound cake, not iced	2" slice (80 g)	300	**140**	45
Strawberry Cheese Danish Coffee Cake	⅛ cake (57 g)	170	**80**	18
Walnut Danish Ring	⅛ cake (55 g)	230	**120**	18

SWEETS

FOOD	AMOUNT	CALORIES TOTAL	FAT	SAT-FAT
Yellow Layer Cake				
with white icing	2½" wedge (80 g)	350	**170**	45
with chocolate icing	2½" wedge (80 g)	320	**140**	36
CAKES AND SNACK CAKES BY BRAND NAME				
Entenmann's				
All Butter Loaf	⅙ cake (57 g)	210	**80**	45
All Butter French Crumbcake	⅛ cake (50 g)	210	**90**	45
Apple Puffs	1 puff (85 g)	260	**110**	27
Cheese Coffee Cake	⅛ cake (48 g)	160	**60**	23
Cheese Crumb Babka	⅛ danish (48 g)	160	**60**	23
Cheese-Filled Crumb Coffee Cake	⅛ cake (57 g)	200	**90**	32
Cinnamon Filbert Ring	⅕ danish (60 g)	260	**150**	27
Cinnamon Rugelach	1 piece (21 g)	100	**60**	NA
Crumb Coffee Cake	⅒ cake (57 g)	250	**110**	27
Raspberry Danish Twist	⅛ danish (53 g)	220	**100**	32
Rugelach				
all flavors	1 piece (19 g)	90	**45**	27
Entenmann's Light				
Apple Spice Cake	⅕ cake (79 g)	200	**0**	0
Banana Crunch Cake	⅕ cake (85 g)	220	**0**	0
Banana Loaf	⅙ cake (76 g)	190	**0**	0
Blueberry Crunch Cake	⅙ cake (76 g)	180	**0**	0
Chocolate Crunch Cake	⅕ cake (79 g)	210	**0**	0
Chocolate Loaf Cake	⅕ cake (85 g)	210	**0**	0
Cinnamon Apple Coffee Cake	⅑ cake (54 g)	120	**0**	0
Cinnamon Apple Twist	⅛ danish (53 g)	140	**0**	0
Cherry Cheese Pastry	⅑ pastry (54 g)	130	**0**	0
Crumb Delight	⅑ cake (57 g)	210	**60**	9
Fudge Brownie	⅒ brownie (40 g)	110	**0**	0
Fudge-Iced Chocolate Cake	⅙ cake (85 g)	190	**0**	0
Golden Chocolatey Chip Loaf	⅕ cake (85 g)	220	**0**	0
Golden Loaf Cake	⅛ cake (48 g)	130	**0**	0

SWEETS

FOOD	AMOUNT	CALORIES TOTAL	CALORIES FAT	CALORIES SAT-FAT
Entenmann's Light (cont.)				
Lemon Twist	⅛ danish (53 g)	130	**0**	0
Louisiana Crunch Cake	⅙ cake (76 g)	210	**0**	0
Marble Loaf	⅛ cake (50 g)	130	**0**	0
Mocha-Iced Chocolate Cake	⅙ cake (85 g)	200	**0**	0
Pineapple Crunch Cake	⅙ cake (76 g)	190	**0**	0
Raspberry Cheese Pastry	⅑ cake (54 g)	140	**0**	0
Raspberry Twist	⅛ danish (53 g)	140	**0**	0
Hostess				
Baseballs	1 cake (46 g)	150	**30**	9
Cinnamon Crumb Cakes, low-fat	1 cake (28 g)	90	**5**	0
Cup Cakes				
Chocolate	1 cake (50 g)	180	**50**	23
low-fat	1 cake (46 g)	140	**15**	5
Orange	1 cake (43 g)	160	**45**	18
HoHos	1 cake (29 g)	130	**55**	36
King Dons	1 cake (40 g)	170	**80**	45
Sno Balls	1 cake (50 g)	180	**50**	23
Suzy Q's	1 cake (58 g)	220	**80**	36
Twinkies	1 cake (43 g)	150	**45**	18
low-fat	1 cake (43 g)	130	**15**	5
Krispy Kreme				
Apple Danish	1 pastry (90 g)	380	**210**	54
Cherry Danish	1 pastry (92 g)	370	**180**	45
Cream Cheese Danish	1 pastry (96 g)	410	**240**	72
Honey Bun	1 bun (96 g)	410	**220**	54
Little Debbie				
Apple Streusel Coffee Cake	2 cakes (60 g)	230	**70**	14
Chocolate Cup Cakes	1 cake (45 g)	180	**80**	18
Coffee Cakes	1 cake (30 g)	115	**30**	7
Devil Squares	1 cake (31 g)	135	**55**	14
Oatmeal Lights	1 wrap (38 g)	130	**50**	5
Snack Cakes	1 cake (36 g)	155	**70**	32
Strawberry Shortcake Rolls	1 roll (61 g)	230	**70**	18
Swiss Cake Rolls	1 cake (31 g)	130	**55**	9
Zebra Cakes	1 cake (37 g)	165	**75**	16

SWEETS

FOOD	AMOUNT	CALORIES TOTAL	FAT	SAT-FAT
Pepperidge Farm				
Cream Cheese Carrot	1/9 cake (80 g)	320	**180**	41
Devil's Food Cake	1/6 slice (80 g)	290	**122**	NA
Three-Layer Cake	1/8 cake (69 g)	250	**110**	27
Sara Lee				
Banana Sundae Layer Cake	1/10 cake (81 g)	270	**120**	90
Butter Streusel Coffee Cake	1/6 cake (54 g)	220	**110**	54
Chocolate Swirl Cake	1/4 cake (83 g)	330	**140**	72
Double Chocolate Layer Cake	1/8 cake (79 g)	260	**120**	99
Flaky Coconut Layer Cake	1/8 cake (81 g)	280	**130**	108
French Cheesecake	1/5 cake (133 g)	410	**230**	144
German Chocolate Layer Cake	1/8 cake (83 g)	280	**130**	99
Mint Chocolate Mousse	1/5 mousse (122 g)	440	**250**	189
Original Cream Cheesecake	1/4 cake (121 g)	350	**160**	81
Pecan Coffee Cake	1/6 cake (54 g)	230	**110**	41
Pound Cake, all butter	1/6 cake (76 g)	320	**150**	81
Strawberry French Cheese Cake	1/6 cake (123 g)	320	**130**	81
Strawberry Shortcake	1/8 cake (71 g)	180	**70**	45
Vanilla Layer Cake	1/8 cake (80 g)	250	**120**	90
Tastykake				
Cinnamon Sweet Rolls	1 cake (60 g)	190	**30**	9
Cup Cakes				
Banana Creamies	1 cake (43 g)	170	**60**	14
Butter Cream–Iced Chocolate	2 cakes (64 g)	250	**70**	18
Chocolate Creamies	1 cake (43 g)	180	**70**	18
Chocolate Cup Cakes	3 cakes (92 g)	310	**80**	14
Creme Filled Chocolate	2 cakes (64 g)	230	**70**	9
Vanilla Creamies	1 cake (43 g)	190	**80**	14
Honey Bun	1 cake (92 g)	360	**150**	27
Juniors				
Chocolate Junior	1 cake (94 g)	330	**110**	18

SWEETS

FOOD	AMOUNT	CALORIES TOTAL	CALORIES FAT	CALORIES SAT-FAT
Tastykake (cont.)				
Coconut Junior	1 cake (94 g)	310	**170**	36
Koffee Kake Junior	1 cake (71 g)	270	**180**	14
Kandy Kakes				
Chocolate Kandy Kakes	3 cakes (57 g)	250	**110**	63
Peanut Butter Kandy Kakes	3 cakes (57 g)	280	**130**	63
Koffee Kakes				
Chocolate	3 cakes (99 g)	380	**130**	23
Creme Filled Koffee Kakes	2 cakes (57 g)	240	**80**	14
Kreme Krimpies	3 cakes (85 g)	340	**110**	14
Krimpets				
Butterscotch Krimpets	3 cakes (85 g)	310	**70**	14
Jelly Krimpets	3 cakes (85 g)	280	**40**	9
Strawberries Krimpets	3 cakes (85 g)	310	**70**	9
Marshmallow Treats	1 cake (57 g)	210	**35**	9
Pound Kake	1 cake (85 g)	320	**120**	45
Tropical Delights				
Coconut Tropical Delights	2 cakes (57 g)	190	**80**	41
Guava Tropical Delights	2 cakes (57 g)	190	**60**	36
Papaya Tropical Delights	2 cakes (57 g)	190	**60**	36
Pineapple Tropical Delights	2 cakes (57 g)	190	**60**	36
Weight Watchers				
Chocolate Eclair	1 eclair (59 g)	150	**35**	9
Chocolate Raspberry Royale	1 dessert (99 g)	190	**30**	9
New York–Style Cheesecake	1 cake (70 g)	150	**45**	23
CAKE MIXES BY BRAND NAME				
Any brand				
Angel Food	1/12 cake (36 g)	140	**0**	0

SWEETS

FOOD	AMOUNT	CALORIES TOTAL	CALORIES FAT	CALORIES SAT-FAT
Betty Crocker				
Angel Food				
fat-free	38 g mix	140	**0**	0
Easy	¼ loaf or 3 cupcakes (44 g mix)	170	**0**	0
Lemon Bars	29 g mix	130	**30**	9
	prep with eggs	140	**40**	9
Pound Cake	57 g mix	250	**60**	27
	prep with eggs	270	**70**	27
Super Moist				
Butter Recipe Chocolate	43 g mix	190	**40**	14
	prep with eggs, butter, frosting	270	**120**	63
Butter Recipe Yellow	43 g mix	170	**20**	9
	prep with eggs, butter, frosting	260	**100**	54
Carrot	51 g mix	200	**30**	9
	prep with eggs, butter, frosting	320	**140**	23
Chocolate Fudge	43 g mix	170	**25**	9
	prep with eggs, butter, frosting	270	**110**	23
Devil's Food	43 g mix	180	**35**	14
	prep with eggs, butter, frosting	240	**100**	36
Double Chocolate Swirl	43 g mix	170	**25**	9
	prep with eggs, butter, frosting	270	**110**	23
French Vanilla	43 g mix	170	**30**	9
	prep with eggs, butter, frosting	240	**90**	18
Lemon	43 g mix	170	**30**	9
	prep with eggs, butter, frosting	240	**90**	18
Yellow	43 g mix	170	**30**	9
	prep with eggs, butter, frosting	250	**90**	23

SWEETS

FOOD	AMOUNT	CALORIES TOTAL	CALORIES FAT	CALORIES SAT-FAT
Betty Crocker (cont.)				
Stir 'n Bake				
Carrot Cake	61 g mix	250	**60**	14
Coffee Cake	47 g mix	200	**50**	14
Devil's Food	58 g mix	240	**70**	18
Super Moist Light				
Devil's Food	52 g mix	210	**30**	14
	prep with eggs & frosting	230	**40**	18
Sweet Rewards reduced fat				
Devil's Food	43 g mix	160	**15**	5
	prep with oil & eggs	200	**45**	10
Yellow	43 g mix	160	**10**	5
	prep with oil & eggs	200	**40**	15
Duncan Hines				
Angel Food	¼ cup mix (38 g)	140	**0**	0
Moist Deluxe				
Butter Recipe Golden	52 g mix	230	**50**	18
	prep with eggs, butter, topping	320	**140**	63
Devil's Food	⅓ cup mix (43 g)	180	**40**	14
	prep with eggs, oil, topping	290	**130**	27
White	⅓ cup mix (43 g)	170	**30**	14
	prep with eggs, oil, topping	190	**50**	9
all other flavors	⅓ cup mix	180	**30**	14
	prep with eggs, oil, topping	250	**100**	18
Pillsbury				
Chocolate Chip Streusel Coffee Cake	50 g mix	210	**50**	18
	prep with oil & eggs	270	**110**	23
Cinnamon Streusel Coffee Cake	47 g mix	190	**45**	14
	prep with oil & eggs	260	**100**	23

SWEETS

FOOD	AMOUNT	CALORIES TOTAL	FAT	SAT-FAT
Pillsbury (cont.)				
Moist Supreme (Jell-O pudding in the mix)				
Butter Recipe	43 g mix	170	**25**	9
	prep w. eggs & butter	260	**110**	54
Devil's Food	43 g mix	180	**35**	14
	prep w. oil & eggs	270	**130**	27
Lemon	52 g mix	210	**35**	14
	prep w. oil & eggs	300	**120**	27
Yellow	43 g mix	180	**35**	14
	prep w. oil & eggs	240	**90**	23
CAKE DECORATIONS				
Dessert Decorations, real chocolate (Betty Crocker)				
Alphabet & Numbers	5 pieces (4 g)	20	**15**	5
Animal Shapes	3 pieces (4 g)	20	**15**	5
Happy Birthday Pieces	2 pieces (4 g)	20	**15**	5
Lace Decors	4 pieces (4 g)	20	**15**	5
Leaves	3 pieces (4 g)	20	**15**	5
CAKE FROSTING BY BRAND NAME				
Betty Crocker Rich & Creamy frosting				
Chocolate	2 tbsp (33 g)	130	**45**	14
Coconut Pecan	2 tbsp	150	**70**	27
Cream Cheese	2 tbsp (34 g)	140	**45**	14
Dark Chocolate	2 tbsp (36 g)	150	**50**	14
French Vanilla	2 tbsp (34 g)	140	**45**	14
Rainbow Chip	2 tbsp (36 g)	160	**60**	27
Vanilla	2 tbsp (34 g)	140	**45**	14
Betty Crocker Soft Whipped frosting				
Chocolate	2 tbsp (24 g)	100	**45**	18
Cream Cheese	2 tbsp (24 g)	100	**40**	14
Fluffy Lemon	2 tbsp (24 g)	110	**45**	14
Fluffy White	2 tbsp (24 g)	100	**40**	14
Strawberry	2 tbsp (24 g)	110	**45**	14
Vanilla	2 tbsp (24 g)	100	**40**	14

SWEETS

FOOD	AMOUNT	CALORIES TOTAL	CALORIES FAT	CALORIES SAT-FAT
Betty Crocker Sweet Rewards reduced-fat frosting				
Chocolate	2 tbsp (33 g)	120	**20**	9
Vanilla	2 tbsp (33 g)	120	**20**	5
Cake Mate Decorating Icing				
all colors	1 tsp (6 g)	30	**10**	0
Duncan Hines Homestyle Frosting				
Vanilla	2 tbsp (32 g)	140	**50**	14
all other flavors	2 tbsp (32 g)	130	**45**	14
Washington Frosting Mix				
Creamy Fudge	¼ cup mix (26 g)	110	**15**	5
	prep w. margarine	150	**50**	10
Creamy White	¼ cup mix (26 g)	110	**15**	5
	prep w. margarine	150	**50**	10

CANDY

FOOD	AMOUNT	CALORIES TOTAL	CALORIES FAT	CALORIES SAT-FAT
3 Musketeers	1 bar (60.4 g)	260	**70**	36
Miniatures	7 pieces (41 g)	170	**45**	23
After Eight	5 pieces (49 g)	170	**50**	32
Almond Joy	1 bar (49 g)	240	**120**	72
Snack size	2 bars (38 g)	190	**90**	63
Andes				
Crème de menthe thins	8 pieces (38 g)	210	**120**	108
Mint Parfait Thins	8 pieces (38 g)	210	**120**	99
Baby Ruth	1 bar (59.5 g)	280	**110**	63
Fun size	2 bars (42 g)	190	**80**	45
Butterfinger	1 bar (60 g)	280	**100**	54
Fun size	2 bars (42 g)	200	**70**	36
Butterscotch				
Brach's	3 pieces (18 g)	70	**0**	0
Callard & Bowser	3 pieces (18 g)	60	**10**	5
Candy Corn	26 pieces (39 g)	140	**0**	0
Caramels				
Chocolate Chew (Riesen)	5 pieces (40 g)	180	**60**	27
Classic (Hershey's)	6 pieces (37 g)	160	**45**	41
Classic (Kraft)	6 pieces (37 g)	160	**50**	41
Creams (Goetze's)	3 pieces (34 g)	130	**30**	9
Milk Maid (Brach's)	4 pieces (39 g)	150	**40**	9
Milkfuls (Storck)	6 pieces (40 g)	170	**25**	14

SWEETS

FOOD	AMOUNT	CALORIES TOTAL	CALORIES FAT	CALORIES SAT-FAT
Carob Peanut Clusters	1 piece (28 g)	150	**90**	36
Carob Raisins	40 pieces (41 g)	160	**45**	36
Chocolate-covered peanuts	15 pieces (40 g)	220	**120**	54
Chocolate-covered peanut clusters	3 pieces (43 g)	230	**130**	63
Chocolate-covered fudge mix	14 pieces (40 g)	190	**70**	36
Chocolate Mint Pattie (Brach's)	3 pieces (36 g)	140	**26**	18
Dots	12 pieces (43 g)	150	**0**	0
Dove Chocolate Bar	1 bar (37 g)	200	**110**	63
Dove Promises, all flavors	7 pieces (42 g)	220–230	**120**	72
Fondant (mints, candy corn, other)	1 oz	105	**0**	0
French Burnt Peanuts	28 pieces (40 g)	190	**80**	9
Fruit Roll-ups, all flavors	2 rolls (28 g)	110	**10**	0
Goldenberg's Peanut Chews	3 pieces (37 g)	270	**120**	27
Good & Plenty	33 pieces (40 g)	130	**0**	0
Gumdrops	1 oz	100	**0**	0
Gummy Bears	10 pieces (40 g)	140	**0**	0
Halvah (Joyva)				
Chocolate-covered	½ bar (57 g)	380	**210**	45
Chocolate-flavored	½ bar (57 g)	390	**230**	36
Marble	½ bar (57 g)	390	**230**	36
Hard candies	1 oz	110	**0**	0
	1 piece (5.6 g)	18	**0**	0
Heath Sensations	⅓ bag (43 g)	220	**120**	63
Hershey's Chocolate				
Cookies 'n Creme	1 bar (43 g)	230	**110**	54
Hugs	8 pieces (38 g)	210	**110**	54
Kisses	8 pieces (39 g)	210	**110**	72
with almonds	8 pieces (38 g)	210	**120**	63
Milk Chocolate	1 bar (43 g)	230	**120**	81
with almonds	1 bar (41 g)	230	**130**	63
Sweet Escapes				
Caramel & Peanut Butter	1 bar (20 g)	70	**25**	9
Chocolate-Toffee	1 bar (39 g)	190	**70**	45
Crispy Caramel Fudge	1 bar (20 g)	80	**20**	9
Miniatures	5 pieces (42 g)	230	**130**	72

SWEETS

FOOD	AMOUNT	CALORIES TOTAL	CALORIES FAT	CALORIES SAT-FAT
Hershey's Chocolate (cont.)				
Symphony	1 bar (42 g)	240	**140**	72
Triple Chocolate Wafer	1 bar (39 g)	160	**45**	23
	1 bar (20 g)	80	**25**	14
Hot Tamales	19 pieces (40 g)	150	**0**	0
Jellybeans	14 pieces (39 g)	140	**0**	0
Jordan Almonds (Brach's)	10 pieces (41 g)	180	**50**	0
Jujyfruits	15 pieces (40 g)	160	**0**	0
Junior Mints	16 pieces (40 g)	160	**25**	18
Kit Kat	1 bar (42 g)	220	**100**	63
Snack	3 bars (47 g)	240	**110**	72
Licorice Sticks	4 pieces (37 g)	120	**5**	0
M&M's				
Almond	¼ cup (42 g)	230	**110**	36
Peanut Butter	¼ cup (42 g)	220	**110**	72
Peanut	¼ cup (42 g)	220	**100**	41
Plain	¼ cup (42 g)	210	**80**	54
Malted Milk Balls	17 pieces (40 g)	180	**60**	60
Maple Nut Goodlies (Brach's)	7 pieces (39 g)	190	**80**	9
Mars Bar	1 bar (50 g)	240	**110**	36
Marshmallows (Kraft)	5 pieces (34 g)	110	**0**	0
Mary Janes	6 pieces (40 g)	160	**40**	0
Mighty Malts malted milk balls	10 pieces (42 g)	200	**60**	60
Mike and Ike	19 pieces (40 g)	150	**0**	0
Milk Chocolate–Covered Jots	39 pieces (40 g)	190	**60**	36
Milk Chocolate Peanut Jots	17 pieces (40 g)	200	**90**	27
Milk Duds	13 pieces (40 g)	170	**50**	36
Milky Way	1 bar (58 g)	270	**90**	45
Fun size	2 bars (40 g)	180	**60**	32
Lite	1 bar (45 g)	170	**50**	23
Miniatures	5 pieces (41 g)	180	**60**	32
Mr. Goodbar	1 bar (49 g)	270	**150**	63
Mounds Bar	1 bar (53 g)	250	**120**	99
Snack size	2 bars (38 g)	180	**90**	72
Necco Mints	2 pieces (6 g)	24	**0**	0
Nerds	1 box (11 g)	40	**0**	0

SWEETS

FOOD	AMOUNT	CALORIES TOTAL	FAT	SAT-FAT
Nestlé				
100 Grand	1 bar (43 g)	200	**70**	45
Nestle Crunch	1 bar (44 g)	230	**100**	63
Fun size	4 bars (40 g)	210	**100**	63
Nips				
Caramel	2 pieces (14 g)	60	**10**	9
Chocolate Parfait	2 pieces (14 g)	60	**15**	9
Coffee	2 pieces (14 g)	50	**10**	14
NutRageous	1 bar (54 g)	290	**150**	45
Nonpareils (dark chocolate)	17 pieces (41 g)	200	**80**	54
Pastel Mints (Petite)	¼ cup (40 g)	210	**100**	18
Payday	1 bar (52 g)	250	**120**	18
Snack size	2 bars (40 g)	190	**90**	32
Peanut brittle	½ cup (40 g)	180	**45**	9
Peanut Butter Cups (Estee)	5 candies (38 g)	200	**110**	63
Pretzel Flipz (Nestle)	9 pieces (28 g)	130	**50**	45
Raisinets	¼ cup (45 g)	200	**70**	41
Raisins, chocolate-covered	34 pieces (40 g)	170	**60**	45
Reese's				
Miniatures	5 pieces (39 g)	210	**110**	41
Peanut Butter Cups	2 cups (45 g)	250	**130**	45
Pieces	50 pieces (39 g)	190	**70**	41
Sticks	1 pkg (42 g)	220	**120**	54
Rolo	7 pieces (42 g)	200	**80**	45
Russell Stover				
Almond Delights	2 pieces (40 g)	210	**110**	45
Assorted Caramels	3 pieces (40 g)	190	**80**	45
Assorted Chocolates	1 box (57 g)	300	**120**	72
	3 pieces (40 g)	190	**70**	45
Assorted Creams	3 pieces (40 g)	180	**60**	36
Assorted Whips	2 pieces (40 g)	210	**60**	36
Bars				
Almond Delight Bar	1 bar (57 g)	290	**150**	63
Caramel Bar	1 bar (46 g)	230	**100**	63
French Chocolate Mint	1 bar (42 g)	240	**140**	63
Mint Dream	1 bar (32 g)	160	**70**	45
Pecan Delight Bar	1 bar (57g)	310	**180**	63
Pecan Roll	1 bar (50 g)	260	**160**	18

SWEETS

FOOD	AMOUNT	CALORIES TOTAL	FAT	SAT-FAT
Russell Stover (cont.)				
Caramels	3 pieces (40 g)	170	**70**	45
Cherry Blimps	2 pieces (40 g)	180	**80**	36
Chocolate Truffle Golf Balls	1 ball (40 g)	200	**110**	72
Dark Chocolate Assortment	3 pieces (40 g)	190	**70**	45
French Chocolate Mints	4 pieces (40 g)	220	**130**	81
Home-Fashioned Favorites	3 pieces (40 g)	170	**60**	45
Milk Chocolate Assortment	3 pieces (40 g)	190	**80**	45
Nut, Chewy and Crispy Centers	3 pieces (40 g)	200	**90**	45
Toffee Sticks	3½ sticks (40 g)	230	**140**	72
Skittles (all flavors)	¼ cup (42 g)	170	**15**	0
	1 bag (61.5 g)	250	**25**	5
Snickers	1 bar (59 g)	280	**130**	45
Fun size	2 bars (40 g)	190	**90**	32
Toffee				
Brach's	3 pieces (18 g)	80	**20**	9
Butter Toffee (Farley's)	3 pieces (16 g)	70	**10**	5
English Toffees (Callard & Bowser)	2 pieces (17 g)	80	**35**	23
Licorice Toffees (Callard & Bowser)	2 pieces (17 g)	80	**30**	31
Tootsie Roll				
Midgies	6 pieces (40 g)	160	**25**	5
Pops	1 pop (17 g)	60	**0**	0
Twix	2 cookies (50 g)	280	**130**	45
Miniatures	3 pieces (29 g)	150	**60**	23
Villa Cherries (Brach's)	2 pieces (38 g)	150	**30**	18
Werther's				
Chocolates	10 pieces (40 g)	220	**120**	63
Original	3 pieces (15 g)	60	**10**	9
Whitman's				
Bars				
Cookies 'N Cream	1 bar (28 g)	150	**80**	63
Sampler				
Assorted Chocolates	4 pieces (50 g)	250	**110**	63
Assorted Creams	3 pieces (40 g)	180	**70**	45
Dark Chocolates	3 pieces (40 g)	200	**90**	54

SWEETS

FOOD	AMOUNT	CALORIES TOTAL	CALORIES FAT	CALORIES SAT-FAT
Whitman's (cont.)				
Truffles	3 pieces (40 g)	200	**100**	63
Whoppers	17 pieces (40 g)	180	**65**	63
Yogurt-covered				
Almonds	11 pieces (42 g)	210	**120**	63
Peanuts	17 pieces (41 g)	210	**120**	54
Pretzels	6 pieces (40 g)	150	**80**	72
Raisins (Harmony)	30 pieces (41 g)	180	**70**	54
York Peppermint Patties	3 patties (41 g)	160	**25**	18
COOKIES AND BARS BY BRAND NAME				
Assorted, no brand name				
Biscotti	1 (30 g)	270	**20**	5
Chocolate-Covered Pretzels	8 pieces (30 g)	140	**45**	27
White Chocolate–Covered Pretzels	8 pieces (30 g)	140	**45**	27
Danish Butter Cookies	4 (30 g)	160	**45**	45
French Roulettes				
Chocolate	4 (45 g)	225	**155**	36
Cinnamon	4 (33 g)	190	**130**	18
Original	4 (33 g)	190	**130**	18
Mundel Bread	2 pieces (30 g)	130	**60**	9
Archway				
Chocolate Chip 'n Toffee	1 (28 g)	130	**60**	18
Frosty Lemon	1 (26 g)	110	**40**	14
Gingersnaps	5 (32 g)	150	**5**	14
Oatmeal	1 (25 g)	110	**35**	9
Apple-Filled	1 (25 g)	100	**30**	5
Date-Filled	1 (28 g)	100	**30**	5
Iced	1 (28 g)	120	**45**	14
Raisin	1 (26 g)	110	**30**	9
Ruth's Golden	1 (28 g)	120	**45**	9
Strawberry-Filled	1 (25 g)	100	**30**	14
Old-Fashioned Molasses	1 (26 g)	100	**25**	5
Old-Fashioned Windmill	1 (20 g)	90	**30**	5
Rocky Road	1 (28 g)	130	**50**	14

SWEETS

FOOD	AMOUNT	CALORIES TOTAL	CALORIES FAT	CALORIES SAT-FAT
Delicious				
Animal Crackers	9 (28 g)	130	**45**	14
Applesauce Oatmeal	3 (306 g)	140	**40**	14
Assorted Sandwich	3 (30 g)	150	**50**	14
Banana Ramas	2 (25 g)	120	**40**	18
Butterfinger	3 (28 g)	130	**50**	23
Butter Thins	10 (29 g)	110	**45**	18
Chocolate Chip	2 (28 g)	130	**40**	9
Chocolate Chip Thins	10 (29 g)	110	**45**	18
Coconut Bars	3 (28 g)	140	**60**	27
Duplex Sandwich	3 (30 g)	150	**50**	14
English Toffee Heath	3 (36 g)	170	**80**	36
Fig Bars, lowfat	2 (37 g)	130	**27**	5
Fruit Bars (all flavors)	2 (29 g)	90	**0**	0
Ginger Snaps	4 (29 g)	130	**30**	5
Graham				
Cinnamon	2 crackers (30 g)	130	**30**	5
Honey	2 crackers (30 g)	130	**30**	5
Jelly Tops	5 (28 g)	140	**60**	14
Land O Lakes Frosted				
Butter	2 (27 g)	120	**50**	36
Lemon Sandwich	3 (30 g)	150	**50**	14
Oatmeal	2 (28 g)	120	**45**	9
Iced	2 (28 g)	130	**40**	9
Shortbread Cookies	4 (29 g)	140	**50**	27
Skippy Peanut Butter	3 (28 g)	150	**90**	18
Strawberry Sandwich	3 (30 g)	150	**50**	14
Sugar	2 (28 g)	140	**50**	9
Sugar Wafers	4 (30 g)	140	**54**	27
Vanilla Sandwich	3 (30 g)	150	**50**	14
Vanilla Wafers	8 (28 g)	130	**40**	9
Entenmann's				
Chocolate Chip	4 (28 g)	130	**60**	18
Chocolate Chip & Pecans	4 (28 g)	140	**70**	18
Chocolate Chunk	1 (16 g)	80	**35**	14
Chocolate Sandwich	3 (33 g)	150	**60**	14
Oatmeal Macaroon	3 (33 g)	150	**60**	18
Oatmeal Raisin	4 (28 g)	130	**45**	9

SWEETS

FOOD	AMOUNT	CALORIES TOTAL	FAT	SAT-FAT
Entenmann's (cont.)				
Peanut Butter Chocolate Chunk	1 (16 g)	80	**40**	14
Vanilla Sandwich	3 (33 g)	160	**60**	14
Entenmann's Light Cookies				
Chocolatey Chip	2 (30 g)	120	**35**	14
Oatmeal Raisin	2 (30 g)	100	**0**	0
Entenmann's Soft Bake Cookies				
Chocolate Chip	1 (20 g)	100	**45**	18
Milk Chocolate Chip	1 (20 g)	100	**45**	18
Original Recipe Chocolate Chip	3 (30 g)	150	**60**	18
White Chocolate Macadamia Nut	1 (20 g)	100	**50**	18
Estee				
Chocolate Chip	4 (31 g)	150	**60**	18
Chocolate Sandwich	3 (34 g)	160	**50**	14
Creme Wafer all flavors	5 (33 g)	155	**75**	14
Original Sandwich	3 (34 g)	160	**50**	9
Peanut Butter Sandwich	3 (34 g)	160	**60**	14
Vanilla Sandwich	3 (34 g)	160	**50**	9
Famous Amos				
Chocolate Chip & Pecans	4 (28 g)	140	**70**	18
Chocolate Chip	4 (28 g)	130	**60**	18
Chocolate Chunk	1 (16 g)	80	**35**	14
Chocolate Sandwich	3 (33 g)	150	**60**	14
Oatmeal Macaroon	3 (33 g)	160	**60**	18
Oatmeal Raisin	4 (28 g)	130	**45**	9
Peanut Butter Chocolate Chunk	1 (16 g)	80	**40**	14
Vanilla Sandwich	3 (33 g)	160	**60**	14
Fifty 50				
Chocolate Chip	4 (32 g)	170	**90**	NA
Coconut	4 (32 g)	160	**90**	27
Duplex Sandwich	3 (36 g)	160	**60**	14
Hearty Oatmeal	4 (32 g)	140	**50**	14

SWEETS

FOOD	AMOUNT	CALORIES TOTAL	FAT	SAT-FAT
Grandma's				
Fudge Chocolate Chip	1 (39 g)	170	**60**	18
Molasses	1 (39 g)	160	**35**	14
Keebler				
Animal				
Iced	1 (32 g)	150	**45**	9
Sprinkled	1 (32 g)	150	**40**	9
Chips Deluxe	1 (15 g)	80	**40**	14
with Peanut Butter	1 (16 g)	80	**40**	18
Chocolate Chewy	2 (28 g)	130	**60**	18
Chocolate Lover's	1 (17 g)	90	**45**	23
Coconut	2 (16 g)	80	**45**	18
Rainbow	1 (16 g)	80	**35**	18
Soft 'n Chewy	1 (16 g)	80	**30**	9
Classic Collection				
Chocolate Fudge Creme	1 (17 g)	80	**30**	9
French Vanilla Creme	1 (17 g)	80	**30**	9
Cookie Stix				
Butter Cookies	4 (27 g)	130	**45**	18
Chocolate Chip	4 (27 g)	130	**45**	14
Sugar Cookies	5 (27 g)	150	**50**	14
E. L. Fudge				
Butter-flavored Sandwich	2 (25 g)	120	**50**	9
Fudge Sandwich	2 (25 g)	120	**50**	9
Summer	2 (25 g)	120	**50**	9
Fudge Shoppe				
Deluxe Grahams	3 (27 g)	140	**60**	41
Fudge Sticks	3 (29 g)	150	**70**	41
Fudge Stripes	3 (32 g)	160	**70**	41
reduced fat	3 (27 g)	130	**40**	18
Grasshopper	4 (30 g)	150	**60**	41
Grahams	1 (16 g)	80	**45**	9
reduced fat	1 (16 g)	80	**30**	5
Chocolate	8 (31 g)	130	**35**	9
Cinnamon Crisp	8 (31 g)	130	**25**	9
low-fat	8 (28 g)	110	**10**	5
Honey	8 (31 g)	140	**40**	9
low-fat	9 (28 g)	120	**15**	5

SWEETS

FOOD	AMOUNT	CALORIES TOTAL	FAT	SAT-FAT
Keebler (cont.)				
Sandies				
Pecan Shortbread	1 (16 g)	80	**45**	9
reduced fat	1 (16 g)	80	**30**	5
Simply Shortbread	1 (15 g)	80	**40**	18
Snackin' Grahams				
Cinnamon	21 (29 g)	130	**25**	9
Honey	33 (30 g)	130	**25**	9
Soft Batch Chocolate Chip	1 (16 g)	80	**35**	9
Vanilla Wafers	8 (31 g)	150	**60**	18
reduced fat	8 (31 g)	130	**30**	5
Little Debbie				
Figaroos				
low-fat	1 (43 g)	150	**23**	5
Fudge Rounds	1 cake (34 g)	140	**50**	14
Fudge Macaroons	1 (29 g)	140	**70**	36
Lemon Creme Wafers	1 (20 g)	100	**45**	9
Marshmallow Pies	1 (39 g)	160	**50**	27
Marshmallow Supremes	1 (32 g)	130	**45**	9
Muffin Loaves	1 (55 g)	220	**90**	14
Nutty Bars	2 (57 g)	310	**160**	27
Oatmeal Creme Pies	1 (38 g)	170	**60**	14
Oatmeal Lights	1 (38 g)	130	**20**	5
Peanut Butter Bars	2 (54 g)	270	**140**	27
Peanut Butter Naturals	1 wrap (44 g)	230	**120**	23
Star Crunch	1 (31 g)	140	**50**	14
Lu				
Fondant	4 (32 g)	170	**80**	72
Le Chocolatier	3 (28 g)	150	**70**	63
Le Pim's Orange	2 (25 g)	90	**25**	9
Le Truffé	4 (33 g)	170	**80**	63
Murray				
Assortment	5 (27 g)	130	**60**	14
Butter	8 (30 g)	130	**40**	9
Chocolate Chips	8 (30 g)	140	**45**	14
Ginger Snaps	5 (30 g)	140	**40**	9

SWEETS

FOOD	AMOUNT	CALORIES TOTAL	FAT	SAT-FAT
Murray (cont.)				
Lemon Cremes	3 (28 g)	140	**50**	14
Sugar Wafers	5 (28 g)	150	**70**	18
Vanilla Cremes	3 (28 g)	140	**50**	14
Vanilla Wafers	8 (28 g)	120	**25**	9
Murray Sugar Free				
Chocolate Chip	3 (31 g)	140	**60**	27
Chocolate Sandwich	3 (28 g)	120	**50**	14
Shortbread	8 (28 g)	120	**40**	9
Nabisco				
Apple Newtons, fat-free	2 (29 g)	100	**0**	0
Barnum's Animal Crackers	12 (31 g)	140	**35**	5
Cameo Creme Sandwich	2 (28 g)	130	**40**	9
Chips Ahoy!	3 (32 g)	160	**70**	23
Chewy	3 (36 g)	170	**70**	23
Chunky	1 (16 g)	80	**35**	14
Munch Size	6 (32 g)	160	**70**	18
Reduced Fat	3 (31 g)	140	**45**	14
Snack Bars	1 (37 g)	150	**45**	18
Chocolate Teddy Grahams	24 (30 g)	140	**40**	9
Cinnamon Teddy Grahams	24 (30 g)	140	**40**	9
Cranberry Newtons, fat-free	2 (29 g)	100	**0**	0
Famous Chocolate Wafers	5 (32 g)	140	**35**	14
Fig Newtons	2 (31 g)	110	**20**	0
Fat-Free	2 (29 g)	100	**0**	0
Fudge Striped Shortbread	3 (32 g)	160	**70**	14
Ginger Snaps	4 (28 g)	120	**25**	5
Grahams	4 (28 g)	120	**25**	5
Honey Maid Grahams				
Chocolate	8 (28 g)	120	**25**	5
Cinnamon	8 (28 g)	120	**20**	0
Honey	8 (28 g)	120	**25**	0
Low-fat	8 (28 g)	110	**15**	0
Oatmeal Crunch	8 (28 g)	120	**20**	0
Lorna Doone Shortbread	4 (29 g)	140	**60**	9
Mallomars	2 (26 g)	120	**45**	23
Marshmallow Twirls	1 (30 g)	130	**50**	9

SWEETS

FOOD	AMOUNT	CALORIES TOTAL	FAT	SAT-FAT
Nabisco (cont.)				
Mystic Mint Sandwich	1 (17 g)	90	**40**	9
Newtons Cobblers	1 (22 g)	70	**0**	0
Nilla Wafers	8 (32 g)	140	**40**	9
Nutter Butter				
Bites	10 (30 g)	150	**60**	9
Chocolate	2 (28 g)	130	**45**	9
Peanut Butter Sandwich	2 (28 g)	130	**50**	9
Peanut Creme Patties	5 (31 g)	160	**80**	14
Oatmeal	1 (17 g)	80	**30**	5
Iced Oatmeal	1 (17 g)	80	**25**	0
Oreos				
Brownie Bars	1 (37 g)	150	**50**	14
Chocolate Sandwich	3 (33 g)	160	**60**	14
Double Stuf Chocolate Sandwich	2 (28 g)	140	**60**	14
Fudge Covered Oreos	1 (21 g)	110	**50**	14
Raspberry, fat-free	2 (29 g)	10	**0**	0
Reduced Fat	3 (32 g)	130	**30**	9
SnackWell's (*see* listing after **Rippin' Good**)				
Social Tea Biscuits	6 (28 g)	120	**30**	5
Vanilla Sandwich	3 (35 g)	160	**45**	9
Pepperidge Farm				
Bordeaux	4 (28 g)	130	**50**	23
Brussels	3 (30 g)	150	**60**	27
Chessmen	3 (26 g)	120	**70**	27
Chocolate Chip Blondie (fat-free)	1 blondie (40 g)	120	**0**	0
Chocolate Chunk Classic Cookies				
Chesapeake	1 (26 g)	140	**70**	23
Chocolate Chunk Pecan	1 (26 g)	140	**70**	23
Milk Chocolate Macadamia	1 (26 g)	130	**60**	23
Montauk Milk Chocolate Chunk	1 (26 g)	130	**60**	23
Nantucket Chocolate Chunk	1 (26 g)	140	**60**	23

SWEETS

FOOD	AMOUNT	CALORIES TOTAL	FAT	SAT-FAT
Pepperidge Farm (cont.)				
Sausalito Milk Chocolate Macadamia Chocolate Chunk	1 (26 g)	140	**70**	23
Tahoe White Chocolate Macadamia Chocolate Chunk	1 (26 g)	140	**70**	27
Chocolate Laced Pirouettes	5 (33 g)	180	**90**	23
Fruitful				
Apricot Raspberry Cup	3 (32 g)	140	**50**	18
Raspberry Tart	2 (30 g)	120	**25**	9
Strawberry Cup	3 (32 g)	140	**50**	18
Fudge Dipped Brownie (reduced fat)	1 (48 g)	190	**40**	9
Fudge Striped Chocolate Chunk Cookies	1 (18 g)	80	**20**	14
Geneva	3 (31 g)	160	**80**	32
Lido	1 (17 g)	90	**45**	14
Milano	3 (34 g)	180	**90**	32
Double Chocolate	2 (27 g)	140	**70**	27
Endless Chocolate	3 (34 g)	180	**90**	45
Milk Chocolate	3 (33 g)	170	**80**	32
Mint	2 (25 g)	130	**70**	32
Orange	2 (25 g)	130	**70**	32
Old-Fashioned				
Chocolate Chip	3 (28 g)	140	**60**	23
Ginger Man	4 (28 g)	130	**35**	9
Hazelnut	3 (32 g)	160	**70**	18
Lemon-Nut Crunch	3 (31 g)	170	**80**	18
Shortbread	2 (26 g)	140	**70**	23
Sugar	3 (30 g)	140	**60**	14
Santa Fe Oatmeal Raisin	1 (26 g)	120	**45**	9
Soft Baked				
Chocolate Chunk	1 (26 g)	130	**60**	23
reduced fat	1 (26 g)	110	**35**	18
Milk Chocolate Macadamia	1 (26 g)	130	**60**	23

SWEETS

FOOD	AMOUNT	CALORIES TOTAL	CALORIES FAT	CALORIES SAT-FAT
Pepperidge Farm (cont.)				
Oatmeal Raisin	1 (26 g)	110	**40**	9
reduced fat	1 (26 g)	100	**25**	5
Rippin' Good				
Assorted Creme Wafers	3 (28 g)	140	**60**	14
Butter Thins	10 (29 g)	110	**45**	18
Chocolate Chip	3 (32 g)	150	**60**	18
Chocolate Chip Sandwich	2 (31 g)	150	**60**	18
Chocolate Chip Thins	10 (29 g)	110	**45**	18
Chocolate Sandwich	3 (34 g)	160	**60**	14
Coconut Bars	3 (26 g)	130	**50**	23
Cookie Jar Assortment	3 (33 g)	150	**60**	18
Duplex Sandwich	3 (34 g)	160	**50**	14
Ginger Snaps	5 (28 g)	130	**40**	9
Granola & Peanut Butter Sandwich	2 (31 g)	150	**50**	14
Iced Spice	3 (32 g)	130	**25**	5
Lemon Crisp	3 (32 g)	160	**70**	14
Lemon Sandwich	3 (34 g)	160	**50**	14
Macaroon Sandwich	2 (31 g)	150	**60**	23
Oatmeal	3 (32 g)	150	**50**	14
Iced	3 (36 g)	150	**40**	9
Striped	2 (28 g)	150	**70**	36
Peanut Butter Sandwich	2 (31 g)	150	**60**	14
Shortbread Cookies	4 (29 g)	140	**50**	27
Strawberry Sandwich	3 (34 g)	160	**50**	14
Striped Dainties	3 (28 g)	150	**70**	36
Sugar	3 (32 g)	150	**60**	14
Toffee 'n Creme Sandwich	2 (31 g)	150	**60**	14
Vanilla Sandwich	3 (34 g)	160	**50**	14
Vanilla Wafers	5 (28 g)	120	**35**	9
SnackWell's (Nabisco)				
Caramel Delights	1 (18 g)	70	**20**	5
Chocolate Chip	13 (29 g)	130	**35**	14
Chocolate Sandwich	2 (26 g)	110	**25**	5
Creme Sandwich	2 (26 g)	110	**25**	5
Devil's Food	1 (16 g)	50	**0**	0
Double Chocolate Chip	13 (30 g)	130	**35**	14

SWEETS

FOOD	AMOUNT	CALORIES TOTAL	FAT	SAT-FAT
SnackWell's (Nabisco) (cont.)				
Golden Devil's Food	1 (16 g)	50	**0**	0
Mint Creme	2 (25 g)	110	**30**	9
Oatmeal Raisin	2 (30 g)	120	**30**	5
Peanut Butter Chip	13 (29 g)	120	**35**	9
Stella D'oro				
Almond Toast Cookie	2 (29 g)	110	**20**	5
Anginetti Cookies	4 (31 g)	140	**35**	5
Anisette Sponge Cookies	2 (27 g)	90	**10**	0
Anisette Toast Cookies	3 (35 g)	130	**10**	0
Biscotti				
Chocolate Chunk	1 (22 g)	90	**25**	9
French Vanilla	1 (22 g)	90	**20**	5
Breakfast Treats Cookies	1 (23 g)	100	**30**	9
Lady Stella Cookie				
Assortment	3 (28 g)	130	**45**	14
Sunshine				
Ginger Snaps	7 (29 g)	130	**40**	9
Golden Fruit				
Cranberry Biscuits	1 (20 g)	80	**20**	0
Raisin Biscuits	1 (20 g)	80	**15**	5
Hydrox Chocolate Sandwich				
Cremes	3 (31 g)	150	**60**	18
Oatmeal, Country Style	2 (24 g)	120	**45**	9
Peanut Butter Sugar Wafers	4 (32 g)	170	**80**	18
Sugar Wafers	3 (26 g)	130	**60**	14
Vanilla Wafers	7 (31 g)	150	**60**	14
Vienna Fingers	2 (29 g)	140	**50**	14
reduced fat	2 (29 g)	130	**40**	9
Twix				
Chocolate Caramel	1 (29 g)	140	**60**	23
Weight Watchers				
Apple Raisin Bar	1 (21 g)	70	**20**	5
Chocolate Chip	2 (30 g)	140	**45**	18
Chocolate Sandwich	3 (31 g)	140	**35**	9
Oatmeal Raisin	2 (30 g)	120	**15**	0
Vanilla Sandwich	3 (31 g)	140	**25**	9

SWEETS

FOOD	AMOUNT	CALORIES TOTAL	FAT	SAT-FAT
COOKIES AND BARS, MIXES				
Betty Crocker				
Chocolate Chip	3 tbsp mix (28 g)	120	**30**	14
	2 cookies	160	**70**	23
Double Chocolate Chunk	3 tbsp mix (28 g)	120	**25**	14
	2 cookies	150	**60**	18
Oatmeal Chocolate Chip	3 tbsp mix (28 g)	120	**30**	14
	2 cookies	160	**70**	18
Peanut Butter	3 tbsp mix (28 g)	120	**35**	5
	2 cookies	160	**70**	14
Sugar	3 tbsp mix (28 g)	120	**25**	9
	2 cookies	170	**70**	18
COOKIES, READY-TO-MAKE				
Pillsbury				
Chocolate Chip	28 g dough	130	**50**	23
reduced fat	28 g dough	110	**25**	14
with walnuts	28 g dough	140	**60**	18
Cookies with M&M's	28 g dough	130	**50**	18
Double Chocolate Chip & Chunk	28 g dough	130	**50**	18
Cookie mixes, refrigerated				
Nestlé Toll House				
Chocolate Chip	2 tbsp (32)	140	**50**	18
DANISH PASTRY				
Almond	1 (100 g)	420	**190**	41
Apple-Filled	1 (110 g)	350	**110**	36
Blueberry-Filled	1 (100 g)	360	**150**	27
Cheese	1 (110 g)	380	**170**	45
Cherry Cheese	1 (100 g)	360	**140**	45
Cherry-Filled	1 (100 g)	320	**100**	27
Coconut	1 (100 g)	400	**150**	45
Custard-Filled	1 (100 g)	350	**140**	36
Danish Ring				
Pecan	⅛ ring (57 g)	230	**100**	23
Walnut	⅛ ring (55 g)	240	**120**	23

SWEETS

FOOD	AMOUNT	CALORIES TOTAL	CALORIES FAT	CALORIES SAT-FAT
Danish Pastry (cont.)				
Danish Twist (Entenmann's)				
Cinnamon	⅙ ring (61 g)	260	**130**	27
Raspberry	⅙ ring (53 g)	220	**100**	27
Lemon-Filled	1 (100 g)	350	**140**	31
Orange Danish (Pillsbury, ready-to-make)	1 (41 g)	140	**50**	NA
Pecan	1 (100 g)	400	**180**	41
DONUTS				
Bakery type				
Apple Raisin Rosebud	1 (55 g)	220	**90**	23
Blueberry Cake	1 (56 g)	250	**100**	23
Blueberry-Filled	1 (65 g)	250	**120**	27
Carrot Cake	1 (80 g)	340	**180**	45
Cherry Cake	1 (56 g)	250	**100**	27
Chocolate Chip	1 (70 g)	310	**140**	41
Chocolate Creme-Filled	1 (65 g)	290	**150**	41
Chocolate-Iced Chocolate Cake	1 (65 g)	270	**110**	31
Chocolate-Iced Yellow Cake	1 (55 g)	230	**90**	27
Cinnamon Rosebud	1 (64 g)	260	**110**	36
Custard-Creme Filled	1 (65 g)	240	**130**	63
Glazed Yeast Raised	1 (55 g)	240	**130**	36
Honey Wheat	1 (80 g)	340	**150**	36
Lemon Custard–Filled	1 (65 g)	190	**140**	36
Old-Fashioned	1 (62 g)	270	**140**	36
Orange Glazed Cake	1 (65 g)	310	**160**	41
Sour Cream	1 (80 g)	340	**160**	36
Spicy Apple–Filled	1 (65 g)	250	**130**	36
Entenmann's				
Crumb-Topped Donuts	1 (60 g)	260	**110**	27
Devil's Food Crumb Donuts	1 (60 g)	250	**120**	36
Rich Frosted Donuts	1 (57 g)	280	**170**	54
Old-Fashioned Donuts	1 (62 g)	270	**140**	36
Sour Cream Donuts	1 (80 g)	340	**160**	36

SWEETS

FOOD	AMOUNT	CALORIES TOTAL	CALORIES FAT	CALORIES SAT-FAT
Hostess				
Cinnamon Donettes	4 (61 g)	240	**90**	36
Frosted Donettes	3 (43 g)	200	**110**	63
Frosted Donuts	1 (47 g)	330	**160**	72
Plain Donuts	1 (47 g)	150	**80**	36
Powdered Donettes	3 (43 g)	180	**70**	27
Powdered Donuts	1 (47 g)	230	**160**	72
Krispy Kreme				
Enrobed Knibbles	5 pieces (54 g)	270	**150**	72
Glazed Donuts	1 (52 g)	200	**100**	27
Glazed Knibbles	3 pieces (49 g)	220	**120**	27
Mini Crullers	3 (56 g)	230	**90**	27
Powdered Sugar Knibbles	3 pieces (48 g)	220	**120**	36
Little Debbie				
Donut Sticks	1 (47 g)	220	**110**	42
TastyKake				
Cinnamon Donut	1 (48 g)	190	**90**	14
Donut Holes				
Chocolate glazed	5 (54 g)	230	**130**	23
Glazed	3 (39 g)	190	**100**	23
Orange glazed	5 (54 g)	230	**130**	18
Sweet Sprinkles	4 (56 g)	250	**120**	27
Plain Donut	1 (41 g)	160	**80**	14
Powdered Sugar	1 (48 g)	190	**80**	14
Rich Frosted Donut	1 (58 g)	250	**130**	54
DUMPLINGS, TURNOVERS, AND STRUDEL				
Apple Dumpling (Pepperidge Farm)	1 dumpling (85 g)	290	**99**	23
Apple-Filled Pastry	1 pastry (64 g)	190	**45**	9
Apple Fritter	1 fritter (70 g)	300	**130**	32
Apple Strudel	1 piece (64 g)	200	**90**	27
Apple Turnover (Pepperidge Farm)	1 turnover (89 g)	330	**130**	27
Apricot-Filled pastry	1 roll (64 g)	200	**50**	9
Apricot Strudel	1 piece (64 g)	220	**100**	27
Custard-Filled Pastry	1 (64 g)	190	**40**	9

SWEETS

FOOD	AMOUNT	CALORIES TOTAL	CALORIES FAT	CALORIES SAT-FAT
Peach Dumpling				
(Pepperidge Farm)	1 dumpling (85 g)	320	**99**	23
Toaster Strudel				
Pillsbury (all flavors)	1 (54 g)	180	**60**	14
FROZEN DESSERTS				
Frozen custard				
Kohr Brothers				
Light, Vanilla & Chocolate	4 fl oz	130	**50**	36
FROZEN NOVELTIES				
Ben & Jerry's Peace Pops				
Cookie Dough	1 (108 g)	420	**230**	126
Dole Juice Bars				
Fruit 'n Juice, all flavors	1	70	**0**	0
Fruit Juice Bar, all flavors	1	45	**0**	0
Dove Bars				
Dark Chocolate Chocolate	1 bar (79 g)	260	**150**	90
French Vanilla Bite-Size	5 bars (92 g)	330	**190**	117
Milk Chocolate & Vanilla	1 bar (77 g)	260	**150**	99
Milk Chocolate with Almonds & Vanilla Ice Cream	1 bar (80 g)	280	**170**	99
Eskimo Pie				
Chocolate-coated Vanilla Ice Cream Bar	1 (49 g)	120	**70**	54
Vanilla Cone with no-sugar Choc & Peanut topping	1 (74 g)	210	**110**	72
Vanilla Ice Cream Sandwich, no sugar	1 (65 g)	160	**40**	18
Good Humor				
Chocolate Eclair	1 bar (59 g)	170	**70**	27
Dark & Milk Chocolate	1 bar (57 g)	180	**110**	72
Strawberry Shortcake	1 bar (59 g)	170	**70**	27
Toasted Almond	1 bar (88 mL)	180	**80**	36
Fudgsicle	1 (43 g)	60	**5**	5
fat-free	1 (43 g)	60	**0**	0
King Cone	1 (136 mL)	300	**90**	54
Original Ice Cream Bar	1 (89 mL)	190	**90**	72

SWEETS

FOOD	AMOUNT	CALORIES TOTAL	FAT	SAT-FAT
Good Humor (cont.)				
Popsicle (Twister)	1 (52 mL)	45	**0**	0
Premium Sundae Cone	1 (136 mL)	290	**120**	63
Strawberry Shortcake	1 bar (88 mL)	160	**70**	27
Toasted Almond	1 bar (88 mL)	180	**80**	36
Häagen-Dazs Ice Cream Bars				
Cookies & Cream Crunch	1 (88 g)	320	**200**	117
Mint Dark Chocolate	1 (84 g)	290	**180**	108
Strawberry & White Chocolate	1 (80 g)	270	**170**	108
Vanilla & Almonds	1 (87 g)	320	**210**	108
Vanilla and Milk Chocolate	1 (100 g)	330	**220**	126
Klondike				
Almond	1 piece (96 g)	310	**190**	117
Big Bear Ice Cream Cookie Sandwich	1 (85 g)	290	**110**	72
Big Bear Vanilla Ice Cream Sandwich	1 (78 g)	200	**60**	27
Caramel Crunch	1 piece (96 g)	290	**150**	117
Choco Taco	1 bar (91 g)	310	**150**	90
Chocolate	1 piece (92 g)	280	**170**	117
Krispy Krunch	1 piece (96 g)	300	**170**	126
Neapolitan	1 piece (91 g)	280	**170**	126
Original	1 piece (93 g)	290	**180**	126
Reduced Fat, no sugar added	1 piece (78 g)	190	**90**	63
Vanilla (Original)	1 piece (93 g)	290	**180**	126
Milky Way				
Ice Cream Bar, Dark	1 (69 g)	220	**120**	81
Low-Fat Milk Shake	1 cup (237 mL)	220	**30**	NA
Mrs. Fields Cookie Ice Cream Sandwich	1 (116 g)	395	**185**	117
Nestlé				
Bon Bons	8 pieces (93 g)	350	**230**	144
Cool Creations Tiger Tails	1 pop (69 g)	60	**0**	0

SWEETS

FOOD	AMOUNT	CALORIES TOTAL	CALORIES FAT	CALORIES SAT-FAT
Nestlé (cont.)				
Crunch				
reduced fat	1 bar (51 g)	150	**80**	54
Vanilla	1 bar (62 g)	200	**130**	99
Drumstick Sundae Cones				
Chocolate	1 (99 g)	320	**150**	90
Vanilla	1 (99 g)	340	**170**	99
reduced fat	1 (97 g)	300	**130**	45
Vanilla Caramel	1 (103 g)	360	**180**	117
Drumstick Supreme Cones				
Cappucccino	1 (89 g)	270	**120**	81
Strawberry	1 (88 g)	230	**90**	54
Triple Chocolate	1 (89 g)	260	**120**	63
Snickers				
Ice Cream Bar	1 (52 g)	200	**110**	72
snack bar	1 (50 g)	180	**100**	54
Ice Cream Cone	1 (87 g)	290	**130**	72
Trix				
Pops	1 (53 g)	40	**0**	0
Tropicana Orange Cream Bar	1 (71 g)	80	**10**	5
Weight Watchers Smart Ones				
Chocolate Chip Cookie Dough Sundae	1 dessert (75 g)	190	**40**	18
Chocolate Mousse	1 dessert (77 g)	190	**45**	18
Chocolate Treat	1 bar (87 g)	100	**5**	0
Double Fudge Brownie Parfait	1 parfait (109 g)	190	**25**	18
Orange Vanilla Treat	2 bars (80 g)	70	**10**	5
Strawberry Parfait Royale	1 parfait (104 g)	180	**20**	9
Vanilla Sandwich	1 (65 g)	150	**25**	9
FROZEN YOGURT—FREEZER COMPARTMENT, GROCERY STORE				
Ben & Jerry's Frozen Yogurt				
Cherry Garcia (low-fat)	½ cup	170	**30**	18
Chocolate Chip Cookie Dough	½ cup	210	**35**	23
Chocolate Fudge Brownie	½ cup	180	**20**	9

SWEETS

FOOD	AMOUNT	CALORIES TOTAL	CALORIES FAT	CALORIES SAT-FAT
Ben & Jerry's Frozen Yogurt (cont.)				
Coffee Almond Fudge	½ cup	180	**40**	9
Vanilla Heath Bar Crunch	½ cup	210	**50**	23
Breyers Frozen Yogurt				
Natural Vanilla	½ cup	120	**25**	14
Colombo Nonfat Shoppe-Style Frozen Yogurt				
chocolate flavors	½ cup	130	**0**	0
all other flavors	½ cup	110	**0**	0
Häagen-Dazs Frozen Yogurt				
Vanilla	½ cup	140	**0**	0
Vanilla Fudge	½ cup	160	**0**	0
Vanilla-Raspberry Swirl	½ cup	130	**0**	0
Fat-Free Frozen Yogurt				
Vanilla	½ cup	140	**0**	0
Vanilla-Raspberry Swirl	½ cup	140	**0**	0
Kemp's Frozen Yogurt				
Chocolate	½ cup	110	**25**	18
Fudge Marble	½ cup	110	**0**	0
Mint Fudge	½ cup	110	**0**	0
Strawberry	½ cup	90	**0**	0
Vanilla	½ cup	110	**25**	18
Safeway Select Fat Free Frozen Yogurt				
all flavors	½ cup	80–90	**0**	**0**

FROZEN YOGURT—YOGURT OR ICE CREAM SHOP

Colombo *(These figures are for 1 fl oz. Ask the server for the number of ounces in your serving.)*

FOOD	AMOUNT	CALORIES TOTAL	CALORIES FAT	CALORIES SAT-FAT
Lite (nonfat)	1 fl oz	25	**0**	0
Low-fat	1 fl oz	28	**4**	2
Peanut butter low-fat	1 fl oz	30	**6**	1
ICBIY				
Nonfat	small (6¾ fl oz)	135	**0**	0
	medium (9⅓ fl oz)	187	**0**	0
	large (12 fl oz)	240	**0**	0
Original	small (6¾ fl oz)	182	**43**	NA
	medium (9⅓ fl oz)	251	**59**	NA
	large (12 fl oz)	324	**76**	NA

SWEETS

FOOD	AMOUNT	CALORIES TOTAL	FAT	SAT-FAT
TCBY				
Nonfat	small (5 fl oz)	138	**0**	0
	medium (7 fl oz)	193	**0**	0
	large (9 fl oz)	248	**0**	0
Original	small (5 fl oz)	163	**38**	23
	medium (7 fl oz)	228	**53**	32
	large (9 fl oz)	293	**68**	41
ICE CREAM, BY BRAND NAMES				
Ben & Jerry's				
Ice Cream				
Butter Pecan	½ cup	310	**220**	99
Cherry Garcia	½ cup	240	**140**	90
Chocolate Chip Cookie Dough	½ cup	270	**150**	81
Chocolate Fudge Brownie	½ cup	260	**110**	63
Chubby Hubby	½ cup	350	**210**	99
Chunky Monkey	½ cup	280	**170**	90
Coffee with Heath Toffee Crunch	½ cup	280	**170**	90
Cool Britannia	½ cup	260	**140**	90
New York Super Fudge Chunk	½ cup	290	**190**	99
Peanut Butter & Jelly	½ cup	280	**150**	81
Wavy Gravy	½ cup	330	**210**	99
Vanilla Caramel Fudge	½ cup	280	**150**	90
All Natural Ice Cream				
Chocolate Chip Cookie Dough	½ cup	300	**140**	90
Chocolate Fudge Brownie	½ cup	280	**130**	90
New York Super Fudge Chunk	½ cup	320	**190**	108
Phish Food	½ cup	300	**120**	90
Totally Nuts	½ cup	310	**190**	99
Low-fat Ice Cream				
Blackberry Cobbler	½ cup	180	**30**	18
Blond Brownie Sundae	½ cup	190	**25**	14
Coconut Cream Pie	½ cup	160	**25**	14

SWEETS

FOOD	AMOUNT	CALORIES TOTAL	FAT	SAT-FAT
Ben & Jerry's (cont.)				
S'mores	½ cup	190	**20**	9
Sweet Cream & Cookies	½ cup	170	**25**	18
Vanilla & Chocolate Mint Patty	½ cup	180	**25**	18
Breyers				
Ice Cream				
Butter Pecan	½ cup	180	**110**	54
Cherry Vanilla	½ cup	150	**70**	45
Chocolate	½ cup	160	**80**	54
Chocolate Chip	½ cup	170	**90**	63
Chocolate Chip Dough	½ cup	180	**90**	54
Coffee	½ cup	150	**80**	54
French Vanilla	½ cup	160	**90**	54
Mint Chocolate Chip	½ cup	170	**90**	63
Peanut Butter & Fudge Swirls	½ cup	180	**80**	41
Strawberry	½ cup	130	**60**	41
Vanilla	½ cup	150	**80**	54
Viennetta				
Cappuccino & Vanilla	1 slice (68 g)	190	**100**	63
Chocolate	1 slice (68 g)	190	**100**	72
Vanilla	1 slice (68 g)	190	**100**	63
Edy's				
Grand Ice Cream				
Butter Pecan	½ cup	160	**90**	45
Cherry Chocolate Chip	½ cup	150	**70**	45
Chocolate Chip Dough	½ cup	170	**80**	45
Cookie Jar	½ cup	180	**80**	54
Double Fudge Brownie	½ cup	170	**80**	45
French Vanilla	½ cup	160	**90**	45
Rocky Road	½ cup	170	**90**	45
Vanilla Bean	½ cup	140	**70**	45
Ice Cream				
Baby Ruth Baseball Sundae	½ cup	170	**70**	45
Girl Scouts Thin Mint	½ cup	170	**90**	54
Godzilla Vanilla	½ cup	160	**70**	45

SWEETS

FOOD	AMOUNT	CALORIES TOTAL	FAT	SAT-FAT
Edy's (cont.)				
Homemade Ice Cream				
Banana Cream Pie	½ cup	130	**50**	27
Vanilla	½ cup	140	**60**	36
Light Ice Cream				
Almond Praline	½ cup	110	**35**	18
Cheesecake Chunk	½ cup	120	**45**	27
Chiquita 'n Chocolate	½ cup	110	**45**	23
Chocolate Fudge Mousse	½ cup	110	**25**	18
French Silk	½ cup	120	**35**	23
Rocky Road	½ cup	120	**40**	23
Vanilla	½ cup	100	**25**	18
Häagen-Dazs				
Ice Cream				
Butter Pecan	½ cup	310	**210**	99
Chocolate	½ cup	270	**160**	99
Coffee	½ cup	270	**160**	99
Dulce de Leche Caramel	½ cup	290	**150**	90
Macadamia Brittle	½ cup	300	**180**	99
Rum Raisin	½ cup	270	**160**	90
Strawberry	½ cup	250	**150**	90
Vanilla	½ cup	270	**160**	99
Vanilla Swiss Almond	½ cup	310	**190**	99
Low-fat Ice Cream				
Chocolate	½ cup	170	**25**	14
Strawberry	½ cup	150	**20**	9
Vanilla	½ cup	170	**25**	14
Healthy Choice Low-Fat Ice Cream				
Cappuccino Chocolate Chunk	½ cup	120	**20**	9
Chocolate Chunk	½ cup	120	**20**	9
Fudge Brownie	½ cup	120	**20**	9
Praline & Caramel	½ cup	130	**20**	5
Rocky Road	½ cup	140	**20**	9
Vanilla	½ cup	100	**20**	9
Lucerne Ice Cream				
Almond & Caramel	½ cup	170	**90**	45

SWEETS

FOOD	AMOUNT	CALORIES TOTAL	CALORIES FAT	CALORIES SAT-FAT
Lucerne Ice Cream (cont.)				
Cappuccino Chip	½ cup	150	**70**	45
Cherry Blossom	½ cup	140	**70**	45
Chocolate	½ cup	140	**80**	45
Chocolate Chip	½ cup	150	**80**	45
Chocolate Marble	½ cup	140	**60**	36
Coffee	½ cup	140	**70**	45
French Vanilla	½ cup	150	**70**	45
Golden Nut Sundae	½ cup	160	**70**	45
Mint Chocolate Chip	½ cup	150	**80**	45
Neapolitan	½ cup	140	**70**	14
Rocky Road	½ cup	160	**80**	45
Swiss Chocolate Cherry	½ cup	140	**60**	36
Tin Roof Sundae	½ cup	160	**70**	45
Lucerne Low-fat Ice Cream				
Neapolitan	½ cup	100	**20**	14
Newman's Own Ice Cream				
Milk Chocolate Mud Bath	½ cup	190	**90**	63
Obscene Vanilla Bean	½ cup	170	**90**	63
Pistol Packin' Praline Pecan	½ cup	200	**100**	63
Safeway Select				
Fat-Free Ice Cream				
all flavors	½ cup	70–90	**0**	0
Health Advantage Low-fat Ice Cream				
Coffee Latte Chocolate Swirl	½ cup	100	**10**	5
Cookies 'n Cream	½ cup	100	**20**	9
Rocky Road	½ cup	110	**20**	5
Vanilla	½ cup	90	**15**	9
Light Ice Cream				
Chocolate Chip Cookie Dough	½ cup	130	**50**	27
Cookies 'n Cream	½ cup	130	**50**	27
Rocky Road	½ cup	130	**45**	23
Vanilla	½ cup	120	**40**	27
Starbucks Coffee				
Biscotti Bliss	½ cup	260	**120**	63
Chocolate Chocolate Fudge	½ cup	290	**150**	72

SWEETS

FOOD	AMOUNT	CALORIES TOTAL	FAT	SAT-FAT
Starbucks Coffee (cont.)				
Coffee Almond	½ cup	250	**117**	63
Coffee Almond Fudge	½ cup	260	**120**	63
Espresso Swirl	½ cup	220	**90**	54
Frappuccino	1 bar (80 g)	110	**20**	9
Italian Roast	½ cup	230	**110**	63
Java Chip	½ cup	250	**117**	63
Starbucks Low-fat Ice Cream				
Latte	½ cup	170	**25**	14
Mocha Mambo	½ cup	170	**25**	14
ICE CREAM CONES				
Cake Cone	1 (4.5 g)	20	**0**	0
Oreo Chocolate Cone	1 (13 g)	50	**5**	0
Sugar Cone	1 (13 g)	50	**0**	0
Waffle Cone	1 (20 g)	80	**10**	0
ICE CREAM TOPPINGS				
Hershey's				
Candy Bar Sprinkles	2 tbsp	140	**45**	27
Chocolate Chips	1 oz	140	**72**	45
Coconut	2 tbsp	58	**37**	33
Mrs. Richardson				
Hot Fudge	2 tbsp	140	**60**	54
Caramel	2 tbsp	130	**0**	0
Parlor Perfect				
Confetti Sprinkles	2 tbsp	110	**20**	5
Cookie 'n Nut Crunch	2 tbsp	80	**30**	9
Ice Cream Critters	2 tbsp	90	**15**	14
Praline Nut Crunch	2 tbsp	100	**40**	18
Shell Topping				
Chocolate	2 tbsp	210	**150**	72
Chocolate Fudge	2 tbsp	200	**120**	72
Heath	2 tbsp	230	**150**	63
Hershey's Krackel	2 tbsp	190	**130**	54
Reese's Chocolate & Peanut Butter	2 tbsp	220	**150**	72

SWEETS

FOOD	AMOUNT	CALORIES TOTAL	FAT	SAT-FAT
Sundae Syrup				
all flavors	2 tbsp	110	**0**	0
Walnut	2 tbsp	170	**80**	9
Syrup				
Chocolate	2 tbsp	100	**0**	0
Smuckers				
Butterscotch, Caramel	2 tbsp	130	**0**	0
Dove Dark Chocolate	2 tbsp	140	**45**	14
Dove Milk Chocolate	2 tbsp	130	**35**	14
Fruit, all flavors	2 tbsp	90–110	**0**	0
Hot Fudge	2 tbsp	140	**35**	9
Fat-Free	2 tbsp	90	**0**	0
SHERBET				
Safeway Select Sherbet				
all flavors	½ cup	120	**15**	9
SORBET				
Ben & Jerry's				
Devil's Food Chocolate	½ cup	160	**15**	9
Doonesberry	½ cup	130	**0**	0
Lemon Swirl	½ cup	120	**0**	0
Purple Passion Fruit	½ cup	120	**0**	0
Edy's Whole Fruit				
all flavors	½ cup	140	**0**	0
Häagen-Dazs				
all flavors	½ cup	120	**0**	0
Chocolate	2 tbsp	100	**0**	0
MUFFINS, BAKERY				
Almond Poppy Seed	1 (127 g)	320	**120**	23
Apple Walnut	1 (127 g)	320	**120**	23
Banana Walnut	1 (113 g)	440	**200**	31
Banana Walnut (Hostess)	3 minis (34 g)	160	**80**	9
Blueberry	1 (127 g)	290	**90**	23
Blueberry (Entenmann's)	1 (57 g)	120	**0**	0
Blueberry (Hostess)	3 minis (34 g)	150	**70**	9
Blueberry (Sara Lee)	1 (64 g)	220	**100**	18

SWEETS

FOOD	AMOUNT	CALORIES TOTAL	CALORIES FAT	CALORIES SAT-FAT
Blueberry (Weight Watchers)	1 (71 g)	160	**0**	0
Bran	1 (127 g)	320	**90**	18
Chocolate Chocolate Chip (Weight Watchers)	1 (71 g)	190	**20**	9
Cinnamon Apple (Hostess)	3 minis (34 g)	160	**80**	14
Corn	1 (127 g)	330	**90**	23
Corn (Sara Lee)	1 (64 g)	260	**130**	27
Country Corn Muffin	1 (113 g)	420	**210**	31
Lemon-Flavored Poppyseed	1 (113 g)	450	**180**	27
Oat Bran	1 (127 g)	330	**100**	31
Strawberry	1 (127 g)	300	**90**	23
Low-fat Muffins				
Blueberry (Entenmann's)	1 (57 g)	120	**0**	0
Muffin Tops				
Blueberry	1 (113 g)	260	**80**	18
Bran	1 (113 g)	270	**70**	18
Corn	1 (113 g)	290	**80**	18

MUFFINS (MIXES BY BRAND NAME)

FOOD	AMOUNT	CALORIES TOTAL	CALORIES FAT	CALORIES SAT-FAT
Betty Crocker				
Banana Nut	3 tbsp mix (37 g)	140	**25**	5
	1 muffin prepared	170	**50**	10
Double Chocolate	¼ cup mix (43 g)	190	**60**	27
	1 muffin prepared	200	**70**	27
Lemon Poppy Seed	¼ cup mix (37 g)	150	**15**	5
	1 muffin prepared	190	**60**	15
Twice the Blueberries (low-fat)	¼ cup mix (39 g)	120	**15**	5
	1 muffin prepared	140	**35**	10
Wild Blueberry	¼ cup mix (40 g)	140	**15**	5
	1 muffin prepared	170	**50**	13
Betty Crocker Corn Bread or Muffin Mix	⅙ pkg (31 g mix)	110	**10**	0
	1 muffin prepared	160	**50**	14

SWEETS

FOOD	AMOUNT	CALORIES TOTAL	CALORIES FAT	CALORIES SAT-FAT
Betty Crocker (cont.)				
Betty Crocker Sweet Rewards Fat-Free Muffin Mix				
Wild Blueberry	3 tbsp mix (36 g)	120	**0**	0
	1 muffin prepared	120	**0**	0
Duncan Hines Muffin Mix				
Chocolate Chip	¼ cup mix (42 g)	180	**70**	27
	1 muffin prepared	190	**70**	27
Cranberry Orange	¼ cup mix (34 g)	150	**40**	9
	1 muffin prepared	150	**45**	9
Jiffy Muffin Mix				
Blueberry	¼ cup mix (38 g)	160	**45**	18
	1 muffin prepared	190	**60**	27
Corn	¼ cup mix (38 g)	160	**35**	14
	1 muffin prepared	180	**50**	14
Raspberry	¼ cup mix (38 g)	170	**60**	23
	1 muffin prepared	180	**70**	25
Mrs. Crutchfield's Fat-Free Muffin Mix				
All flavors	¼ cup mix (40 g)	150	**0**	0
Raga Muffin Mix				
Blueberry	¼ cup mix (40 g)	160	**40**	18
	1 muffin prepared	170	**50**	20
All other flavors	¼ cup mix (40 g)	160	**40**	18
	1 muffin prepared	170	**40**	18
PIES—BAKERY, HOMEMADE, OR RESTAURANT				
Apple	⅙ pie (104 g)	280	**120**	36
Boston Cream Pie	1/12 pie	370	**153**	83
Cherry	⅙ pie (104 g)	310	**120**	32
Cherry Beehive Pie (Entenmann's)	⅕ pie (130 g)	270	**0**	0
Coconut Custard	¼ pie (139 g)	410	**200**	81
Dutch Apple	⅙ pie (104 g)	290	**110**	27
Key Lime	⅛ pie (113 g)	390	**150**	54
Meringues				
Chocolate	3" wedge (130 g)	320	**110**	27
Coconut	3" wedge (130 g)	340	**130**	36
Lemon	⅙ pie (113 g)	290	**120**	36
Peach	⅙ pie (104 g)	300	**100**	23

SWEETS

FOOD	AMOUNT	CALORIES TOTAL	FAT	SAT-FAT
Pecan	⅙ pie (113 g)	440	**200**	45
Pumpkin	⅙ pie (109 g)	290	**100**	31
Sweet Potato Pie	⅙ pie (104 g)	270	**80**	23
SNACK PIES, BY BRAND NAME				
Hostess Fruit Pies				
Apple	1 pie (128 g)	480	**200**	81
Blueberry	1 pie (128 g)	480	**190**	90
Cherry	1 pie (128 g)	470	**200**	99
Lemon	1 pie (128 g)	500	**220**	99
Tastykake				
Apple Pie	1 pie (113 g)	270	**100**	9
Blueberry Pie	1 pie (113 g)	320	**100**	23
Cherry Pie	1 pie (113 g)	290	**100**	9
Coconut Creme Pie	1 pie (113 g)	370	**190**	36
French Apple Pie	1 pie (120 g)	360	**110**	27
Lemon Pie	1 pie (113 g)	300	**120**	14
Peach Pie	1 pie (113 g)	300	**100**	22
Strawberry Pie	1 pie (106 g)	310	**100**	27
Tasty-Klair Pie	1 pie (113 g)	410	**180**	45
PIES, FROZEN				
Marie Callender's				
Apple Cobbler	¼ cobbler (120 g)	350	**160**	36
Berry Cobbler	¼ cobbler (120 g)	390	**170**	45
Peach Cobbler	¼ cobbler (120 g)	370	**160**	27
Mrs. Smith's				
Blackberry Cobbler	⅛ cobbler (113 g)	250	**80**	36
Oven Bake & Serve Pie				
Apple	⅛ pie (131 g)	310	**130**	23
reduced fat	⅙ pie (123 g)	210	**70**	14
Cherry	⅛ pie (131 g)	310	**120**	23
Coconut Custard	⅕ pie (142 g)	280	**110**	45
Dutch Apple Crumb	⅛ pie (131 g)	350	**130**	27
Hearty Pumpkin	⅛ pie (131 g)	240	**70**	14
Pumpkin Custard	⅛ pie (131 g)	230	**70**	18

SWEETS

FOOD	AMOUNT	CALORIES TOTAL	FAT	SAT-FAT
Mrs. Smith's (cont.)				
Restaurant Classics				
Cookies & Cream	1/9 pie (120 g)	390	**180**	117
French Silk Chocolate	1/9 pie (126 g)	560	**360**	216
Key Lime, Authentic	1/9 pie (123 g)	420	**170**	108
Special Recipe				
Deep Dish Apple	1/10 pie (139 g)	370	**150**	27
Deep Dish Cherry	1/10 pie (139 g)	340	**120**	23
Deep Dish Cherry-Berry	1/10 pie (139 g)	360	**140**	27
Deep Dish Peach	1/10 pie (139 g)	330	**120**	23
Thaw-and-Serve Pie				
Banana Cream	1/4 pie (108 g)	290	**130**	36
Boston Cream	1/8 pie (69 g)	180	**60**	18
Chocolate Cream	1/4 pie (108 g)	330	**150**	36
Coconut Cream	1/4 pie (114 g)	340	**170**	45
Lemon Cream	1/4 pie (108 g)	300	**140**	36
Sara Lee				
Apple	1/8 pie (131 g)	340	**140**	32
Chocolate Cream	1/4 pie (108 g)	330	**150**	36
Coconut Cream	1/5 pie (136 g)	480	**280**	NA
Lemon Meringue	1/6 pie (142 g)	350	**100**	23
PIE CRUSTS				
Graham cracker crumbs	3 tbsp (18 g)	80	**18**	5
Graham Cracker (Keebler Ready Crust)	1/8 9" crust (21 g)	110	**45**	9
reduced fat	1/8 9" crust (21 g)	100	**30**	9
2 Extra Servings	1/10 10" crust	130	**60**	14
Hershey's Chocolate (Keebler Ready)	1/8 9" crust (21 g)	110	**45**	9
Oreo (Nabisco)	1/6 9" crust	140	**60**	14
2 Regular 9" Crusts				
Bel-Air	1/8 9" crust (18 g)	80	**45**	23
Mrs. Smith's	1/8 9" crust (18 g)	80	**45**	9
2 Deep-Dish 9" Crusts				
Bel-Air	1/8 9" crust (21 g)	100	**60**	32
Mrs. Smith's	1/8 9" crust (25 g)	110	**60**	14

SWEETS

FOOD	AMOUNT	CALORIES TOTAL	CALORIES FAT	CALORIES SAT-FAT
PUDDINGS—REFRIGERATOR CASE OR SHELF				
Banana				
Kozy Shack	4 oz (113 g)	130	**30**	18
Butterscotch				
Hunt's Snack Pack	3.5 oz (99 g)	130	**40**	9
Swiss Miss	3.5 oz (99 g)	130	**45**	9
Cheesecake (Jell-O)				
Blueberry	3.5 oz (99 g)	140	**40**	23
Strawberry	3.5 oz (99 g)	150	**40**	23
Chocolate				
Hunt's Snack Pack	3.5 oz (99 g)	140	**50**	14
fat-free	3.5 oz (99 g)	90	**0**	0
Jell-O	4 oz (113 g)	160	**45**	18
fat-free	4 oz (113 g)	100	**0**	0
Kozy Shack	4 oz (113 g)	140	**30**	18
Kraft Handi-Snacks	3.5 oz (99 g)	130	**45**	14
fat-free	3.5 oz (99 g)	90	**0**	0
Swiss Miss	3.5 oz (99 g)	150	**60**	14
Chocolate/Vanilla Swirls				
(Jell-O)	4 oz (113 g)	160	**45**	18
fat-free	1 snack (113 g)	100	**0**	0
Double Chocolate Fudge				
Healthy Choice	3.5 oz (99 g)	110	**20**	5
Flan				
Kozy Shack	4 oz (113 g)	150	**35**	18
French Vanilla				
Healthy Choice	3.5 oz (99 g)	110	**20**	5
Rice				
Kozy Shack	4 oz (113 g)	130	**30**	18
Tapioca				
Hunt's Snack Pack	3.5 oz (99 g)	130	**35**	14
Swiss Miss	4 oz (113 g)	140	**30**	18
fat-free	3.5 oz (99 g)	90	**0**	0
Vanilla				
Hunt's Snack Pack	3.5 oz (99 g)	130	**45**	14
fat-free	3.5 oz (99 g)	80	**0**	0
Kozy Shack	6 oz (170 g)	195	**45**	27

SWEETS

FOOD	AMOUNT	CALORIES TOTAL	CALORIES FAT	CALORIES SAT-FAT
Vanilla (cont.)				
Kraft Handi-Snacks	3.5 oz (99 g)	140	**40**	14
Swiss Miss	3.5 oz (99 g)	140	**45**	9
PUDDINGS—RESTAURANT				
Caramel Bavarian Cream	½ cup	246	**128**	73
Chocolate Mousse	½ cup	324	**199**	115
Creme Caramel	1 cup	303	**125**	27
Custard, baked	1 cup	305	**125**	61
SPREADS				
Nutella	2 tbsp	160	**80**	9
Milky Way				
Chocolate & Hazelnut	2 tbsp	170	**90**	27
SUGARS, SYRUPS, ETC.				
Honey	1 tbsp	65	**0**	0
Jams, Jellies, and Preserves	1 tbsp	55	**0**	0
Molasses	1 tbsp	43	**0**	0
Sugar				
Brown, firmly packed	1 tbsp	51	**0**	0
	½ cup	410	**0**	0
White				
Granulated	1 tsp	16	**0**	0
	½ cup	385	**0**	0
Powdered, sifted	1 cup	385	**0**	0
Syrups (*see also* **ICE CREAM TOPPINGS**)				
Chocolate-flavored (Hershey's)	2 tbsp	100	**0**	0
Corn	1 tbsp	61	**0**	0
Maple	1 tbsp	61	**0**	0
MISCELLANEOUS SWEETS				
Apple Fritter	1 (70 g)	300	**130**	31
Apple Tart (French)	⅛ tart	265	**100**	65
Baklava (Apollo)	4½ pieces (125 g)	540	**280**	45
Baklava	1 piece (28 g)	140	**72**	0
Carob Chips	1 oz (2⅔ tbsp)	140	**63**	52

SWEETS

FOOD	AMOUNT	CALORIES TOTAL	FAT	SAT-FAT
Chocolate				
Baking, unsweetened	1 oz	145	**135**	81
Chocolate Chips	1 oz	140	**72**	45
semi-sweet	30 chips (15 g)	70	**35**	23
Mini Baking Bits (M&M's)	1 tbsp (14 g)	70	**35**	18
Hershey's				
Milk chocolate chips	1 tbsp (15 g)	80	**40**	23
Reduced-fat baking chips	1 tbsp (16 g)	60	**20**	31
Reese's Peanut Butter Chips	1 tbsp (15 g)	80	**35**	36
Nestlé				
Milk chocolate morsels	1 tbsp (14 g)	70	**40**	23
Chocolate Eclairs	1 (57 g)	220	**110**	90
Custard cream puffs	1	303	**163**	63
Escalloped Apples (Stouffer's)	⅔ cup (158 g)	180	**25**	0
Gelatin dessert	½ cup	70	**0**	0
Jell-O, all flavors	½ cup	90	**0**	0
Madeleines	1	90	**50**	30
Puff Pastry, frozen (Pepperidge Farm)				
Sheets	⅙ sheet (41 g)	200	**100**	23
Shells	1 shell (47 g)	230	**130**	27
Tiramisu	2" × 2" piece	395	**220**	130

VEGETABLES AND VEGETABLE PRODUCTS

FOOD	AMOUNT	CALORIES TOTAL	FAT	SAT-FAT
See also **FROZEN, MICROWAVE, AND REFRIGERATED FOODS.**				
Alfalfa seeds, sprouted, fresh	1 cup	10	**0**	0
Artichoke				
fresh	1 medium	65	**0**	0
	1 large	83	**0**	0
cooked	1 medium	53	**0**	0
hearts				
canned in water	½ cup	35	**0**	0
marinated, undrained	½ cup	190	**135**	18

VEGETABLES AND VEGETABLE PRODUCTS

FOOD	AMOUNT	CALORIES TOTAL	CALORIES FAT	CALORIES SAT-FAT
Asparagus				
fresh	½ cup	15	**0**	0
	4 spears	13	**0**	0
cooked	½ cup	22	**0**	0
	4 spears	15	**0**	0
Baked beans, canned				
Brown Sugar & Bacon (Campbell's)	½ cup	170	**25**	9
Plain or vegetarian	½ cup	117	**5**	2
in tomato sauce (Campbell's)	½ cup	130	**20**	9
Pork 'n Beans (Hanover)	½ cup	120	**15**	5
with beef	½ cup	160	**42**	20
with franks	½ cup	184	**76**	27
with pork	½ cup	134	**18**	7
and sweet sauce	½ cup	141	**17**	7
and tomato sauce	½ cup	124	**12**	5
Bamboo shoots				
fresh	1 cup	41	**0**	0
cooked	1 cup	15	**0**	0
canned	1 cup	25	**0**	0
Beans				
Black				
dry	1 cup	661	**25**	6
boiled	1 cup	227	**8**	2
Fermented Black	1 tbsp	35	**10**	0
Great Northern				
dry	1 cup	621	**19**	6
boiled	1 cup	210	**7**	2
canned	1 cup	300	**9**	3
Kidney				
dry	1 cup	613	**14**	2
boiled	1 cup	225	**8**	1
canned	1 cup	208	**7**	1
Kidney, California red				
dry	1 cup	607	**0**	0
boiled	1 cup	219	**0**	0

VEGETABLES AND VEGETABLE PRODUCTS

FOOD	AMOUNT	CALORIES TOTAL	CALORIES FAT	CALORIES SAT-FAT
Beans (cont.)				
Kidney, Red				
dry	1 cup	619	**18**	3
boiled	1 cup	225	**8**	1
canned	1 cup	216	**8**	1
Kidney, Royal Red				
dry	1 cup	605	**7**	1
boiled	1 cup	218	**3**	0
Lima, Baby				
fresh	1 cup	216	**6**	1
boiled	1 cup	188	**5**	1
Lima, Large				
fresh	1 cup	176	**11**	3
boiled	1 cup	208	**6**	1
canned	1 cup	186	**4**	1
Navy				
dry	1 cup	697	**24**	6
boiled	1 cup	259	**9**	2
canned	1 cup	296	**10**	3
Pink				
dry	1 cup	721	**21**	6
boiled	1 cup	252	**7**	2
Pinto				
dry	1 cup	656	**20**	4
boiled	1 cup	235	**8**	2
canned	1 cup	186	**7**	1
Refried				
canned (Del Monte)	1 cup	260	**32**	10
Mexican restaurant	¾ cup	375	**146**	60
Snap (includes green, Italian, and yellow)				
fresh	1 cup	34	**0**	0
cooked	1 cup	44	**0**	0
canned	1 cup	36	**0**	0
Soy				
dry	1 cup	774	**334**	48
boiled	1 cup	298	**139**	20

VEGETABLES AND VEGETABLE PRODUCTS

FOOD	AMOUNT	CALORIES TOTAL	CALORIES FAT	CALORIES SAT-FAT
Beans (cont.)				
Soy products				
Aka Miso	1 tbsp	30	**10**	0
Miso	1 cup	565	**150**	22
Shiro Miso	1 tbsp	30	**10**	0
Tofu	1 piece (2½ x 2¾ x 1″)	88	**50**	7
White, small				
dry	1 cup	723	**23**	4
boiled	1 cup	253	**10**	1
Yellow				
fresh	1 cup	676	**46**	12
boiled	1 cup	254	**17**	4
Beets				
fresh	1 cup slices	60	**0**	0
	2 beets	71	**0**	0
cooked	1 cup slices	52	**0**	0
	2 beets	31	**0**	0
canned, drained	1 cup	54	**0**	0
Black-eyed or cowpeas, cooked	1 cup	190	**0**	0
Broadbeans				
fresh	1 cup	511	**21**	3
boiled	1 cup	186	**6**	1
canned	1 cup	183	**5**	0
Broccoli				
fresh	1 cup chopped	24	**0**	0
	1 spear	42	**0**	0
cooked	1 cup chopped	46	**0**	0
	1 spear	53	**0**	0
Brussels sprouts, cooked	1 cup	60	**0**	0
	1 sprout	8	**0**	0
Cabbage				
fresh	1 cup shredded	16	**0**	0
	1 head	215	**0**	0
cooked	1 cup shredded	32	**0**	0
	1 head	270	**0**	0

VEGETABLES AND VEGETABLE PRODUCTS

FOOD	AMOUNT	CALORIES TOTAL	CALORIES FAT	CALORIES SAT-FAT
Cabbage, Chinese				
fresh	1 cup shredded	9	**0**	0
cooked	1 cup shredded	20	**0**	0
Cabbage, red				
fresh	1 cup shredded	19	**0**	0
cooked	1 cup shredded	32	**0**	0
Cabbage, Savoy				
fresh	1 cup shredded	19	**0**	0
cooked	1 cup shredded	35	**0**	0
Carob flour	1 tbsp	14	**0**	0
	1 cup	185	**0**	0
Carrots				
fresh	1	31	**0**	0
	1 cup shredded	48	**0**	0
cooked	1 cup sliced	70	**0**	0
Carrot juice	1 cup	98	**0**	0
Cauliflower, cooked or				
fresh	3 flowerets	13	**0**	0
	1 cup pieces	24	**0**	0
Celery, fresh	1 stalk	6	**0**	0
	1 cup diced	18	**0**	0
Chard, Swiss				
fresh	1 cup chopped	6	**0**	0
	1 leaf	9	**0**	0
cooked	1 cup chopped	35	**0**	0
Chickpeas or garbanzos				
dry	1 cup	729	**109**	11
boiled	1 cup	269	**38**	4
canned	½ cup (120 g)	110	**18**	3
Chili with beans, canned	1 cup	286	**126**	54
Coleslaw (*see also* **FAST FOODS**)				
made with mayonnaise	1 cup	171	**161**	34
Collard greens, cooked	1 cup chopped	27	**0**	0
Corn				
Cooked	1 ear	70	**7**	1
	1 cup kernels	130	**15**	3

VEGETABLES AND VEGETABLE PRODUCTS

FOOD	AMOUNT	CALORIES TOTAL	FAT	SAT-FAT
Corn (cont.)				
baby corn nuggets (Haddon House)	½ cup (132 g)	30	**13**	9
canned, cream style	1 cup	186	**10**	1
popcorn (*see* **SNACKS**)				
Cucumber, fresh	1 cucumber	39	**0**	0
	1 cup slices	14	**0**	0
Dandelion greens				
fresh	1 cup chopped	25	**0**	0
cooked	1 cup chopped	35	**0**	0
Eggplant				
fresh	1 eggplant	27	**0**	0
cooked	1 cup cubes	27	**0**	0
Endive, fresh	1 cup chopped	8	**0**	0
	1 head	86	**0**	0
Garlic, fresh	1 clove	4	**0**	0
Kale				
fresh	1 cup chopped	33	**0**	0
cooked	1 cup chopped	41	**0**	0
Kohlrabi				
fresh, diced	1 cup	41	**0**	0
cooked, drained	1 cup	40	**0**	0
Leeks				
fresh	1	76	**0**	0
	¼ cup chopped	16	**0**	0
cooked	1	38	**0**	0
	¼ cup chopped	8	**0**	0
Lentils				
dry	1 cup	649	**17**	2
boiled	1 cup	231	**7**	1
Lettuce, fresh				
butterhead, Boston	1 head (5-inch)	21	**0**	0
	1 outer or 2 inner leaves	2	**0**	0
crisphead, iceberg	1 head (6-inch)	70	**0**	0
	1 wedge (¼ head)	20	**0**	0
	1 cup chopped	5	**0**	0
loose leaf, romaine	1 cup chopped	10	**0**	0

VEGETABLES AND VEGETABLE PRODUCTS

FOOD	AMOUNT	CALORIES TOTAL	CALORIES FAT	CALORIES SAT-FAT
Mushrooms				
fresh, sliced, chopped	1 cup, pieces	20	**0**	0
	1 lb	127	**0**	0
cooked	1 cup, pieces	42	**0**	0
canned	1 cup, pieces	38	**0**	0
Mushrooms, shiitake				
dried	4 mushrooms	44	**0**	0
cooked	4 mushrooms	40	**0**	0
	1 cup pieces	80	**0**	0
Mustard Greens	1 cup chopped	14	**0**	0
Okra				
fresh	8 pods	36	**0**	0
	1 cup slices	38	**0**	0
cooked	8 pods	27	**0**	0
	1 cup slices	50	**0**	0
Onions				
fresh	1 cup chopped	54	**0**	0
cooked	1 cup chopped	58	**0**	0
fried onion rings, frozen	7 rings	285	**168**	54
Onions, green, fresh	1 cup chopped	26	**0**	0
Parsley, fresh	10 sprigs	3	**0**	0
Parsnips				
fresh	1 cup slices	100	**0**	0
cooked	1 cup slices	126	**0**	0
Peas, green				
fresh	1 cup	118	**4**	0
cooked	1 cup	134	**0**	0
Peas, split				
fresh	1 cup	671	**21**	0
boiled	1 cup	231	**7**	0
Peas and carrots, canned	1 cup	96	**6**	0
Peas and onions, canned	1 cup	122	**8**	0
Peppers				
hot chili, fresh	1	18	**0**	0
	½ cup chopped	30	**0**	0
jalapeño	½ cup chopped	17	**0**	0

VEGETABLES AND VEGETABLE PRODUCTS

FOOD	AMOUNT	CALORIES TOTAL	CALORIES FAT	CALORIES SAT-FAT
Peppers (cont.)				
sweet, raw	1	18	**0**	0
	½ cup chopped	12	**0**	0
sweet, cooked	1	18	**0**	0
	½ cup chopped	12	**0**	0
Potatoes				
Au gratin	1 cup	320	**167**	104
Baked in skin	8 oz	145	**0**	0
	1 lb	325	**0**	0
	1 (2⅓ x 4¾ inch)	173	**0**	0
Boiled in skin	5 oz	104	**0**	0
	1 cup diced or sliced	118	**0**	0
	1 lb	345	**0**	0
Boiled, pared before cooking	8 oz	146	**0**	0
	5 oz	88	**0**	0
	1 cup diced or sliced	101	**0**	0
	1 lb	295	**0**	0
Raw, without skin	4 oz	88	**0**	0
Fried, Frozen (*see also* **FROZEN, MICROWAVE, AND REFRIGERATED FOODS**)				
French fries (*see also* **FAST FOODS**)				
Act II Microwave	1 box (88 g)	240	**110**	23
Frozen, other preparations				
Hash brown	½ cup	170	**81**	38
Mashed				
with whole milk	½ cup	81	**6**	3
with whole milk and butter	½ cup	111	**40**	22
Potato Chips (*see* **SNACK FOODS**)				
Potato Pancakes	1 pancake	495	**113**	31
Potato Salad	½ cup	180	**92**	16
Potato Sticks (*see* **SNACK FOODS**)				
Scalloped, from dry mix	1 cup	230	**99**	25

VEGETABLES AND VEGETABLE PRODUCTS

FOOD	AMOUNT	CALORIES TOTAL	CALORIES FAT	CALORIES SAT-FAT
Pumpkin, cooked	1 cup mashed	49	**0**	0
Radishes, fresh	10	7	**0**	0
	½ cup slices	10	**0**	0
Sauerkraut, canned	1 cup	44	**0**	0
Shallots, fresh	1 tbsp chopped	7	**0**	0
Spinach				
fresh	1 cup chopped	6	**0**	0
	10 oz pkg	46	**0**	0
cooked	1 cup	41	**0**	0
Spinach souffle, made with whole milk, eggs, cheese, butter	1 cup	218	**165**	64
Squash				
acorn				
raw	1 cup cubes	56	**0**	0
	1 lb	180	**0**	0
cooked	1 cup cubes	115	**0**	0
butternut				
raw	1 cup cubes	63	**0**	0
	1 lb	203	**0**	0
cooked	1 cup cubes	83	**0**	0
spaghetti				
cooked	1 cup	45	**0**	0
Succotash, cooked	1 cup	222	**14**	1
Sweet potatoes				
raw with skin	1 lb	343	**0**	0
baked in skin	1 (5" long, 2" diameter)	118	**0**	0
	½ cup mashed	103	**0**	0
boiled without skin	1 cup mashed	344	**0**	0
Tomatoes				
fresh	1	24	**0**	0
	1 cup chopped	35	**0**	0
cooked	1 cup	60	**0**	0
canned in tomato juice	1 cup	67	**0**	0
crushed	½ cup	30	**0**	0
Tomato juice, canned	1 cup	42	**0**	0

VEGETABLES & VEGETABLE PRODUCTS

FOOD	AMOUNT	CALORIES TOTAL	FAT	SAT-FAT
Tomato products, canned				
Marinara sauce	1 cup	171	**75**	11
Paste	1 tbsp	14	**0**	0
Puree	1 cup	102	**0**	0
Sauce (*see also* **SAUCES, GRAVIES, AND DIPS**)	1 cup	74	**0**	0
Spaghetti sauce (*see* **SAUCES, GRAVIES, AND DIPS**)				
Turnips, cooked	1 cup cubes	28	**0**	0
Vegetable juice cocktail, canned	1 cup	44	**0**	0
Water chestnuts, canned	1 cup slices	70	**0**	0
Watercress				
whole	1 cup (10 sprigs)	7	**0**	0
chopped fine	1 cup	24	**0**	0
Yam, cooked	1 cup cubes	158	**0**	0
Zucchini				
raw	1 cup slices	19	**0**	0
cooked	1 cup slices	28	**0**	0

MISCELLANEOUS

FOOD	AMOUNT	CALORIES TOTAL	FAT	SAT-FAT
Baking powder	1 tbsp	5	**0**	0
Carob flour	1 tsp	14	**0**	0
Cocoa powder	1 tsp	5	**0**	0
Cocoa powder	1 tbsp	14	**5**	0
Curry powder	1 tsp	5	**0**	0
Garlic powder	1 tsp	10	**0**	0
Gelatin, dry	1 envelope	25	**0**	0
Ketchup	1 tbsp	15	**0**	0
Mustard	1 tsp	5	**0**	0
Olives, green	4 medium	15	**15**	2
Olives, ripe	3 small or 2 large	15	**15**	3
Oregano	1 tsp	5	**0**	0

MISCELLANEOUS

FOOD	AMOUNT	CALORIES TOTAL	FAT	SAT-FAT
Paprika	1 tsp	5	**0**	0
Pickles				
dill	1 medium	5	**0**	0
sweet	1	20	**0**	0
gherkin	1	20	**0**	0
Vinegar	1 tbsp	0	**0**	0
Yeast, all types	1 tbsp	20	**0**	0

Food Tables Index

GLOSSARY
APPENDIX
REFERENCES
INDEX
TABLE OF EQUIVALENT MEASURES

Glossary

Adipose tissue. Tissue in which fat is stored.

Aerobic exercise. Steady, repetitive exercise that uses the large muscles and requires a steady supply of oxygen — in contrast to exercise that requires bursts of activity separated by periods of rest. Examples of aerobic exercises include walking, swimming, running, and biking. Aerobic exercise burns more fat than active sports, which burn more carbohydrate.

Basal metabolic rate (BMR). The rate at which energy is used when the body is completely at rest to maintain such vital functions as breathing, heartbeat, and digestion.

Burning or oxidation. The chemical process of combining substances (carbohydrates, fats, proteins in foods) with oxygen, resulting in the release of stored energy.

Calorie. A unit of heat or energy produced by burning (oxidizing) nutrients. Carbohydrates and proteins contain 4 calories per gram; fats, 9 calories per gram; and alcohol, 7 calories per gram.

Carbohydrate. One of the three major energy-containing nutrients in foods; the others are protein and fat. Carbohydrates are simple (sugars) or complex (starches); each type contains 4 calories per gram.

Cholesterol. A fatlike substance found in the cell membranes of all animals, including humans. Cholesterol is transported in the bloodstream. Some of it is manufactured by the body and some comes from the foods of animal origin that we eat. A healthy level of cholesterol for adults is below 200 mg/dL. A higher level is often associated with increased risk of heart disease.

Complex carbohydrate. One of the two major types of carbohydrates, which include starches. They are found in whole-grain and cereal products and vegetables. In their natural state complex carbohydrates are accompanied by dietary fiber.

Dietetic. A term used to describe a food that has been nutritionally altered in some way. "Dietetic" can mean less sodium, less fat, less sugar, or fewer calories. If a food is intended for weight loss, it must meet requirements for low-calorie or reduced-calorie claims. If not for weight loss, its label must state its special dietary purpose.

Energy. Power to do work. The energy in foods is measured in calories. A high-energy food is a high-calorie food. Energy expenditure, as in exercise, is quantified in terms of calories expended or burned.

Fat. One of the three major energy-containing nutrients in foods; the others are carbohydrates and proteins. Fat is an oily substance that is found in many foods, especially oils, dairy products, and meat products. Fat is the major form of storing energy in the body. Fat stores in the adipose tissue total 140,000 or more calories. Fat contains 9 calories per gram.

Fat Budget. The number of calories from ingested fat that are allowed per day to reach and maintain a person's desirable and healthy weight. Fat Budget is based on a percentage of minimum total calorie intake (BMR).

Fattening. A term commonly used to describe any substance that contributes to making a person fat. Many foods (often starchy ones) have been mislabeled "fattening." Truly, the most fattening substance is fat.

Fiber, dietary. A nondigestible substance found in plant products. It can be insoluble, like wheat fiber, or soluble, like oat bran, pectin (in fruits), and guar gum (in beans). Both types of fiber provide bulk and moderate the absorption of nutrients. Insoluble fibers help regularity, whereas soluble fibers reduce blood cholesterol.

Glucose. A simple carbohydrate or sugar found in foods and in the body; the preferred fuel for quick energy and the only fuel used by the brain.

Glycogen. The storage form of carbohydrate in the body; long chains of glucose linked end to end. Glycogen stores, which total about 800 calories, occur in liver, muscle, and other tissues.

Gram. A metric measure of weight. One ounce equals 28.35 grams. Nutrients are listed in grams on food labels: a gram of fat contains 9 calories; a gram of protein or carbohydrate contains 4 calories.

Hidden fat. Fat in foods that is not visible. Hidden fats include oils used in frying or baking and fat naturally present in foods, such as butterfat in cheese and whole milk, fat marbled throughout beef, or fat in the skin of poultry.

Ideal or desirable weight. Weight associated with general good health and lowest mortality rates.

Lean body mass. The metabolically active tissue in the body, primarily composed of muscle. The greater a person's lean body mass, the higher the metabolic rate and the greater protection against weight gain.

Lite or light. A product advertised as light must contain one-third fewer calories or half as much fat as the regular product. The nutrition information on the label should be used to determine how much fat the product contains and thus how it can be fit into your Fat Budget.

Low-fat. A term used to imply that a food is acceptable for a weight reduction diet. Foods labeled low-fat, such as 2% low-fat milk, may actually contain substantial amounts of fat. To be sure a product can fit in your Fat Budget, check the fat content on the food label.

Metabolism. The sum of the chemical changes that occur to substances in the body. Much of metabolism is the conversion of food into living tissue and energy.

Monounsaturated fat. One of the three types of fat commonly found in foods. Monounsaturated fats help to reduce blood cholesterol levels. The richest source of monounsaturated fat is olive oil. Like all other fats, monounsaturated fat has 9 calories per gram.

Nitrites. Substances used to preserve, color, and flavor meat products. In the body, nitrites can be converted into nitrosamines, which have been shown to cause cancer.

Nondairy. A term commonly used for imitation dairy foods that contain no dairy products, such as imitation creamers, sour creams, and whipped toppings. While these products do not contain cholesterol, they do contain fat, often highly saturated fat like coconut oil. Check the nutrition information on labels to determine how much fat is present.

Nutrients. Substances in foods (carbohydrates, proteins, and fats) that contain energy and are building blocks for making living tissue. Also includes substances needed for normal bodily functioning, such as minerals and vitamins.

Obesity. Excess accumulation of body fat. Obese is defined as 20 percent to 40 percent over ideal weight, massively obese as greater than 40 percent over ideal weight.

Oxidation. See burning.

Polyunsaturated fat. One of the three types of fat commonly found in foods. Polyunsaturated fat helps to lower blood cholesterol, but in excess can lower the "good" cholesterol in the blood and has been associated with an increased risk of cancer. The most common sources of polyunsaturated fats are corn, sunflower, and safflower oils. Like all other fats, polyunsaturated fat has 9 calories per gram.

Protein. One of the nutrients in foods that provides energy and building blocks for making essential body constituents such as muscle, enzymes, and cell membranes.

Reduced-calorie. A term regulated by the Food and Drug Administration that means a product is at least one-quarter lower in calories than the food with which it is being compared. It does not necessarily mean that the product is low in fat. Consult the nutrition information on labels for fat content.

Saturated fat. One of the three types of fat commonly found in foods, saturated fat has a powerful effect on raising blood cholesterol levels. The most common sources of saturated fats are butterfat; beef, veal, lamb, and chicken fat; cocoa butter; hydrogenated vegetable oil; and the tropical oils (coconut, palm kernel, and palm). Like all other fats, saturated fat has 9 calories per gram.

Simple carbohydrate. One of the two major types of carbohydrate, also known as sugar. In contrast to complex carbohydrates, simple carbohydrates are not usually associated with any nutritionally beneficial substances and are often said to contain empty calories.

Sugar. Any carbohydrate with a sweet taste.

Thermogenesis or thermogenic effect of food. The process of producing heat. Carbohydrates in foods we eat increase the metabolic rate and produce a thermogenic effect.

Vegetable oil. The fat from plant products. Some vegetable oils are mostly unsaturated and are liquid at room temperature (olive, corn, sunflower) and some are mostly saturated and are solid at room temperature (coconut, palm kernel, palm). All vegetable oils are 100 percent fat and have 9 calories per gram.

Appendix: The Nitty-Gritty of How to Keep a Food Record

Now that you are ready to get started on the road to leanness, your first activity is to discover which foods have been making you fat. If you are like most people who want to lose weight, you know that you are eating some fattening foods. What you don't know is exactly how much fat is in the foods you eat or which foods are loaded with fat. That is why you are going to keep a food record — an activity you will find rewarding, insightful, and maybe, even fun.

These records are for YOU — not your spouse, your mother, your brother, your doctor, your dietitian — just for you, and they must be exact and accurate. They must reflect *exactly* what you eat, not what you think you *should* have eaten. Beautiful imaginary food records are useless. If you eat 3 chunky chocolate chip cookies, record 3 cookies, not 1 or 2. Okay, you regret your overindulgence, but being sorry won't put the cookies back in the package or take the fat out of your fat stores. If you don't record what you eat, you fool only yourself, not your body. **Your body accurately and honestly records everything you put into it.** On the other hand, if you write down the 120 fat calories you consumed, you can balance your fat intake for the rest of the day or the next by eating less fat.

Don't become frantic about your food diary. Your first records may reflect what got you into this mess in the first place. Don't feel guilty. Everyone used to wolf down hot dogs and French fries with no thought. We called whole milk a calcium-rich food and steak a high-protein food. Who knew?

More than likely, cutting fat won't be a problem, but eating above your minimum total caloric intake may be a struggle. As the days and weeks progress and you ease into *Choose to Lose,* your records will reflect a diet lower in fat and higher in complex carbohydrates. But it takes time. Don't expect to change your entire diet overnight. Experience shows that changes made more gradually are more apt to last.

KEEPING GOOD RECORDS

Time

Record what you eat when you eat it. Write down what you eat when you eat it. Don't fill in your record at the end of the day. You're bound to forget what you don't want to remember. Recording the time helps you see patterns in your eating. The fact that you ate breakfast at 8 A.M. and then no food until 3 P.M. may give you a clue as to why you blew your Fat Budget on a late lunch of a double cheeseburger, milk shake, and fries.

Food and Amount

List all foods separately and list amounts accurately. Measure everything. Be accurate. Don't *just* write down cream cheese. Write down the amount you ate — cream cheese, 2 tablespoons. Don't *simply* write down chicken breast. Write down chicken breast with skin, batter-fried. Don't *merely* write down roast beef sub. Ask the sandwich man how much roast beef and how much mayonnaise he gave you. (He'll know — his boss makes sure he gives everyone the same amount.) Write down 6-inch sub roll, 4 ounces roast beef, 1 tablespoon mayonnaise. These details are important. If you eat out, ask the waiter what's in the cream sauce and how much butter is on the swordfish, the potato, vegetables, or toast.

> If you are interested in purchasing a *Choose to Lose* Passbook to keep track of your food intake, see the order form on the last page.

Sample Food Record

TIME	FOOD	AMOUNT	TOTAL CALORIES	TOTAL CAL SUBTOTAL	FAT CALORIES	FAT SUBTOTAL
9 A.M.	Cream cheese	2 tbsp				
1 P.M.	Chicken breast with skin, batter-fried	1 breast				
6 P.M.	Roast beef sub					
	roast beef	4 oz				
	6-inch sub roll	1				
	mayonnaise	1 tbsp				

For "Hints on Measuring and Recording What You Measured," see page 594.

Total Calories

When people cut down on fat, they often cut down on total calories. This is unhealthy. When you reduce your fat intake, you must be sure to increase your intake of complex carbohydrates, fruits, and low- or nonfat dairy products. If you don't eat enough of these low-fat foods, you will miss the vitamins, minerals, and fiber they provide, your basal metabolic rate will slow down, and you will be hungry.

On the other hand, grossly overeating your minimum total caloric intake indicates that you are also probably overeating your Fat Budget or overdosing on empty, fiber-deficient nonfat foods.

You only know if you are eating too many or too few total calories if you keep track.

Your total caloric intake should be greater than the number you determined in Chapter 1. Remember: that number is a floor. It is the number of calories needed to sustain your body if you are completely at rest. Since you are active, you need more calories to fuel your activity. How much more depends on how much physical activity you do. For most people 300 to 500 total calories more than the minimum total caloric level will suffice. If you are eating fiber-rich foods, the bulk will protect you from eating too much. But don't worry. If you eat a few hundred more calories in addition as fiber-rich carbohydrates, they will be burned off. They will not be stored as fat. You need to eat to lose. Fat calories — not total calories — make you fat.

Use the Food Tables (look under the heading Total Calories) and labels on packages to determine the total calories of the foods you have eaten. Make sure that you adjust the total calories for the amount you eat relative to the serving size listed. For example, the portion size for cheddar cheese is 1 ounce. The total calories for 1 ounce of cheddar cheese are 114. If you eat 3 ounces of cheddar cheese you record 3 × 114 = 342 total calories.

Fat Calories

Use the Food Tables (look under the heading Fat Calories) and labels on packages to determine the fat calories of the food you have eaten. Make sure that you adjust the fat calories for the amount you eat relative to the serving size listed. For example, the portion size listed for chuck steak is 1 ounce. The fat calories for 1 ounce of chuck, lean and fat, are 78. No one eats 1 ounce of meat. If you ate 5 ounces of chuck, you would record 5 × 78 = 390 fat calories.

If you cannot find the exact food you have eaten in the Food Tables (first try using the Food Table Index on page 567), find a similar food and record its fat calories. For example, if you ate veal parmigiana in a restaurant, you should first check out the Italian Restaurant listing in the **RESTAURANT FOODS** section of the Food Tables. If you come up empty handed, find a high-fat veal parmigiana entry in the **FROZEN, MICROWAVE, AND REFRIGERATED FOODS** section of the Food Tables.

If you have eaten a mixed dish such as chicken stir-fry, you should estimate the sum of the fat calories for the individual ingredients. For example, chicken stir-fry might contain one chicken breast without skin (13 fat calories), a tablespoon of corn oil (120 fat calories), green pepper, onion, and garlic (0 fat calories).

Look in a cookbook (not low-fat) for the dish you ate and figure out the fat calories per serving (see page 174).

Food Labels. Read Chapter 7, "Decoding Food Labels." Take heed. Not only must you determine the calories of fat per serving from the nutrition label, you must adjust for the number of servings you eat. For example, according to the label, 2 tablespoons of cream cheese contain 90 fat calories. If you eat only 1, you would record 45 fat calories.

Entering the Information. If you ate cereal with 1% milk for breakfast, start by entering the amount of cereal you ate. Read the nutrition label on the cereal box to determine the total calories and fat calories for that amount. Your first food entry would look like this.

TIME	FOOD	AMOUNT	TOTAL CALORIES	TOTAL CAL SUBTOTAL	FAT CALORIES	FAT SUBTOTAL
8 A.M.	Cheerios	1 cup	110		15	

Fat Subtotal. You should keep a running tally of your fat intake. This makes you aware of the amount of fat you have "spent" and the amount you have left in your Fat Budget. You can look at your fat subtotal and easily judge the fat impact of any food you are planning to eat.

For your first entry of the day, in the column labeled Fat Subtotal, repeat the number of fat calories you entered in the Fat Calories column.

TIME	FOOD	AMOUNT	TOTALCAL ORIES	TOTAL CAL SUBTOTAL	FAT CALORIES	FAT SUBTOTAL
8 A.M.	Cheerios	1 cup	110		15	15

Total Calorie Subtotal

You will also want to keep a running tally of your total calories to make sure you are eating *enough* total calories.

For your first entry of the day, in the column labeled Total Cal Subtotal, repeat the number of total calories you entered in the Total Calories column.

TIME	FOOD	AMOUNT	TOTAL CALORIES	TOTAL CAL SUBTOTAL	FAT CALORIES	FAT SUBTOTAL
8 A.M.	Cheerios	1 cup	110	110	15	15

For your second entry, record food as 1% milk, amount as 1 cup, and fat calories as 20. Add the fat calories for your second entry to the previous Fat Subtotal to calculate the new Fat Subtotal (20 + 15 = 35).

TIME	FOOD	AMOUNT	TOTAL CALORIES	TOTAL CAL SUBTOTAL	FAT CALORIES	FAT SUBTOTAL
8 A.M.	Cheerios	1 cup	110	110	15	15
	1% milk	1 cup	100		20	35

Add the total calories for the milk to the previous total calorie subtotal to calculate the new total calorie subtotal (100 + 110 = 210).

TIME	FOOD	AMOUNT	TOTAL CALORIES	TOTAL CAL SUBTOTAL	FAT CALORIES	FAT SUBTOTAL
8 A.M.	Cheerios	1 cup	110	110	15	15
	1% milk	1 cup	100	210	20	35

Final Subtotals

At the end of the day compare your fat intake (last entry of your Fat Subtotal) for the day with your Fat Budget. Is your fat consumption way over budget? But is it closer to budget than yesterday? What foods contributed the most fat to the day's intake? Can you eat these in smaller amounts, less often, not at all, or make low-fat substitutions? Look at your total caloric intake for the day (last entry of your Total Cal Subtotal). Have you been eating enough total calories? Are you at least 300 total calories over your minimum total caloric intake?

Food Guide Pyramid Recommendations

You want to be sure that not only are you reducing fat and eating enough total calories, but that you are eating a balanced diet (see Chapter 13, "Ensuring a Balanced Diet"). To guarantee you are satisfying the minimum basic nutritional requirements, copy these recommended numbers of servings for each food category at the bottom of your food record each day. (They are printed in the *Choose to Lose* Passbook):

Fruits (2–4): Vegetables (3–5): Grains (6–11): Dairy (2–3):

You will find definitions for these servings on page 223. Check over your food record and record your intake in the space provided. If you are not meeting your minimal basic nutritional requirements, make an effort to add the necessary foods to your diet. This is supremely important. **Not only do these foods supply the vitamins, minerals, and fiber you need, they make you feel full and thus less likely to binge on high-fat or empty-calorie foods.**

HINTS ON MEASURING AND RECORDING WHAT YOU MEASURED

Liquids

Use measuring cups and spoons to measure the amount of a liquid you use. This table of equivalent measures should help you record liquids*:

1 cup	= 8 fluid ounces
½ cup	= 4 fluid ounces
¼ cup	= 2 fluid ounces
2 tablespoons	= 1 fluid ounce

Remember, we are dealing here with fluid ounces, with volume. One cup holds 8 fluid ounces. A cup does not weigh 8 ounces. A cup of feathers does not weigh the same as a cup of marbles.

Solid Food

Meat: Read the meat package label or use a kitchen scale to measure the weight (in ounces) of *beef, lamb, pork, veal, poultry,* and *fish*. Record the weight of meat in ounces. Indicate if the meat was raw or cooked when it was weighed and if the fat is included or has been trimmed. If necessary, use the following estimates:

1/4 cup meat = 1 ounce
Chicken breast half = 3 ounces
4 ounces raw meat without bone = 3 ounces cooked
6 ounces raw meat with bone = 3 ounces cooked

Bread should be recorded by slice.

---

*See page 616 for a more complete table of equivalent measures.

Crackers and cookies should be recorded by unit (for example, 5 crackers).

Cereals*, cottage cheese†, creams, fats, frozen desserts, canned fish, sliced fruit, grains, milks, nuts‡, pasta, puddings (not canned), rice, salad dressings, sauces and gravies, snacks, soups, vegetables, yogurt. Use measuring cups and spoons to measure and record these foods.

Candy, cheese, cold cuts, nuts. Read the label on the package or use a scale to measure the weight (in ounces). Save packages with nutrition labeling for use later. Record weight of food in ounces or grams.

Cakes and pies should be recorded by fraction of whole dessert (for example, 1/10 cake).

Frozen food should be recorded by package or the fraction of the package you eat. If possible, save packages with labels for future reference.

Fast foods should be recorded by unit, piece, serving, order, or slice, except for shakes, which are measured in fluid ounces.

Fish. If necessary, use the following estimates:

¼ cup = 1 ounce
4 ounces raw = 3 ounces cooked

Commercial packages often use gram measurements. 28 grams = 1 ounce. Here's a summary chart that lists foods and the unit used to measure them.

SUMMARY CHART

FOOD/BEVERAGE	UNIT	EXAMPLES
DAIRY/EGGS		
Butter	pat, teaspoon, cup	½ cup butter
Cheese, hard, soft	ounce	2 oz cheddar

---

*If you have a scale, you may also weigh and record cereal in ounces.

†Cottage cheese and ricotta should be measured by measuring cup rather than weighed.

‡You may also weigh and record nuts in ounces or count and record the number of nuts you eat.

SUMMARY CHART

FOOD/BEVERAGE	UNIT	EXAMPLES
DAIRY, EGGS (cont.)		
Cheese, curd type	cup	½ cup ricotta
Cream	tablespoon	1 tbsp table cream
Milk	cup	1 cup skim
	fluid ounce	8 fl oz skim milk
Yogurt	cup	1 cup nonfat plain yogurt
	fluid ounce	8 fl oz vanilla yogurt
Egg	1 egg	
	1 egg white	
	1 egg yolk	
FAST FOODS	unit	1 cheeseburger
	order	1 order fried shrimp
	ounce	12 oz shake
	piece	1 chicken breast
	serving	1 serving coleslaw
	slice or fraction of whole*	3 slices 12-inch pepperoni pizza
FATS AND OILS		
Margarine	teaspoon, cup	2 tbsp Promise margarine
Oils	teaspoon, cup	1 tsp olive oil
Salad dressings	teaspoon, cup	¼ cup French dressing
FISH AND SHELLFISH	ounce	4 oz sole
	number	8 shrimp
FROZEN AND MICROWAVE	Depends on product	Use information on food label
FRUITS AND FRUIT JUICES	piece	1 apple
	cup	½ cup applesauce
GRAINS AND PASTA		
Bagel	number	1 bagel
Bread	slice	2 slices rye

*Specify diameter of whole (i.e., pizza, 14 inch diameter)

SUMMARY CHART

FOOD/BEVERAGE	UNIT	EXAMPLES
GRAINS AND PASTA (cont.)		
Cereals	cup	⅔ cup Wheatena
	ounce or gram	1 oz Cheerios
Crackers	number	1 Wheat Thin
Pasta/rice	cup	¾ cup cooked rice
Popcorn	cup	5½ cups air-popped
Rolls	number	1 poppy seed roll
MEATS	ounce	5 oz sirloin steak, with fat
NUTS AND SEEDS	ounce	1 oz pecans
	tablespoon	2 tbsp peanut butter
	number of kernels	12 cashews
	cup	½ cup sesame seeds
POULTRY	piece	½ chicken breast
	ounce	3 oz turkey breast
SAUCES, GRAVIES, & DIPS	tablespoon	3 tbsp brown gravy
	cup	½ cup cream sauce
SAUSAGES & COLD CUTS	ounce or gram	4 oz turkey bologna
	slice	3 slices salami
	link	1 link smoked sausage
SOUPS	cup	¾ cup bean soup
SWEETS		
Cakes	fraction of whole	1/12 chocolate cake
Cookies, bars	number	1 sugar cookie
Pie	fraction of whole	⅙ pumpkin pie
Sugar, jelly	cup/teaspoon	1 tsp grape jam
VEGETABLES	cup	½ cup spinach
	number	1 carrot

References

Chapter 1: Fat's the One

Acheson, K. J., Y. Schutz, T. Bessard, K. Anantharaman, J. P. Flatt, and E. Jéquier. "Glycogen Storage Capacity and de Novo Lipogenesis During Massive Carbohydrate Overfeeding in Man." *American Journal of Clinical Nutrition* 48 (1988):240–247.

Acheson, K. J., Y. Schutz, T. Bessard, J. P. Flatt, and E. Jéquier. "Carbohydrate Metabolism and De Novo Lipogenesis in Human Obesity." *American Journal of Clinical Nutrition* 45 (1987):78–85.

Astrup, A., S. Toubro, A. Raben, and A. R. Skov. "The Role of Low-fat Diets and Fat Substitutes in Body Weight Management: What Have We Learned from Clinical Studies?" *Journal of the American Dietetic Association* 97, suppl. (1997):S82–S87.

Barrows, K., and J. T. Snook. "Effect of a High-Protein, Very-Low-Calorie Diet on Resting Metabolism, Thyroid Hormones, and Energy Expenditure in Obese Middle-Aged Women." *American Journal of Clinical Nutrition* 45 (1987):391–398.

Bray, G. A. "Obesity — A Disease of Nutrient or Energy Balance?" *Nutrition Reviews* 45 (1987):33–43.

"Can Eating the 'Right' Food Cut Your Risk of Cancer?" *Tufts University Diet and Nutrition Letter* 6 (1988):2–6.

Donato, K., and D. M. Hegsted. "Efficiency of Utilization of Various Sources of Energy for Growth." *Proceedings of the National Academy of Sciences* 82 (1985):4866–4870.

Dougherty, R. M., A. K. H. Fong, and J. M. Iacono. "Nutrient Content of the Diet When the Fat Is Reduced." *American Journal of Clinical Nutrition* 48 (1988):970–979.

Dreon, D. M., B. Frey-Hewitt, N. Ellsworth, et al. "Dietary Fat: Carbohydrate Ratio and Obesity in Middle-Aged Men." *American Journal of Clinical Nutrition* 47 (1988):995–1000.

Elliot, D. L., L. Goldberg, K. S. Kuehl, and W. M. Bennett. "Sustained Depressions of the Resting Metabolic Rate after Massive Weight Loss." *American Journal of Clinical Nutrition* 49 (1989):93–96.

Flatt, J. P. "Dietary Fat, Carbohydrate Balance, and Weight Maintenance: Effects of Exercise." *American Journal of Clinical Nutrition* 45 (1987):296–306.

———. "Differences in the Regulation of Carbohydrate and Fat Metabolism and

Their Implications for Body Weight Maintenance." *Hormones, Thermogenesis, and Obesity*. Edited by Henry Lardy and Frederick Stratman. New York: Elsevier Science Publishing, 1989.

———. "Effect of Carbohydrate and Fat Intake on Postprandial Substrate Oxidation and Storage." *Topics in Clinical Nutrition* 2(2) (1987):15–27.

———. "Importance of Nutrient Balance in Body Weight Regulation." *Diabetes/Metabolism Reviews* 4(6) (1988):571–581.

———. "Metabolic Feedback on Food Intake Among ad Libitum Fed Mice." *International Journal of Obesity* 9 (1985):A33.

Flatt, J. P., E. Ravussin, K. J. Acheson, and E. Jéquier. "Effects of Dietary Fat on Postprandial Substrate Oxidation and on Carbohydrate and Fat Balances." *Journal of Clinical Investigation* 76 (1985):1019–1024.

Gray, D. S., J. S. Fisler, and G. A. Bray. "Effects of Repeated Weight Loss and Regain on Body Composition in Obese Rats." *American Journal of Clinical Nutrition* 47 (1988):393–399.

Hammer, R. L., C. A. Barrier, E. S. Roundy, J. M. Bradford, and A. G. Fisher. "Calorie Restricted Low-Fat Diet and Exercise in Obese Women." *American Journal of Clinical Nutrition* 49 (1989):77–85.

Hellerstein, M. K., et al. "Measurement of De Novo Hepatic Lipogenesis in Humans Using Stable Isotopes." *Journal of Clinical Investigations* 87 (1991):1841–52.

Hultman, E., and L. H. Nilsson. "Factors Influencing Carbohydrate Metabolism in Man." *Nutrition Metabolism* 18 (Suppl. 1) (1975):45–64.

Jéquier, E. "Carbohydrates as a Source of Energy." *American Journal of Clinical Nutrition* 59, suppl. (1994):682S–685S.

Katch, F., and W. D. McArdle. *Nutrition, Weight Control, and Exercise*. Philadelphia: Lea & Febiger, 1988.

Lissner, L., D. A. Levitsky, B. J. Strupp, et al. "Dietary Fat and the Regulation of Energy Intake in Human Subjects." *American Journal of Clinical Nutrition* 46 (1987):886–892.

Manson, J. E., M. J. Stampfer, C. H. Hennekens, and W. C. Willett. "Body Weight and Longevity." *Journal of the American Medical Association* 257 (1987):353–358.

Mattes, R. D. "Fat Preference and Adherence to a Reduced-Fat Diet." *American Journal of Clinical Nutrition* 57 (1993):373–381.

Mattes, R. D., C. B. Pierce, and M. I. Friedman. "Daily Caloric Intake of Normal-Weight Adults: Response to Changes in Dietary Energy Density of a Luncheon Meal." *American Journal of Clinical Nutrition* 48 (1988):214–219.

Ravussin, E., and P. A. Tataranni. "Dietary Fat and Human Obesity." *Journal of the American Dietetic Association* 97, suppl. (1997):S42–S46.

Romieu, I., W. C. Willett, M. J. Stampfer, G. A. Colditz, et al. "Energy Intake and Other Determinants of Relative Weight." *American Journal of Clinical Nutrition* 47 (1988):406–412.

Schutz, Y., J. P. Flatt, and Eric Jéquier. "Failure of Dietary Fat Intake to Promote Fat Oxidation: A Factor Favoring the Development of Obesity." *American Journal of Clinical Nutrition* 50 (1989):307–14.

Steen, S. N., R. A. Oppliger, and K. D. Brownell. "Metabolic Effects of Repeated Weight Loss and Regain in Adolescent Wrestlers." *Journal of the American Medical Association* 260 (1988):47–50.

Weiss, L., G. E. Hoffmann, R. Schreiber, H. Andres, E. Fuchs, E. Korber, and H. J. Kolb. "Fatty-Acid Biosynthesis in Man, a Pathway of Minor Importance." *Biological Chemical Hoppe-Seyler* 367 (1986):905–12.

Chapter 2: The *Choose to Lose* Plan

Dennison, D. *The DINE System: The Nutritional Plan for Better Health.* St. Louis: C. V. Mosby, 1982.

Chapter 4: Where's the Fat?

National Center for Health Statistics. *Anthropometric Reference Data and Prevalence of Overweight, United States 1976–1980.* DHHS Publication no. 87-1688. Washington, D.C.: Government Printing Office, 1987.

U.S. Department of Health and Human Services. "Health Implications of Obesity." *National Institutes of Health Consensus Development Conference Statement.* Vol. 5, no. 9, 1985. National Institutes of Health, Office of Medical Applications of Research, Building 1, Room 216, Bethesda, MD 20205.

U.S. Department of Health and Human Services. *The Surgeon General's Report on Nutrition and Health.* DHHS (PHS) Publication No. 88-50210. Washington, D.C.: Government Printing Office, 1988.

The fat tables are based on data from the following sources:

U.S. Department of Agriculture. *Composition of Foods.* Agriculture Handbook No. 8. Washington, D.C.: Government Printing Office, sec. 1–16, rev. 1976–1989.

U.S. Department of Agriculture. *Nutritive Value of American Foods in Common Units.* Agriculture Handbook No. 456. Washington, D.C.: Government Printing Office, November 1975.

Chapter 10: Taking Control: Eating Out

Stunkard, A. J., and H. C. Berthold. "What Is Behavior Therapy? A Very Short Description of Behavioral Weight Control." *American Journal of Clinical Nutrition* 41 (1985):821–823.

Stunkard, A. J., T. T. Foch, and Z. Hrubec. "A Twin Study of Human Obesity." *Journal of the American Medical Association* 256 (1986):51–54.

Stunkard, A. J., T. I. A. Sorenson, C. Hanis, et al. "An Adoption Study of Human Obesity." *New England Journal of Medicine* 314 (1986):193–198.

Chapter 11: Exercise: Is It Necessary?

Ballor, D. L., V. L. Katch, M. D. Becque, and C. R. Marks. "Resistance Weight Training during Caloric Restriction Enhances Lean Body Weight Maintenance." *American Journal of Clinical Nutrition* 47 (1988):19–25.

Cooper, Kenneth. *The New Aerobics.* New York: Bantam Books, 1983.

Cooper, Kenneth, and Mildred Cooper. *The New Aerobics for Women.* New York: Bantam Books, 1988.

Hammer, R. L., and C. A. Barrier, E. S. Roundy, J. M. Bradford, and A. G. Fisher. "Calorie Restricted Low-Fat Diet and Exercise in Obese Women." *American Journal of Clinical Nutrition* 49 (1989):77–85.

Hill, J. O., P. B. Sparling, T. W. Shields, and P. A. Heller. "Effects of Exercise and Food Restriction on Body Composition and Metabolic Rate in Obese Women." *American Journal of Clinical Nutrition* 46 (1987):622–630.

Katch, F., and W. D. McArdle. *Nutrition, Weight Control, and Exercise.* Philadelphia: Lea & Febiger, 1988.

Rippe, J. M., A. Ward, J. P. Porcari, and P. S. Freedson. "Walking for Health and Fitness." *Journal of the American Medical Association* 259 (1988):2720–2724.

U.S. Department of Health and Human Services, Public Health Service. *Exercise and Your Heart.* National Institutes of Health Publication No. 18-1677, 1981.

Chapter 12: It's Up to You

Dougherty, R. M., A. K. H. Fong, and J. M. Iacono. "Nutrient Content of the Diet When the Fat Is Reduced." *American Journal of Clinical Nutrition* 48 (1988):970–979.

Goor, R., and N. Goor. *Eater's Choice: A Food Lover's Guide to Lower Cholesterol,* 5th ed. Boston: Houghton Mifflin, 1999.

James, W. P. T., M. E. J. Lean, and G. McNeill. "Dietary Recommendations after Weight Loss: How to Avoid Relapse of Obesity." *American Journal of Clinical Nutrition* 45 (1987):1135–1141.

Chapter 14: *Choose to Lose* for Children

Mossberg, Hans-Olof. "40-Year Follow-up of Overweight Children." *The Lancet* (August 26, 1989):491–93.

Index

Table of Equivalent Measures

Volume Measures

1 gallon	4 quarts
1 quart	4 cups 2 pints
1 pint	2 cups
1 cup	8 fluid ounces 16 tablespoons
½ cup	4 fluid ounces 8 tablespoons
⅓ cup	5 tablespoons + 1 teaspoon
¼ cup	2 fluid ounces 4 tablespoons
2 tablespoons	1 fluid ounce
1 tablespoon	3 teaspoons ½ fluid ounce

Weight Measures

1 pound	16 ounces 454 grams
3.5 ounces	100 grams
1 ounce	28.35 grams